Coccidiosis in Poultry: Current Thinking on *Eimeria* spp., Host–Parasite Relationships, and Potential for Advancements in Control

Coccidiosis in Poultry: Current Thinking on *Eimeria* spp., Host–Parasite Relationships, and Potential for Advancements in Control

Guest Editor

Kenneth Bafundo

MDPI

Basel • Beijing • Wuhan • Barcelona • Belgrade • Novi Sad • Cluj • Manchester

Guest Editor
Kenneth Bafundo
Poultry Diagnostic and
Research Center
College of Veterinary Medicine
University of Georgia
Athens
USA

Editorial Office
MDPI AG
Grosspeteranlage 5
4052 Basel, Switzerland

This is a reprint of the Special Issue, published open access by the journal *Animals* (ISSN 2076-2615), freely accessible at: https://www.mdpi.com/journal/animals/special_issues/ATL1KY0253.

For citation purposes, cite each article independently as indicated on the article page online and as indicated below:

Lastname, A.A.; Lastname, B.B. Article Title. *Journal Name* **Year**, *Volume Number*, Page Range.

ISBN 978-3-7258-3581-2 (Hbk)
ISBN 978-3-7258-3582-9 (PDF)
https://doi.org/10.3390/books978-3-7258-3582-9

Contents

animals

Editorial

Current Thinking on *Eimeria* spp. in Poultry and Potential for Advancements in Control

Kenneth Bafundo

Poultry Diagnostic and Research Center, College of Veterinary Medicine, University of Georgia, Athens, GA 30602, USA; ken.bafundo@gmail.com

1. Introduction

The significance of *Eimeria* infections in chicken production cannot be understated. The infections are strikingly ubiquitous, with the strong likelihood that every commercial flock of chickens encounters one or more infections caused by these parasites. Some flocks, no doubt, experience only mild effects such as slower growth and compromised efficiency, while others undergo severe disease symptoms characterized by performance losses, hemorrhage, and elevated mortality. Parasite development within the intestine often leads to complications brought about by the presence of other pathogens, and in turn, complicated disease symptoms and animal welfare issues are often initiated by coccidial infections [1]. From a financial perspective, estimates show that annual global losses in the range of USD 14 billion routinely occur [2]. Moreover, global events such as the COVID-19 pandemic and the war in Ukraine have likely exacerbated these losses by more than 20% [3]. In order to meet the world's growing demand for high-quality protein, the years ahead will likely see expansive growth in both broiler and egg production. Commensurate with this growth will be the ever-present economic burden of *Eimeria* infections.

Since the 1940s, commercial chicken producers have relied on chemotherapy as the primary means of managing coccidial infections. Early on, chemically synthesized coccidiostats were commonly used to limit mortality, as broiler producers of the 1960s focused on the control of the hemorrhagic coccidia *E. tenella* and *E. necatrix*. During the 1970s and 1980s, several polyether ionophores were introduced, providing broad, effective control of *Eimeria* spp., allowing the industry to benefit from better intestinal health and, along with it, improved growth and feed efficiency. However, beginning in the mid-1990s, research on new anticoccidials waned, and since that point in time, only one anticoccidial product has been developed for coccidiosis in broiler chickens [4].

In the past two decades, the availability and use of anticoccidial products has also been affected by regional public demand for animals reared in the absence of chemotherapeutic agents [1,5]. In the EU, for example, several anticoccidial products have been banned, and the No Antibiotics Ever (NAE) movement in the United States prohibits the use of ionophore anticoccidials in broilers reared for that market segment [1,6]. Together, lack of research and consumer pressure on existing anticoccidial agents has forced the use of older, less resilient anticoccidials in today's production systems, thereby requiring additional support in order for optimal growth and efficiency to be achieved [6].

In order to break the cycle of continuous drug usage, live coccidiosis vaccines are often applied. Successful vaccine usage relies upon specialized management considerations (vaccine storage and application, litter and house management, etc.) that foster oocyst survival and allow for the re-exposure of birds to these infective forms. When these details are effectively addressed, vaccination has the potential to become a reasonable alternative to

Received: 28 January 2025
Accepted: 31 January 2025
Published: 3 February 2025

Citation: Bafundo, K. Current Thinking on *Eimeria* spp. in Poultry and Potential for Advancements in Control. *Animals* **2025**, *15*, 424. https://doi.org/10.3390/ani15030424

chemotherapy. Because vaccines contain drug-sensitive coccidia, modification of the drug sensitivity profiles of coccidia from vaccinated houses is known to occur. This approach is often used to improve the performance of many anticoccidial drugs once the return to chemotherapy occurs.

Of course, live vaccines have well-recognized effects on intestinal integrity, and producers often express concerns about negative performance results during vaccine usage [7]. The "bio-shuttle program", utilizing both vaccine application and short-term anticoccidial use (during approximately 10 days of peak coccidiosis challenge), limits the negative effects of vaccinal *Eimeria* replication during critical growth periods [8], thereby allowing vaccination to be more cost-competitive with standard chemotherapeutic approaches.

Although interest in a non-living vaccine remains very high, progress toward this objective has been difficult. Several laboratories have carried out pen and field trials with a variety of *Eimeria* antigens or genes for various antigens. A number of different application techniques has also been evaluated [4]. To date, however, none of these approaches has exceeded live vaccination in terms of efficacy, bird performance, ease of application, and/or cost effectiveness. It therefore appears that vaccination with live parasites will remain the most practical means of immunological protection for the foreseeable future.

Given this scenario, it should be clear that many poultry veterinarians express concern about the current and future status of coccidial infection [9]. These specialists recognize the need for technological advances for improved *Eimeria* control, a fact which underscores the need for a better understanding of these parasites and the infections that they produce. The enclosed Special Issue is devoted to current research on coccidiosis in chickens, with the hope that it will address some of the needs and information gaps described above.

The papers presented herein deal with several different aspects of *Eimeria* infections. The topics range from studies on the nutritional effects of *Eimeria* infections in broilers and layers to new methods of *Eimeria* control that diminish intestinal inflammation. Novel methods of oocyst speciation using artificial intelligence are also presented, along with reports of the new *Eimeria* species identified in Romania. The effects of coccidial infection on the intestinal microflora, a new and significant area of research, are also addressed in three separate contributions. Additionally, studies on the identification and testing of newly characterized antigens for inclusion in future vaccines are also included.

2. An Overview of Published Articles

Traditionally, the broiler chicken has been the standard focus of investigations dealing with *Eimeria* infections and host nutrition. In fact, nutritional studies in broilers have become one of the more commonly addressed topics in recent years. Varying greatly from this usual approach, the review of Sharma and Kim (Contribution 1) evaluated the effects of coccidial infections in laying fowl, and, in particular, outlined the significance of *Eimeria* infections in the developing pullet, in the laying hen, as well as the effects on egg production. In addition to describing the disease in the pullet/layer and noting the standard methods of control, the comprehensive nature of this work references basic nutritional research gleaned from studies in several avian species, but also presents new results found in production hens. The work clearly emphasizes the *Eimeria*-related effects that influence hen physiology and egg production and offers recommendations for nutritional strategies that could potentially limit these effects. This is an exceptionally cogent review that provides insights into layer health and egg production that have been lacking in the published literature.

Given the growing significance of live oocyst vaccination in broiler production, the work of Myers and Rochell (Contribution 2) examined the effects of fully pathogenic *Eimeria* vaccination on energy utilization when compared to birds receiving anticoccidial medication. When compared to medicated treatments in a 43-day trial, vaccination impaired nutrient and energy digestibility, body weight, and carcass yield when measured at days 18 and 31. In addition, the effects of vaccination on feed conversion were dependent upon energy level in starter feeds, and these effects were evident both at 18 days and in all subsequent feeding periods. The greatest negative effects were recorded for treatments receiving the highest energy levels in starter feed. Consequently, the authors explain that compensating for impaired energy utilization during vaccination by increasing dietary energy content in starter feed is ineffective. Clearly, additional research is required in order to understand the mechanisms of *Eimeria*-induced lipid malabsorption that extend beyond the limitations of energy availability.

The global emphasis on the reduction in antibiotic usage in poultry has led to the investigation and application of several new classes of compounds that influence the development of *Eimeria* and the adverse effects that these infections produce. The studies of Fellici et al. (Contribution 3) applied novel in vitro techniques to determine whether certain essential oils (thyme, oregano, and garlic) affect host cell penetration by *E. tenella* sporozoites and the subsequent development of *E. tenella* schizonts. Their work demonstrated that these oils decrease sporozoite invasion in vitro, thereby limiting the development of *E. tenella* schizonts. In some cases, both the size and number of schizonts, factors that are strongly associated with the pathogenicity of *E. tenella*, were significantly reduced. It was suggested that these in vitro techniques could be applied to the identification and development of new anticoccidial entities.

Sadoris et al. (Contribution 4) evaluated combinations of yucca (*Yucca schidigera*) and quillaja (*Quillaja saponaria*) in *Eimeria*-challenged broilers to determine their effects on intestinal permeability and immune responsiveness. While oocyst production and lesion scores were significantly reduced by oral administration of this combination, the saponin formula also maintained the integrity of the intestinal epithelium. *Eimeria*-induced cytokines (IFN-γ, IL-17, and IL-10) were reduced in the infected treatment group receiving saponins when compared to infected controls, and these levels did not differ from those recorded for the non-infected control group. This observation helps to explain the reported effects of saponins on the limitations of inflammation during and following coccidial infections.

Villar-Patino et al. (Contribution 5) studied the effects of extracts of Alliaceae (*Allium*) during coccidial infection by demonstrating improvements in feed conversion and changes in the microbial composition of the gut during feeding.

To better understand the effects of *Eimeria* infections on the intestinal microflora, Wang et al. (Contribution 6) repeatedly infected broiler chicks with low doses of *E. mitis*, producing birds that were immunologically protected from *E. mitis* challenge. Following this procedure, birds were monitored for changes in intestinal health and the composition of the intestinal microflora. The results indicated that multiple exposures to *E. mitis* increased the numbers of opportunistic bacterial pathogens in the intestine; a generalized decrease in non-pathogenic bacteria also occurred. Both findings were associated with increased levels of dysbacteriosis. Thus, in addition to previously reported impairments of growth performance, *E. mitis* infections, even when produced by repeated low-dose exposures, cause meaningful changes in the intestinal microbiota. The work illustrates the importance of *E. mitis*, a common coccidial parasite of broilers that is often unrecognized during routine *Eimeria* surveillance.

Miska et al. (Contribution 7) studied the effects of *E. maxima* infections on both luminal and mucosal bacterial populations of the intestine over a 14-day period following parasite exposure. The results demonstrated that *E. maxima* affected both alpha and beta diversity in intestinal bacteria, and that these effects commonly occurred at the height of infection. By day 14 post-infection, differences between the infected and control populations were minimal. While *E. maxima* infection increased the levels of *Lactobacillus*, butyrate-producing bacteria, thought to be beneficial in mucosal repair, were more commonly associated with non-infected broilers. The study also showed that luminal and mucosal bacterial populations differed markedly, thus requiring separate analyses of these two populations so that a more complete picture of the ecology of the intestinal microbiota could be presented.

Studies have shown that proteins isolated from the rhoptries of *E. tenella* display characteristics of protective antigens. The work of Li et al. (Contribution 8) investigated *E. tenella* ROP27 (EtROP27) as a potential component of a subunit coccidiosis vaccine. This rhoptry protein is known to be expressed at all stages of *E. tenella* development. The protein was produced in a procaryotic expression system, and after purification, it was tested in broilers challenged with *E. tenella*. Administration of EtROP27 was found to increase Ig-Y titers in serum; improvements in body weight gain and reductions in oocyst production, lesion scores, and bloody feces were also observed. While in vivo testing of this protein was limited in scope, these findings illustrate the potential of EtROP27 as a component of a future vaccine.

In most diagnostic situations, speciation of the coccidia involved is required. Traditionally, this has been accomplished through oocyst identification and lesion scoring, but there are inherent shortcomings and time commitments associated with these methodologies [10]. Kellogg et al. (Contribution 9) developed an artificial intelligence-based technique for the accurate identification of *Eimeria* oocysts in fecal samples. The method also has the capability of differentiating sporulated from non-sporulated oocysts of the species examined. The technique relies upon automated image analysis of samples that are then compared to examples of oocysts that have been previously identified as representative of the species. The reported results suggest that this image analysis technique is approximately equivalent to standard methods in terms of the accuracy of identification, but can markedly reduce analysis time compared to manual procedures. Further development of the methodology is in progress so that the minor species and those with smaller oocysts can be more accurately identified. Once completed, these additions will enhance reliability and make the technique an important tool for diagnosticians and researchers in the field.

Using PCR techniques, Coroian et al. (Contribution 10) examined the coccidial species parasitizing free-range chickens in Romania. *Eimeria acervulina* and *E. tenella* were the most commonly identified species, with *E. praecox*, *E. brunetti*, and *E. mitis* also recognized in several flocks. Surprisingly, *E. maxima* was not identified. Of importance was the recognition of OTUy and OTUz (*E. nagambie* and *E. zaria*, newly recognized species from the chicken) in a number of Romanian flocks. While *E. zaria* has previously been identified in Europe, this work is the first to report OTUy (*E. nagambie*) in a European country. In addition, common helminth infections were also reported, with *Ascaridia*, *Heterakis*, and *Capillaria* frequently recognized.

3. Concluding Comments

As noted above, it is hoped that the papers published in this Special Issue address some of the challenges presented by *Eimeria* infections and potentially fill some of gaps in our knowledge of these important parasites. It is also hoped that this volume will stimulate other researchers to focus their attention on the growing significance of coccidial infections in chickens and direct their efforts toward a better understanding of and potential solutions for this malady. In either case, readers are encouraged to examine these 10 contributions, since each has a place in the modern literature devoted to *Eimeria* in chickens.

Conflicts of Interest: The author declares no conflict of interest.

List of Contributions

1. Sharma, M.; Kim, W.K. Coccidiosis in egg-laying hens and potential nutritional strategies to modulate performance, gut health, and immune response. *Animals* **2024**, *14*, 1015. https://doi.org/10.3390/ani14071015.
2. Myers, A.G.; Rochell, S.J. Effects of starter diet energy concentration on nutrient digestibility and subsequent growth performance and meat yields of broilers under two coccidiosis control programs. *Animals* **2024**, *14*, 1524. https://doi.org/10.3390/ani14111524.
3. Felici, M.; Tugnoli, B.; De Hoest-Thompson, C.; Piva, A.; Grilli, E.; Marugán-Hernández, V. Thyme, Oregano, and Garlic Essential Oils and Their Main Active Compounds Influence Eimeria tenella Intracellular Development. *Animals* **2024**, *14*, 77. https://doi.org/10.3390/ani14010077
4. Saddoris-Clemons, K.; Osho, S.; Garcia, M.; Humphrey, B. Effects of dietary inclusion of a proprietary combination of *Quillaja saponaria* and *Yucca schidigera* on intestinal permeability and immune response in broiler chickens during a coccidia challenge. *Animals* **2024**, *14*, 1737. https://doi.org/10.3390/ani14121737.
5. Villar-Patiño, G.; Camacho-Rea, M.C.; Olvera-García, M.E.; Baltazar-Vázquez, J.C.; Gómez-Verduzco, G.; Téllez, G.; Labastida, A.; Ramírez-Pérez, A.H. Effect of Alliaceae encapsulated extract on growth performance, gut health, and intestinal microbiota in broiler chickens challenged with *Eimeria* spp. *Animals* **2023**, *13*, 3884. https://doi.org/10.3390/ani13243884.
6. Wang, Z.; Shang, P.; Song, X.; Wu, M.; Zhang, T.; Zhao, Q.; Zhu, S.; Qiao, Y.; Zhao, F.; Zhang, R.; et al. Alterations in ileal microbiota and fecal metabolite profiles of chickens with immunity to *Eimeria mitis*. *Animals* **2024**, *14*, 3515. https://doi.org/10.3390/ani14233515.
7. Miska, K.B.; Campos, P.M.; Cloft, S.E.; Jenkins, M.C.; Proszkowiec-Weglarz, M. Temporal changes in jejunal and cecal microbiota of broiler chickens with clinical coccidiosis (*Eimeria maxima*). *Animals* **2024**, *14*, 2976. https://doi.org/10.3390/ani14202976.
8. Li, M.; Lv, X.; Zheng, M.; Wei, Y. The immunoprotective analysis of ROP 27 protein in *Eimeria tenella*. *Animals* **2023**, *13*, 3500. https://doi.org/10.3390/ani13223500.
9. Kellogg, I.; Roberts, D.L.; Crespo, R. Automated image analysis for detection of coccidia in poultry. *Animals* **2024**, *14*, 212. https://doi.org/10.3390/ani14020212.
10. Coroian, M.; Fábián-Ravasz, T.-Z.; Dobrin, P.R.; Györke, A. Occurrence of *Eimeria* spp. and intestinal helminths in free range chickens from Northwest and Central Romania. *Animals* **2024**, *14*, 563. https://doi.org/10.3390/ani14040563.

References

1. Hofacre, C.L.; Smith, J.A.; Mathis, G.F. An optimist's view on limiting necrotic enteritis and maintaining broiler gut health and performance in today's marketing, food safety and regulatory climate. *Poult. Sci.* **2018**, *97*, 1929–1933. [CrossRef] [PubMed]
2. Blake, D.P.; Knox, J.; Dehaeck, B.; Huntington, B.; Rathinam, T.; Ravipati, V.; Ayoade, S.; Gilbert, W.; Adebambo, A.O.; Jatau, I.D.; et al. Re-calculating the cost of coccidiosis in chickens. *Vet. Res.* **2020**, *51*, 115. [CrossRef] [PubMed]
3. Blake, D.P. *Eimeria* of chickens: The changing face of an old foe. *Avian Pathol.* **2024**, *53*, 1–12. [CrossRef] [PubMed]
4. Blake, D.P.; Marugan-Hernandez, V.; Tomley, F.M. Spotlight on avian pathology: *Eimeria* and the disease coccidiosis. *Avian Pathol.* **2021**, *50*, 209–213. [CrossRef] [PubMed]
5. Cervantes, H.M. Antibiotic-free poultry production: Is it sustainable? *J. Appl. Poult. Res.* **2015**, *24*, 91–97. [CrossRef]
6. Bafundo, K.W.; Duerr, I.; McNaughton, J.L.; Gomez, L. The influence of a quillaja and yucca combination on growth performance and lesion scores of broilers administered chemical anticoccidials or a live coccidiosis vaccine. *Int. J. Poult. Sci.* **2022**, *21*, 50–56. [CrossRef]
7. Lee, J.T.; Eckert, N.H.; Ameiss, K.A.; Stevens, S.M.; Anderson, P.N.; Anderson, S.M.; Barri, A.; McElroy, A.P.; Danforth, H.D.; Caldwell, D.J. The effect of dietary protein level on performance characteristics of coccidiosis vaccinated and nonvaccinated broilers following mixed-species *Eimeria* challenge. *Poult. Sci.* **2011**, *90*, 1916–1925. [CrossRef] [PubMed]
8. Mathis, G.; Schaeffer, J.; Cookson, K.; Dickson, J.; LaVorgna, M.; Waldrip, D. Effect of lasalocid or salinomycin administration on performance and immunity following coccidia vaccination of commercial broilers. *J. Appl. Poult. Res.* **2014**, *23*, 577–585. [CrossRef]

9. United States Animal Health Association. Report of the USAHA Committee on Poultry and Other Avian Species. 2019. Available online: https://www.usaha.org/upload/Committee/2019Reports/Poultry_and_Other_Avian_Species_.pdf (accessed on 19 October 2021).
10. Long, P.L.; Joyner, L.P. Problems in the identification of the species of *Eimeria*. *J. Protozool.* **1984**, *31*, 535–541. [CrossRef] [PubMed]

animals

Review

Coccidiosis in Egg-Laying Hens and Potential Nutritional Strategies to Modulate Performance, Gut Health, and Immune Response

Milan Kumar Sharma and Woo Kyun Kim *

Department of Poultry Science, University of Georgia, Athens, GA 30602, USA; milan.sharma@uga.edu
* Correspondence: wkkim@uga.edu

Simple Summary: Coccidiosis is one of the most prevalent diseases in poultry production, inflicting substantial economic losses exceeding USD 15 billion to the poultry industry. Coccidiosis is most common in young broilers but is not limited to them, and pullets and egg-laying hens are equally susceptible. Unfortunately, unlike broilers, low emphasis has been given to laying hens. This review aims to summarize the effect of coccidiosis in laying hens while exploring potential nutritional interventions to fight coccidiosis.

Abstract: Avian coccidiosis, despite advancements in management, nutrition, genetics, and immunology, still remains the most impactful disease, imposing substantial economic losses to the poultry industry. Coccidiosis may strike any avian species, and it may be mild to severe, depending on the pathogenicity of *Eimeria* spp. and the number of oocysts ingested by the bird. Unlike broilers, low emphasis has been given to laying hens. Coccidiosis in laying hens damages the gastrointestinal tract and causes physiological changes, including oxidative stress, immunosuppression, and inflammatory changes, leading to reduced feed intake and a drastic drop in egg production. Several countries around the world have large numbers of hens raised in cage-free/free-range facilities, and coccidiosis has already become one of the many problems that producers have to face in the future. However, limited research has been conducted on egg-laying hens, and our understanding of the physiological changes following coccidiosis in hens relies heavily on studies conducted on broilers. The aim of this review is to summarize the effect of coccidiosis in laying hens to an extent and correlate it with the physiological changes that occur in broilers following coccidiosis. Additionally, this review tries to explore the nutritional strategies successfully used in broilers to mitigate the negative effects of coccidiosis in improving the gut health and performance of broilers and if they can be used in laying hens.

Keywords: coccidiosis; laying hens; gut health; oxidative stress; nutritional strategies; immune response; egg production

Citation: Sharma, M.K.; Kim, W.K. Coccidiosis in Egg-Laying Hens and Potential Nutritional Strategies to Modulate Performance, Gut Health, and Immune Response. *Animals* **2024**, *14*, 1015. https://doi.org/10.3390/ani14071015

Academic Editor: Kenneth Bafundo

Received: 6 March 2024
Revised: 21 March 2024
Accepted: 25 March 2024
Published: 27 March 2024

1. Introduction

Avian coccidiosis, caused by various species of the protozoan *Eimeria*, poses a significant threat to the poultry industry. Despite advancements in management, nutrition, genetics, chemotherapy, and immunology, it remains the most prevalent disease in the poultry industry, imposing substantial economic losses of more than USD 15 billion [1]. The economic loss due to coccidiosis is not only due to clinical outbreaks resulting in low production and high mortality but also subclinical coccidiosis without visible clinical signs; even subclinical coccidiosis contributes to economic losses through reduced nutrient absorption and utilization, thereby compromising optimal performance and the feed conversion ratio (FCR). Reduced nutrient absorption during subclinical coccidiosis may result from intestinal enterocyte and morphometric damages, compromised gastrointestinal integrity, decreased ileal digestibility, and increased susceptibility to secondary diseases like necrotic enteritis [2–7].

Coccidiosis is common in young broilers or laying hen pullets ranging from 3 to 18 weeks of age but not limited to this age group. Newly hatched chicks have high maternal antibodies but have little protection against coccidiosis [8]. Chickens usually become infected when maternal antibodies deplete and the oocyst counts in litter peak at 3–6 weeks of age; however, immunity develops shortly after mild infections [2,5,8]. Older flocks without acquired immunity against *Eimeria* spp. might get the disease [2,8]. In addition, there is no cross-immunity among *Eimeria* species; thus, more than one outbreak is possible in the same flock with different species involved in each episode [2,8]. Generally, anticoccidial drugs are used in laying hens' pullets and are withdrawn before they start laying eggs. Therefore, the acquired immunity against *Eimeria* during the pullet phase might not be enough to prevent infection throughout the production period [8–10]. Coccidiosis may strike any avian species, and it may be mild to severe, depending on the pathogenicity of the *Eimeria* spp. and the number of oocysts ingested by the bird [2,5].

Unlike broilers, limited attention has been given to laying hens regarding coccidiosis. One of the leading causes of laying hen mortality was reported as coccidiosis in cage-free systems [11–13]. A study conducted in Sweden observed 19% coccidiosis in laying hens raised on a litter-based cage-free system [14]. Laying hens recovered from coccidiosis but did not have optimal egg production and had a high FCR. Fossum et al. (2009) [12] observed higher incidences of coccidiosis in laying hens raised in cage-free systems (aviaries and free-range) compared with conventional cages in Sweden from 2001 to 2004, after the ban on conventional cages. Although there are several variations among cage-free systems (aviaries, free-range, or barn), the litter-based floor is typical in all housing systems that allow fecal–oral transmission of the oocyst, leading to disease outbreaks, unlike conventional cages. Therefore, preventing coccidiosis in these systems might become one of the many challenges for egg producers while transitioning from traditional cages to cage-free systems.

As the egg industry worldwide is steadily observing an increase in cage-free facilities, the broad topic of "gut health" associated with "coccidiosis" will have one of the most significant economic impacts on laying hen health and performance. Many countries in Europe have already transitioned to cage-free systems, whereas the United States is slowly transitioning to cage-free systems, and by 2023, more than 35% of laying hens in the U.S. were raised in cage-free environments, and this is still increasing (UEP, 2023). The evolution of these new production systems (aviaries; free-range) with large numbers of hens on the floor/range will influence the occurrence of coccidiosis. Laying hens raised in a litter-based aviary or free-range systems may provide a perfect environment for transmitting coccidiosis, and they have a higher chance of getting the disease than those raised in traditional cage systems. However, laying hens raised in conventional cages are also susceptible to coccidiosis, which contrasts with the traditional view that raising laying hens in conventional cages is unlikely to result in coccidiosis infection [10,15]. Hens raised in a cage system can still ingest enough oocysts from manure belts or edges of cages to induce clinical coccidiosis; however, the incidences are less [10,15]. Yet, there remains a dearth of knowledge regarding the effects of coccidiosis on laying hen performance, gastrointestinal health, oxidative stress, immune response, and skeletal health. However, broilers and laying hens have been intensively selected for many generations for different purposes (rapid growth and meat production vs. egg production), which might have changed their gastrointestinal physiology, immune response, and skeletal development. In this review, we aim to correlate findings from the broiler studies to best explain what might happen in laying hens infected with coccidiosis, particularly concerning gastrointestinal physiology, immune response, and potential mitigation strategies.

2. Avian Coccidiosis

Avian coccidiosis is caused by multiple species of the genus *Eimeria* belonging to the subkingdom protozoa and phylum Apicomplexa [5]. *Eimeria* is an intracellular parasite that infiltrates intestinal enterocytes at different locations, depending on the species. Only seven have been recognized to infect chickens among the hundreds of *Eimeria* species including the following: *E. acervulina, E. brunetti, E. maxima, E. necatrix, E. mitis, E. praecox,* and *E. tenella* [2,5]. Among these *Eimeria* species known to infect chickens, *E. brunetti, E. maxima, E. necatrix, and E. tenella* are considered highly pathogenic, whereas *E. acervulina, E. mitis,* and *E. praecox* exhibit low pathogenicity (Table 1) [2,5]. In the past, there were sporadic reports of outbreaks of *E. hagani* and *E. mivati* in chickens. However, their existence was doubted, and they are now recognized as nomina dubia [16].

Table 1. List of *Eimeria* species important in laying hens with their predilection sites, characteristic lesions, pathogenicity, and immunogenicity.

Species	Pathogenicity	Predilection Site in the Intestine	Characteristic Lesions	Immunogenicity
E. mitis	Low	Mid-intestine (Meckel's diverticulum to cecal junction)	Mucoid exudate and no distinct lesions	Moderate
E. praecox	Low	Anterior intestine (duodenum)	Mucoid exudate and no distinct lesions	Moderate
E. acervulina	Low to moderate	Anterior intestine (duodenum)	Whitish round lesions	Moderate
E. maxima	Moderate to high	Mid-intestine (part of jejunum and ileum)	Petechiae hemorrhage and blood-tinged exudate	High
E. brunetti	Moderate to high	Posterior intestine (Meckel's diverticulum to cecal junction)	Blood clot and coagulated necrosis	High
E. necatrix	High	Mid-intestine (part of jejunum and ileum)	Petechiae hemorrhage and blood-filled exudate	Low
E. tenella	High	Ceca	Hemorrhagic lumen and blood clot	Low

Avian coccidiosis is a common and widespread disease of chickens around the world and is transmitted by the fecal–oral route. Outbreaks of coccidiosis are reported wherever chickens are raised [2,5]. *Eimeria* oocysts can remain viable for a long period of time, and the disease can become endemic to the areas where environmental and managemental conditions favor year-round survival and multiplication [2,5]. The incidences of coccidiosis are usually higher in litter-based intensive systems, which favor the survival and accumulation of oocysts [4].

Eimeria's life cycle is short and has high reproductive potential with self-limiting infection [2,5]. The life cycle of the *Eimeria* consists of an exogenous phase involving the sporulation of the oocyst in the environment and an endogenous phase involving asexual reproduction (schizogony) followed by sexual reproduction (gametogony) within the avian intestine. Infection initiates with the ingestion of sporulated oocysts from the environment [2,5,17]. Each sporulated oocyst contains four sporocysts, each containing two sporozoites. Upon ingestion, mechanical and biochemical forces break down the oocyst, releasing the sporozoites in the digestive tract [2,5]. Subsequently, sporozoites migrate to specific sites depending on the species involved and invade the intestine's epithelial enterocytes. Within enterocytes, sporozoites transform into trophozoites and begin multiple asexual multiplications, forming schizonts, known as schizogony. Mature schizonts release merozoites, which again invade epithelial enterocytes, repeat the asexual cycle, and complete the second schizogonous cycle. Depending on the species, they might complete another schizogonous cycle before forming male (microgametocytes) and female (macrogametocytes) gametocytes [2,5]. Mature gametocytes release microgametes that fertilize mature macrogametocytes, resulting in the formation of oocysts. Upon maturation, oocysts rupture epithelial cells, which are excreted in feces and, under favorable conditions, sporulate within 48–72 h, becoming infective [2,5]. With each successive cycle, the number of sporulated oocysts increases exponentially in the environment, which retains their infectivity for extended periods [18]. The development of *Eimeria* in a host results in the destruction of the host's intestinal tissues, leading to clinical manifestation observed in a disease outbreak.

3. Host–Pathogen Interaction and Immune Responses against Coccidiosis

Eimeria species are highly host-specific and, once infected, can produce protective immunity against that particular species [19–21]. Notably, no cross-immunity is observed among *Eimeria* species. Generally, a large number of *Eimeria* oocysts are required to induce complete protective immunity, except for *E. maxima*, which is considered more immunogenic than other species [22]. The asexual life cycle of *Eimeria* is more immunogenic than the sexual life cycle. However, for the development of complete immunity and protection, both sexual and asexual phases of the life cycle are equally essential, especially for adaptive immunity [23,24].

Gut-associated lymphoid tissue (GALT) is an integral component of mucosal-associated lymphoid tissues and plays a vital role in the immune response as a first line of defense, preventing the progression of diseases and destroying infectious agents at an early stage [18,22]. GALT is a multilayered tissue with antigen-presenting cells, immunoregulatory cells, and effector cell types different from their systemic counterpart [19]. In poultry, during coccidiosis, GALT serves as antigen recognition and presentation, followed by the release of intestinal antibodies and activation of the cell-mediated immune response [18,19,25]. Various lymphoid tissues (Peyer's patches, bursa of Fabricius, and cecal tonsils) have evolved in the GALT to produce verities of immune cells (epithelial, lymphoid, antigen-presenting, and natural killer cells) to protect against invading pathogens [18,19,22]. Peyer's patches serve as antigen recognition and immune response activation, along with facilitating gastrointestinal IgA secretion [19,22,25]. Following this, activated B and T cells migrate to lamina propria, which serves as the effector site for immune responses, initiating antigen-specific/non-specific responses involving antibody production, leukocyte accumulation, and locally produced cytokines [19,25]. The activation of antigen-presenting cells, such as macrophages and dendritic cells, by coccidia upregulates the production of pro-inflammatory cytokines and chemokines from innate immune cells, which are essential for the development of the adaptive immune response. T lymphocytes play a crucial role in the response to coccidia infection, with various T cell subpopulations capable of recognizing multiple antigens and modulating humoral and cell-mediated immunity [18,19,25].

Cell-mediated immune responses are an important effector mechanism that includes antigen-specific and non-specific activation of various cell populations, including T lymphocytes, CD4+ helper T cells and CD8+ cytotoxic T cells, natural killer cells, and macrophages. T cell activation is associated with major histocompatibility complex (MHC), with cytotoxic T cells recognizing antigens presented in the context of MHC-I and helper T cells recognizing antigens associated with MHC-II molecules. Antigen-presenting and dendritic cells help activate naïve CD4+ T cells into different subsets such as Th1, Th2, Treg, and Th17 [19,26,27]. Activated helper T cells are involved in the humoral immune response, cytotoxic activity activation, macrophages, and natural killer cells [21]. Following coccidiosis, cytotoxic T cells, along with macrophages and natural killer cells, identify coccidia-infected host cells and modulate mononuclear cells to produce interferon-γ, activating proinflammatory pathways to inhibit the development of intracellular *Eimeria* within host cells [18,19,28]. Cytokines produced by CD4+ Th1 cells (IL-2, IL-12, IFN-γ, and TNF-α) aid in clearing the infection by promoting the activation of macrophages and other immune cells capable of eliminating intracellular coccidia [26,27]. The importance of cytotoxic T cells during coccidiosis has been demonstrated previously, as evidenced by the presence of intestinal intraepithelial lymphocytes expressing more than 75% CD8+ T cells, the presence of CD8+ cells in GALT within 24 h of infection, and the presence and activation of mono and polynuclear cells [19,22,29,30].

4. Coccidiosis and Its Effects on Pullets and Laying Hens

4.1. Gut Health and Oxidative Stress

The negative effect of coccidiosis on the parameters associated with gastrointestinal health, such as increased intestinal lesion scores, gastrointestinal permeability, a breach in tight junction integrity, damage to intestinal morphology, inflammation, and oxidative stress, are related to *Eimeria* multiplication and release in the gut lumen [2,5–7,31]. Continuous intestinal epithelial cell turnover occurs throughout the lifetime of the chickens. However, coccidiosis infection increases epithelial cell turnover by two times compared with noninfected birds [32]. It has been reported that the highest turnover of epithelial cells was observed between 5 and 7 days post-*Eimeria* inoculation due to the release of large numbers of first- and second-generation merozoites into the lumen. As a result, the mRNA and protein expression of a cell proliferative marker, the proliferating cell nuclear antigen, in the duodenum epithelium was upregulated, thus increasing crypt depth and resulting in a lower villus height to crypt depth ratio [33]. The difference between *Eimeria*-infected and non-infected birds in terms of villus height might range from 10 to 40% and is also dependent on the *Eimeria* species and level of infection [7,34–40]. This epithelial cell turnover during *Eimeria* infection might be a host defense mechanism to limit the multiplication of the *Eimeria* by the expulsion of infected cells [41]. Moreover, the tight junction proteins maintaining the tight junction integrity of the small intestine are composed of claudin, occludin, junctional adhesion molecules, and zonula occludens families [42]. These tight junction proteins were reported to be upregulated in *Eimeria*-infected birds along with increased permeability by more than 100%, supporting the observation of a rapid turnover rate in the intestinal epithelium during *Eimeria* infection [6,7,37,38,43]. Similar results were observed in pullets and laying hens of different age groups and at various stages of egg production [34–36].

During coccidiosis, the equilibrium between reactive oxygen species (ROS) and the host's ability to neutralize ROS is disrupted, leading to oxidative stress. These ROS have a high affinity for the phospholipid bilayer of the cell, initiating lipid peroxidation and cytotoxic changes, which in turn damages the intestine, altering the gastrointestinal integrity and causing abnormal changes in the intestinal morphology [44,45]. Under normal circumstances, the production of ROS and its effects on cellular mechanisms are controlled by the host defense mechanism consisting of enzymes such as glutathione (GSH), glutathione peroxidases (GPx), and catalase [44]. During *Eimeria* infection, although oxidative defense mechanisms are active, the production of ROS exceeds the production of enzymes suppressing ROS, aggravating the situation. Previous studies have reported that *Eimeria* infection increased the markers associated with oxidative stress, such as reduced concentration of total antioxidant capacity (TAC) from 15 to 40% [31,46] and reduced GPx and GSH activity [31,47]. Furthermore, *Eimeria* infection also increased the markers of radical-induced damage, such as superoxide dismutase (SOD) and malondialdehyde (MDA) [34–36,44,48,49].

4.2. Growth Performance

The effect of coccidiosis on the growth performance of birds depends on the *Eimeria* species and the severity of the infection. Previous studies have reported that broilers challenged with 10^5 oocysts of *Eimeria acervulina* reduced their body weight gain (BWG) by 10.3%, whereas 10^6 oocysts per bird decreased BWG by 26.7% [50]. Likewise, similar results were observed with *E. maxima* and *E. tenella* [50]. On average, coccidiosis reduced the body weight (BW) of the chickens by 10%, irrespective of the species involved in the infection [51]. *E. maxima*-induced coccidiosis was observed to have the most severe negative impact on the BW of broilers, ranging from 23% to 37% [6,51], followed by *E. tenella*, ranging from 15% to 27% [52,53], and *E. acervulina*, from 16% to 19% [54,55]. However, Choi et al. (2021) [56] did not observe any reduction in the BW of broilers when they were inoculated with sporulated oocysts of *E. tenella* ranging from 6250 to 50,000 [56]. In cases of mixed *Eimeria* (*E. maxima*, *E. tenella*, and *E. acervulina*) infections, the reduction in BW can range from 13% to 50%

with an increase in *Eimeria* dosage from 6250 to 50,000 *E. maxima*, 6250 to 50,000 *E. tenella*, and 31,250 to 250,000 *E. acervulina* [7,37,46]. However, mixed *Eimeria* infection has been associated with higher mortality rates, reaching up to 47%, and reductions in BW ranging from 9% to 28% in laying hen pullets aged two weeks, 8% to 15% in pullets aged 15 weeks, and 9% to 11% in hens aged 25 weeks during the early phase of infection [34–36]. Most studies have reported reductions in BW during the early phase of infection (1–8 days post-*Eimeria* inoculation) but not during the recovery phase (8–14 days post-*Eimeria* inoculation). During the early phase of infection, the reduction in feed intake and damage to the intestinal linings are more pronounced compared with those in the recovery phase, where feed intake and intestinal damage recover [7,35]. Additionally, during the early phase, nutritional redistribution occurs towards the immune response to subside the infection [7,35,36]. This reduction in feed intake and nutritional redistribution are associated with reduced growth or body weight loss in pullets and laying hens.

4.3. Production Performance

Coccidiosis rarely occurs in laying hens because of previous infection and the resulting immunity [10]. However, pullets reared in cages/on floors with coccidiostat drugs might not have enough exposure to coccidia to stimulate immunity. Outbreaks can occur after moving to layer facilities, either conventional cages or alternative housing systems [5,8,10]. Coccidiosis in laying hens has been reported during the early egg production period, around 23–24 weeks of age, and is characterized by high morbidity and mortality with a dramatic reduction in egg production [4,10,57–59]. Temporary cessation of egg production has been observed in laying hens infected with *E. maxima*, *E. acervulina*, and *E. tenella*. Hegde and Reid (1969) [57] reported that total egg production dropped to less than 20% when susceptible laying hens were independently challenged with different *Eimeria* species (*E. acervulina*, *E. maxima*, *E. brunetti*, *E. necatrix*, or *E. tenella*), with a significant drop observed two weeks post-challenge. It took 4–6 weeks to recover from infection, which varied among species. The mortality was more significant with *E. necatrix* and *E. maxima* following infection; however, the number of culled birds was not different from the control unchallenged group. Single-comb white leghorns infected with *E. mitis* experienced significant reductions in egg production and eggshell quality, undergoing complete molting before resuming egg production [58]. Hens recovered from coccidiosis may not reach their full egg-laying potential [60]. An outbreak of *E. necatrix* in a breeder hen facility resulted in significant mortality among breeder hens and roosters, as well as a 4.3% reduction in egg production following the recovery period compared with hens of the same age [59]. Recent studies in our lab have observed that coccidiosis at the pre-lay stage of growth delayed the onset of maturity and egg production by two weeks, and infection at peak egg production drastically dropped egg production and temporarily ceased egg production in some hens [34,36]. Reduced egg production during a coccidiosis outbreak might be because of the malabsorption of nutrients, including amino acids, carbohydrates, and minerals, which are essential during the laying period [6–8,61]. The possible mechanism of how coccidiosis affects the performance of laying hens is shown in Figure 1. Reduced feed intake and damage to intestinal linings following coccidiosis are evident. This damage caused by *Eimeria* to the intestinal tract affects the digestion and absorption of the nutrients required for egg formation [36]. Furthermore, it has been observed that plasma amino acids are lowered, and amino acid imbalance occurs during coccidiosis [55]. The limited availability of nutrients and amino acids in laying hens might reduce the synthesis of albumin, an important component of an egg, thus reducing egg production. Increased oxidative stress during coccidiosis might damage the developing oocytes and granulosa cells, leading to follicular atresia, production, and deposition of yolk precursors in selected ovaries, or occurrence of apoptosis in the oviducts and ovarian follicles [36,62,63]. These disruptions to normal physiological and reproductive functions might contribute to a decline in egg production. Additionally, the redistribution of energy during coccidiosis for tissue repair, inflammation, and maintaining oxidative status might also be responsible for the decline in egg production [36]. These studies showed that laying hens are susceptible to coccidiosis during the laying phase, with a dramatic reduction in egg production posing substantial challenges to the egg industry. The economic loss to the industry is mostly from reduced egg production, reduced feed intake, and an increased FCR due to malabsorption of nutrients and mortality.

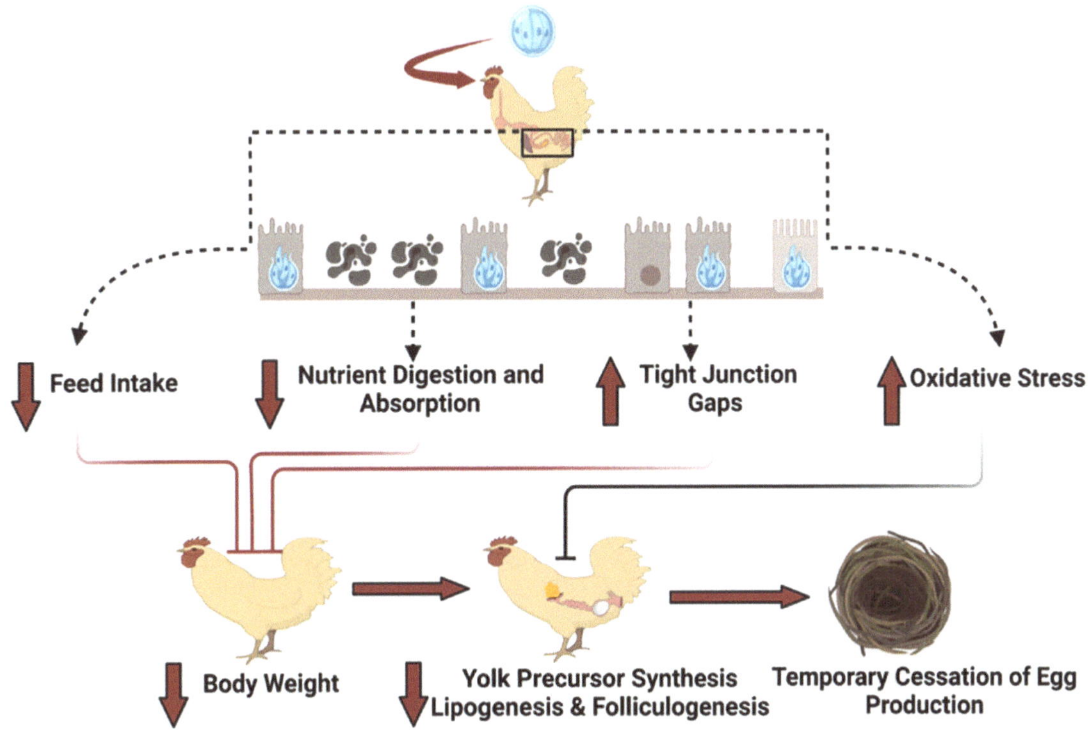

Figure 1. Potential mechanism of action showing how coccidiosis affects egg production in laying hens.

5. Prevention and Control of Coccidiosis

5.1. Biosecurity and Management

Biosecurity is an important tool in poultry production to prevent disease outbreaks on farms. Once a coccidiosis outbreak occurs on a farm and the house becomes contaminated with oocysts, it is virtually impossible to completely decontaminate the environment [2,5]. Strict biosecurity measures should be implemented between farms and chicken houses to minimize the risk of cross-contamination and outbreaks. In litter-based intensive housing systems, *Eimeria* oocysts accumulate in large amounts and can remain viable for longer periods of time [2]. In those systems, management strategies should focus on maintaining the optimum level of moisture in the litter so that it can maintain the viable number of sporulated oocysts sufficient to generate protective immunity but not high enough to induce clinical disease. Furthermore, when employing vaccination as a preventive measure, it is also critical to maintain optimum litter moisture, as oocyst recycling is important in inducing a protective immune response against specific *Eimeria* species [64,65]. Additionally, immunosuppression from stressors such as heat, cold, or high stocking density in chicken houses might increase susceptibility to the disease; therefore, it is critical to maintain optimum environments in the houses [66].

5.2. Chemotherapy

Traditionally, chemotherapeutic (anticoccidial) drugs have been used in the poultry industry for years all around the world to control coccidiosis outbreaks in farms; however, the emergence of drug resistance among the zoonotic pathogens and concerns raised by the World Health Organization and consumers led to a ban of antibiotics use in food-producing animals including poultry as growth promoters and disease prevention [67,68]. In poultry, coccidiostat (inhibits the replication of *Eimeria*) or coccidiocidal (kills and destroys the *Eimeria*) drugs are used, which can be either synthetic or ionophores [64,68]. Synthetic anticoccidial drugs (chemicals) are produced by chemical synthesis from other chemical compounds that have diverse modes of action in controlling *Eimeria*, from inhibiting mitochondrial respiration and disrupting energy production (clopidol or decoquinate) to competitive inhibition of thiamine uptake (amprolium) or inhibiting folic acid pathways (sulphonamides). Some of the synthesized chemicals approved to be used in the poultry industry are diclazuril, halofuginone, clopidol, robenidine, decoquinate, nequinate, amprolium, and zoalene [68,69]. These chemicals act on the intracellular stage of *Eimeria* after they invade the host intestine and interfere with one or more stages of the life cycle [69]. However, while using these chemicals, it is important to consider that they are more susceptible to the development of resistance [68]. Additionally, these chemicals vary in their mode of action, efficacy, dose range, and susceptibility to resistance; it is important to rotate between them at certain intervals to minimize the development of resistance [68]. On the other hand, ionophores are fermented products from bacteria such as *Streptomyces* spp. and *Actinomadura* spp. [64,68]. Ionophores facilitate the transportation of ions across cell membranes, increasing the concentration of intracellular ions and disrupting the normal ion balance across cell membranes [70]. The common ionophores used in the poultry industry are monensin, narasin, salinomycin, maduramicin, semduramicin, and lasalocid. Even though these ionophores are not of human importance, the ban on antibiotics in animal production may include these, which have been the backbone of coccidiosis prevention programs for many years, forcing the poultry industry to look for alternatives [68]. Although the poultry industry has used shuttle or rotational programs to minimize resistance to these drugs in the past, most field strains of *Eimeria* have shown varying levels of resistance to more than one drug [64,71].

5.3. Vaccination

Vaccination against *Eimeria* species has become one of the most widely used methods in controlling coccidiosis outbreaks in the United States. The discovery of the self-limiting infection of *Eimeria* species and the development of resistance to reinfection by the same *Eimeria* species by Beach and Corl. (1925) [72] paved the way for the development and use of a live vaccine to control coccidiosis in the USA by 1952. In addition to self-limiting infection, *Eimeria* species are less susceptible to younger chicks, and the generation of acquired immunity against *Eimeria* species with limited to no pathogenesis made vaccination an efficient strategy to control coccidiosis in commercial poultry production [73]. The objective of immunization is to expose the birds to *Eimeria* species to elicit protective immunity against them, which is adequate to prevent future infections [9,15,65,66,74]. However, the major disadvantage of vaccination is the absence of cross-immunity among different *Eimeria* species; thus, it is more labor-intensive to prepare and has a high cost because of including multiple species in the vaccine [75]. When vaccinating meat birds against coccidiosis, the emphasis is placed on *E. acervulina*, *E. maxima*, and *E. tenella*. In contrast, for egg-laying hens, the focus shifts to *E. tenella*, *E. acervulina*, *E. maxima*, *E. brunetti*, *E. necatrix*, and *E. praecox*, primarily because of variations in the life spans of the respective hosts (Table 2) [15]. To ensure birds build strong immunity against *Eimeria* species, they must undergo oocyst recycling through three to four successive re-infections so that they have complete immunity against *Eimeria* by 3–4 weeks of age when most coccidiosis outbreaks occur [17,65,75,76]. In the United States, two types of live vaccines are available based on the pathogenicity of the parasites, including (i) nonattenuated and (ii) attenuated live oocyst vaccines [65,75]. Nonattenuated vaccines contain laboratory or field strains of *Eimeria* oocysts that have their virulence preserved to elicit protective immunity. With a nonattenuated live oocyst vaccine, immunity against *Eimeria* is produced by completing the life cycle and boosted by recycling oocysts from the litter for reinfections [9,17,65,73,75–77]. However, the significant disadvantages of nonattenuated vaccines are their short shelf life, high cost of production due to the inclusion of all *Eimeria* species, occasional outbreaks of coccidiosis on farms, halted growth, and a possible lack of compensation for growth after recovery [9,15,65,75,77]. On the other hand, live attenuated oocyst vaccines are selected via passage through embryonated hen eggs or by selection for precociousness and have demonstrated reduced pathogenicity while maintaining the ability to induce protective immunity [15,65,75,78]. This selection for precocious has decreased pathogenicity while still having the capacity to stimulate the immune response against selective *Eimeria* species [65,73–75,77]. Recombinant or subunit vaccines are composed of immunogenic antigens capable of producing protective immunity against *Eimeria*. However, identifying the antigens that elicit protective immunity and limited to no cross-immunity among *Eimeria* species are limiting factors for the development of recombinant vaccines [65,74,77]. Thus, for complete protection against the prevalent *Eimeria* species, it is essential to identify antigens against several isolates of the same species and for multiple species to ensure complete protection [65,77].

Table 2. Commercial vaccines available in the U.S. for use in laying hens; adapted from Price (2014) [15] and Cervantes (2023) [68].

Trade Name	*Eimeria* Species Present in Vaccine	Vaccine Type	Route of Administration	Age of Birds at Vaccination	Manufacturer
Coccivac-D2	*E. acervulina, E. brunetti, E. maxima, E. mivati, E. necatrix,* and *E. tenella*	Nonattenuated	Hatchery spray, ocular, drinking water, feed spray	Single dose at 1 to 4 d	Merck
Immucox C	*E. acervulina, E. brunetti, E. maxima, E. necatrix, E. tenella,* and *E. praecox*	Nonattenuated	Drinking water, oral gel	Single dose 1 to 4 d	Ceva
Paracox	*E. acervulina, E. brunetti, E. maxima, E. mitis, E. necatrix, E. praecox,* and *E. tenella*	Attenuated	Drinking water, feed spray	Single dose at 1 to 9 d	Merck
Immunocox5	*E. acervulina, E. brunetti, E. maxima, E. necatrix,* and *E. tenella*	Nonattenuated	Gel	Single dose 1 to 4 d	Ceva

6. Nutritional Intervention for Coccidiosis

6.1. Role of Vitamins

6.1.1. Vitamin D

Vitamin D is a fat-soluble vitamin, and the active metabolites of vitamin D play an intrinsic role in calcium and phosphorus homeostasis and bone mineralization [79–82]. More recently, the immunomodulatory roles of vitamin D in improving the host defense against invading pathogens have been described [83–86]. The absorption of vitamin D is fat-dependent and undergoes hydroxylation into 25-hydroxyvitamin D3 [25(OH)D_3] in the liver [86,87]. Subsequently, 25(OH)D_3 is converted into its active form, 1,25-dihydroxy vitamin D [1,25$(OH)_2D_3$], in the kidney by the enzyme 25-hydroxyvitamin D-1α-hydroxylase. The hydroxylation of 25(OH)D_3 to 1,25$(OH)_2D_3$ is not limited only to the kidney but also occurs in the bone, breast, and thigh muscles, small intestine, and immune cells [88].

The inclusion of 25(OH)D_3 at varying levels in either layer pullets or turkey poults challenged with *Eimeria* reduced BW loss and fecal oocyst shedding, increased macrophage nitric oxide (NO) production, stimulated the activity of innate immune cells, and modulated adaptive immunity (CD4+, CD8+ and CD4+CD25+ T cells) and inflammatory cytokines (IL-1β and IL-10), as well as maintained tight junction [85,86,89]. Furthermore, supplementation of either vitamin D or 25(OH)D_3 or in combination has been shown to improve the FCR and BWG in broilers vaccinated against *Eimeria* spp. at high doses; however, no beneficial effects were observed in reducing intestinal lesions and improving morphometry [90–92]. In broilers challenged with either *E. maxima* or mixed *Eimeria* spp. and fed reduced calcium/phosphorus diets, supplementation of 25(OH)D_3 at the rate of either 3000 IU/kg or 4000 IU/kg has been shown to improve bone mineralization [83,84,93]. Furthermore, in laying hens challenged with lipopolysaccharides, 25(OH)D_3 supplementation improved egg production by improving follicular development and oocyte maturation by increasing the expression of VDR in oocytes, maintaining the plasma estradiol/progesterone and luteinizing hormone levels [94–97]. Moreover, the lipopolysaccharide challenge increased the immune response (CD4+CD25+ T cells, proinflammatory cytokines, and plasma IgM levels) and oxidative stress (SOD, TAC, and GPx) in laying hens, which were normalized by the addition of 25(OH)D_3 [97,98]. Previous studies have reported that inflammation and oxidative stress associated with coccidiosis are responsible for at least a 10% reduction in bone volume, bone mineral content and density, and osteoclastic activity [46,48,99]. These results suggest that vitamin D and its metabolites can potentially reduce the negative impacts of coccidiosis related to growth performance, egg production, oxidative stress, and bone mineralization.

6.1.2. Vitamin E

Similar to vitamin D, vitamin E is a fat-soluble vitamin with a potent antioxidant capacity to protect cells and tissues from lipoperoxidation damage induced by reactive oxygen species [100–102]. Vitamin E from diets gets absorbed and incorporated into cell membranes to protect unsaturated fatty acids inside and outside the cells from reactive oxygen species. Furthermore, vitamin E prevents the excessive generation of reactive oxygen species from respiratory processes [103,104]. In addition, vitamin E is also involved in reducing inflammation, production of cytokine (TNF-α and IL-8) protecting cells of immune responses (lymphocytes, macrophages, and plasma cells), and enhancing the proliferation and functions of immune cells [100,105].

A previous study by Colnago et al. (1984) [106] observed that supplementation of 100 IU/kg vitamin E (DL-α-tocopheryl acetate) in a broiler diet and subsequent challenge by *E. tenella* or *E. maxima* reduced mortality, improved weight gain and the FCR, and reduced lesion scores. Similar results were observed when broiler chicks were challenged twice at 10 and 38 days of age and supplemented with both vitamin A (8 g/kg) and E (300 mg/kg) [107]. However, Allen and Fetterer (2002) [101] conducted two consecutive studies with increasing levels of vitamin E (13–200 IU/kg) in broilers challenged with *E. maxima* and did not observe significant differences in performance except for a slight reduction in lesion score. In broilers fed vitamin E at either 40 or 80 IU/kg alone or in combination with arginine and challenged with mixed *Eimeria* spp., heterophil and monocyte oxidative bursts (ROS production) and NO production decreased at seven days post-inoculation. They also observed lower lesions in challenged birds fed vitamin E, and birds fed a high level of vitamin E (80 IU/kg) in combination with arginine had higher levels of humoral antibodies IgG, IgM, and IGA [108]. Increased humoral and cell-mediated immune response with lower inflammatory mediators were observed in laying hens challenged with *Salmonella* enteritidis and fed diets supplemented with 30 IU/kg vitamin E [109]. Furthermore, da Silva et al. (2011) [110] observed that the inclusion of vitamin E at the rate of 65 mg/kg of diet increased the cell-mediated immune response of chickens vaccinated against coccidiosis and New Castle disease as measured by the cutaneous basophil hypersensitivity test. The antioxidant status (TAC, MDA, GPx activity, or SOD) and plasma lipid peroxidation of the broilers were improved following the *Eimeria* challenge in broilers, but the effects were not enough to improve performance compared with non-challenged broilers [111,112]. Furthermore, vitamin E has been shown to enhance the phagocytic activity of macrophages in chickens against pathogens [100].

In the case of egg layers, vitamin E inclusion has increased egg production by enhancing follicular development by increasing the concentration of reproductive hormones such as follicle-stimulating hormone, luteinizing hormone, estrogens, and progesterone [102].

These results from previous studies showed that vitamin E can be used in laying hens to improve their performance and boost the immune response; however, further studies are needed to confirm the effective dose.

6.1.3. Other Vitamins

Vitamin A is a fat-soluble nutrient constituting a broad range of retinoid compounds and plays a crucial role in various physiological processes, including vision, growth, cell growth and differentiation, reproduction, immunity, skeletal development, and antioxidant properties [113–115]. The association between vitamin A deficiency and the increased incidence and severity of coccidiosis was noted as early as 1945 [116]. Dietary supplementation of vitamin A above the 1960 NRC recommendation (8000 IU/lb) has been shown to enhance broiler performance and the recovery of chicks infected with *E. tenella*, *acervulina*, or *necartix* and results in an improved performance after recovery [117,118]. Furthermore, Dalloul et al. (2002) [113] reported that vitamin A deficiency compromised the intestinal defense mechanism against *E. acervulina* infection, as evidenced by a reduced population of intraepithelial lymphocytes (CD4+ and CD8+ T cells) through alterations in concanavalin A-induced spleen lymphocyte proliferation. Moreover, vitamin A supplementation (12,000 IU/kg) enhanced intestinal morphometry and tight junction integrity in broiler chickens co-infected with *Clostridium* and *Eimeria* [119].

Unlike vitamins A, D, and E, vitamin C is a water-soluble vitamin with a strong antioxidant capacity and acts as a cofactor for collagen biosynthesis, thus maintaining epithelial barrier function and stimulating wound healing [120]. Moreover, it enhances innate immunity by increasing the phagocytic activity of mononuclear cells and adaptive immune responses by differentiation and proliferation of B- and T-cells [121]. Supplementing vitamin C (110–220 ppm) improved weight gain, increased feed intake, reduced mortality, and lowered the corticosterone level and heterophil: lymphocyte ratio in the blood [122,123]. Additionally, concurrent supplementation of vitamin C and protease improved mucin and NO production [124], and vitamin C and arginine or vitamin C and vitamin E in different experiments improved oxidative status but were not able to improve performance in *Eimeria*-infected broilers [112]. The effect of supplementing different vitamins on minimizing the effect of coccidiosis is summarized in Table 3. A hypothesis for using vitamins during coccidiosis could posit that both fat-soluble and water-soluble vitamins may be readily absorbed by the body, potentially exerting beneficial effects on modulating immune responses despite reduced feed intake and nutrient utilization. While there is a notable gap in the available literature concerning the utilization of vitamins to support laying hens during disease conditions, the positive outcomes observed in broilers suggest that similar benefits might extend to laying hens as well. However, further studies are necessary to confirm the optimal dosage for this beneficial effect in laying hens.

Table 3. Effects of vitamin supplementation on performance and intestinal health of chickens infected with *Eimeria* species.

Nutritional Interventions	Breed	*Eimeria* Infection	Dosage	Impacts on Performance	Impact on Health	Reference
25-hydroxycholecalciferol	White Leghorn	*E. acervulina*, *E. maxima*, and *E. tenella* (Inovocox)	4000 IU/kg	Improved performance	• ↓ CD8+ cells. • ↑ CD4+CD25+ cells. • ↓ IL-1β expression. • ↑ IL-10 expression.	[85]
Vitamin D_3 + 25-hydroxycholecalciferol	Cobb 700	7× live vaccine (*E. acervulina*, *E. maxima*, and *E. tenella*)	2000–8000 IU/Kg	Improved BW, BWG, and FCR	• ↑ Duodenal morphology. • ↑ Tibia breaking strength. • ↑ Tibial bone mineralization.	[91]
Vitamin D_3 or 25-hydroxycholecalciferol	Ross 308	*E. maxima* (7000)	4000 IU/Kg	Improved BW, BWG, and FCR	• ↑ Bone breaking strength. • ↑ Bone mineralization. • ↓ Jejunal morphology.	[84]
Vitamin D_3 or 25-hydroxycholecalciferol	Ross 308	*E. maxima* (7000)	4000 IU/kg	No effect on performance	• ↑ Bone breaking strength. • ↑ Ash content and percentage.	[83]
Vitamin D_3 or 25-hydroxycholecalciferol	Cobb 500	2× live vaccine	1375 or 2750 IU/kg	Improvement in BW and FCR	• ↓ Intestinal morphology. • ↑ Tibia breaking strength. • ↑ Bone ash %.	[92]
25-hydroxycholecalciferol	Cobb 500	*E. acervulina* (62,500), *E. tenella* (12,500), and *E. maxima* (12,500)	3000 IU/kg	No effect on performance	• ↑ Bone ash. • ↑ Bone mineral content of the birds. • ↑ Tissue and bone volume of the femur.	[93]
DL-alpha tocopheryl acetate	Hubbard	*E. tenella* (150,000) or *E. maxima* (50,000)	100 IU/Kg	Improved BWG, FI, and FCR	• ↓ Mortality. • ↓ Lesion scores. • ↑ Packed cell volume.	[106]
DL-alpha tocopheryl acetate	Ross	*E. maxima* (175,000 or 40,000)	13–200 ppm	No effect on performance	• ↓ Lesion scores. • ↑ Plasma NO.	[101]
DL-alpha tocopheryl acetate	Cobb 500	*E. acervulina* (100,000), *E. maxima* (60,000), and *E. tenella* (40,000): Coccivac-B	40 or 80 IU/kg	-	• ↑ Heterophil and monocyte oxidative burst. • ↑ IgG, IgM, and IgA production. • ↓ Lesion scores.	[108]
D-α-tocopherol	Ross 308	Livacox Q vaccination	65 mg/kg	-	• ↑ Cell-mediated immunity.	[110]
D-α-tocopherol	Ross 308	*E. tenella*	100, 316, or 562 mg/kg	-	• ↓ MDA concentration.	[111]
D-α-tocopherol and L-ascorbic acid	Cobb 500	100× Advent vaccine	80 mg/kg VE + 1 g/kg VC	Improved BWG	• ↑ Nitric oxide production. • ↑ GPx activity. • ↓ Lesion scores.	[112]
L-ascorbic acid	Vantress-White Plymouth Rock cross	*E. maxima* (30,000), *E. brunetti* (100,000), *E. necatrix* (30,000), or *E. tenella* (30,000)	110 mg/kg	No effect on performance	• Increased mortality.	[123]
L-ascorbic acid	Hubbard	*E. tenella* (300,000)	150 or 300 mg/kg	Improved BW, FI, and FCR	• ↓ Plasma corticosterone. • ↓ Heterophil/lymphocyte ratio.	[122]

Table 3. *Cont.*

Nutritional Interventions	Breed	*Eimeria* Infection	Dosage	Impacts on Performance	Impact on Health	Reference
Vitamin A	Plymouth Rock		8000 IU/lb	Improved BWG	• ↑ Recovery.	[117]
Vitamin A	Ross 308	*E. acervlina* (10,000)	VE deficiency	-	• ↓ CD4+, CD8+ cells. • ↑ Oocyst shedding. • ↓ IFN-γ. • ↓ Lymphocyte proliferation.	[113]
Vitamin A	Broilers	*E. maxima* + *Clostridium perfringens*	12,000 IU/kg	Improved BW, BWG, and FI	• ↑ Intestinal morphology. • ↑ Expression of tight junction proteins.	[119]
Menadione sodium bisulfite	Lancaster X New Hampshire	*E. tenella* (500,000)	1.06 or 9.72 g/ton	-	• ↓ Mortality.	[125]

IU: international unit; BW: body weight; BWG: body weight gain; FI: feed intake; FCR: feed conversion ratio; NO: nitric oxide; IL-1β: Interleukin 1β; IL-10: Interleukin 10; IFN-γ: Interferon γ; IgA: immunoglobin A; IgG: immunoglobin G; IgM: immunoglobin M; ppm: parts per million; VE: vitamin E; VC: vitamin C; MDA: malondialdehyde; GPx: glutathione peroxidase.

6.2. Role of Functional Amino Acids

6.2.1. Arginine

Arginine, an essential functional amino acid, is vital in protein synthesis and accretion in chickens. Additionally, it contributes to secondary functions such as immune modulation, wound healing, and antioxidant activity [126–129]. In response to inflammation, arginine is metabolized to nitric oxide (NO) by nitric oxide synthase, which serves as an immune modulator, antioxidant defense, and cytotoxic mediator for non-specific host defenses [126–128]. Additionally, metabolites derived from arginine, such as proline, hydroxyproline, and polyamines, have been demonstrated to promote cell proliferation (Figure 2). Furthermore, ornithine and proline are primary amino acids found within collagen that are essential for wound healing and skeletal growth [126–128,130,131].

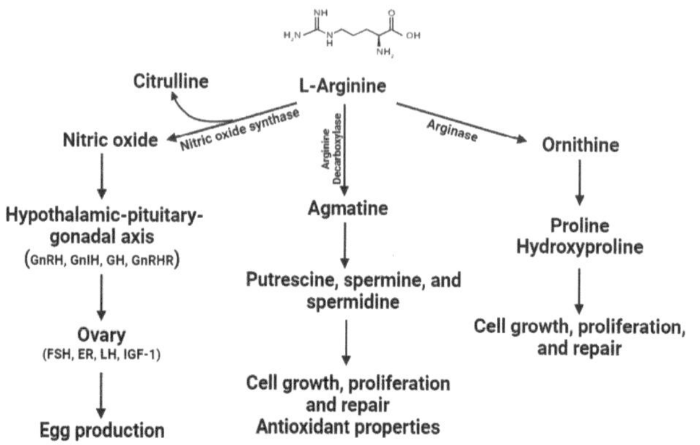

Figure 2. Effect of L-arginine and its metabolites on gut health and reproductive functions in laying hens. FSH: follicular-stimulating hormone; ER: estrogen; LH: luteinizing hormone; IGF: insulin-like growth factor 1.

The secondary functions of arginine during coccidiosis have been extensively investigated in recent years and have shown beneficial effects in chickens. An early study by Allen and Fetterer (2000) [132] reported that *E. acervulina* infection reduced the plasma concentration of arginine, whereas its metabolite NO was significantly increased. The hypothesis for using arginine above the requirement could be that additional arginine might aid in balancing amino acids in plasma and provide a surplus for secondary functions. However, supplementation of arginine (500 mg/kg) did not reverse the negative effect associated with *E. acervulina*, *maxima*, or *tenella* infection except for a reduction in oocyst shedding of *E. tenella* [133]. Increasing dietary arginine concentration above the NRC requirement improved gut health by increasing villus height and goblet cell count and increasing mTOR pathways for healing [134,135]. The intestinal integrity of infected birds was improved when supplementing arginine, ranging from 1.24 to 1.48%, as observed by the upregulation of tight junction protein, permeability, and intestinal morphology [136,137]. Furthermore, arginine supplementation also reduced the expression of TLR-4 and IL-1β in broilers along with modulating the humoral immune response (secretory IgG and IgA) in broilers challenged with 20× Coccivac-B vaccine [134]. In addition, Yazdanabadi et al. (2020a; b) [138,139] reported that arginine supplemented 25–50% over the requirement increased the improved BWG, NO, and proinflammatory concentration along with gut morphometry. Furthermore, supplementation of arginine above the requirements has been found to improve the oxidative status of chickens following *Eimeria* challenge by GPx activity, reducing MDA and maintaining the levels of GSH and TAC [136,137,140].

In the case of laying hens, there is a lack of evidence for arginine supplementation and *Eimeria* challenge; however, in lipopolysaccharide-induced immune suppression, arginine supplementation above the requirements reduced the expression of inflammatory cytokines such as IL-1β, IL-10, TLR4, and NF-κB [141]. Furthermore, NO, a metabolite of arginine, plays a significant role in the hypothalamus–ovarian axis and plays a vital role in laying hen performance as well [142–144]. Based on previous studies in broilers and laying hens, supplementation of L-arginine above the requirements can be used in laying hens challenged with *Eimeria* to alleviate the symptoms associated with it.

6.2.2. Methionine

Methionine is considered the first limiting amino acid in corn- and soy-based diets for laying hens and broilers and the first amino acid to be incorporated into the polypeptide chain during translation [126,145]. Aside from protein synthesis, methionine has antioxidant functions, acting as a precursor of cysteine, which is essential for GSH synthesis [128,146]. Both GSH and cysteine protect the cells against ROS [146]. Because of its role in balancing oxidative status through GSH and cysteine, its efficacy has been tested in broilers exposed to coccidiosis. A study by Pourali et al. (2014) [147] reported that supplementing 150% of total sulfur amino acids improved BWG, oocyst shedding, and hepatic malondialdehyde concentration; however, no effect was observed in GPx in broilers challenged with mixed *Eimeria* species. A study by Lai et al. (2018, 2023) [148,149] reported that supplementing extra methionine above the requirement in vaccinated birds did not have any beneficial effect on the performance, immune response, or oxidative status of Partridge shank broilers. However, reducing the level of methionine in the diets during *Eimeria* infection significantly affects the performance, oxidative status, and immune responses in challenged broilers [31,47,150,151], and supplementing 0.75% methionine in low-crude protein diets increased the detrimental effect of *Eimeria* infection on broiler health and performance [6]. In laying hens under heat stress, similar results were observed as those for *Eimeria*-challenged broilers, where reducing the methionine level reduced the production performance and bone mineralization and increased the oxidative stress of laying hens [152]. The role of TSAA is evident in maintaining performance, immune response, and oxidative status during *Eimeria* infection; however, it is still unclear if supplementing methionine above the requirement has a beneficial effect on laying hens under coccidiosis.

6.2.3. Other Amino Acids

Glutamine, a conditionally essential amino acid, is the primary fuel for immune and intestinal epithelial cells and a precursor for GSH synthesis [153]. Supplementation of glutamine (0.5% or 1%) in coccidia-challenged broilers has been shown to reduce the expression of inflammatory cytokines IL-10 and IFN-γ, whereas it improves intestinal integrity and morphometry [154]. Furthermore, supplementing 0.75% glutamine in low-crude protein diets improved BWG during the recovery phase and maintained the expression of Claudin-1 compared to that of a normal protein diet in *Eimeria*-infected broilers [6].

Threonine is another essential amino acid and is essential for mucin and antibody (IgA, IgM) production in the gastrointestinal tract [155]. The intestinal mucus layer acts as the first line of defense against invading pathogens [155]. Therefore, factors increasing mucin production or intestinal secretions, as observed in coccidiosis, might increase threonine requirements during infection [156]. Supplementing 124% of threonine to its requirement in broilers challenged with mixed *Eimeria* species improved performance to the level of non-challenged birds, intestinal morphometry, oocyst shedding, and humoral immune response as measured by higher antibody production [157]. In contrast, threonine deficiency worsened the effects of coccidiosis in broilers challenged with coccidiosis by impairing intestinal morphometry and integrity, inflammatory responses, and lymphocyte population [158]. The effect of supplementing functional amino acids on the performance and intestinal health of chickens under coccidiosis infection is summarized in Table 4. Although these functional amino acids have shown beneficial effects in broilers under coccidiosis and in laying hens under normal conditions, their beneficial effect in laying hens under diseased conditions has not been fully explored. These amino acids have shown potential in mitigating the negative effect of coccidiosis in broilers and might have the same functions in laying hens, which need to be explored.

Table 4. Effects of functional amino acids supplementation on the performance and intestinal health of chickens infected with *Eimeria* species.

AA	Breed	*Eimeria* Challenge	AA Dosage	Impact on Performance	Impact on Health	Reference
Arginine	SexSal	*E. maxima* (30,000), or *E. tenella* (50,000), or *E. acervulina* (500,000)	+500 or +1000 mg/kg	No effect on BWG	• ↓ Oocyst shedding of *E. tenella*.	[108]
	Cobb 500	*E. acervulina* (100,000), *E. maxima* (60,000), and *E. tenella* (40,000): Coccivac-B	Added 0.3% or 0.6% to make 1.74% and 2.04% (NRC, 1994)		• ↑ Heterophil and monocyte oxidative burst. • ↑ IgG, IgM.	[108]
	Ross 708	20× Coccivac-B	106·4% and 160·8% of (NRC, 1994)	Improved BWG and FCR	• ↑ Villus height and crypt depth. • ↑ Goblet cell counts and density. • ↓ Expression of TLR4, IL-1β.	[141]
	Cobb 500	100× Advent coccidiosis vaccine	1.80% (NRC 1994)	No effect on BWG	• ↑ NO production. • ↑ GPx activity. • ↓ Intestinal lesions.	[135]
	Ross 308	*E. acervulina* (150,000), *E. tenella* (15,000), *E. maxima* (15,000), and *E. necatrix* (15,000)	105 and 110% (breeder recommendation)	No effect on BWG or FCR	• ↓ Crypt depth and oocyst shedding.	[135]
	Ross 308	*E. acervulina* (3.5×10^5)	Low arginine (0.74%)	Reduced BWG, FI, and FCR	• ↓ Plasma Arg: Lys ratio.	[129]
			High arginine (1.23%)	Improved BWG, FI, and FCR	• ↑ Plasma Arg: Lys ratio. • ↑ NO production.	
	Cobb 500	*E. acervulina* (62,500), *E. tenella* (12,500), and *E. maxima* (12,500)	100%, 108%, and 116% (breeder recommendation)	Improved BWG, FI, and FCR	• ↓ Gut permeability. • ↑ Tight junction proteins. • ↑ VH and reduced CD. • ↑ NO production. • ↓ SOD activity.	[136]
	Ross 308	*E. necatrix* (15,000), *E. maxima* (20,000), *E. acervulina* (15,000), and *E. tenella* (170,000)	125 and 150% (breeder recommendation)	Improved BWG, FI, and FCR	• ↓ IL-1β, IL-2, IL-6, IFN-γ, and TNF-α. • ↑ NO production. • ↓ Oocyst count.	[139]
	Cobb 500	*E. acervulina* (62,500), *E. tenella* (12,500), and *E. maxima* (12,500)	16% CP plus 0.75% ARG	No effect on performance	• ↓ Gut permeability. • ↑ VH and AA digestibility.	[159]
	Cobb 500	*E. acervulina* (62,500), *E. tenella* (12,500), and *E. maxima* (12,500)	150% (breeder recommendatio)		• ↓ Gut permeability. • ↑ Tight junction proteins. • ↓CD8+:CD4+. • ↑ NO production.	[140]

Table 4. *Cont.*

AA	Breed	*Eimeria* Challenge	AA Dosage	Impact on Performance	Impact on Health	Reference
TSAA	Ross 308	750 *E. acervulina*, *E. tenella*, and *E. maxima*	150% TSAA (breeder recommendation)	Improved BWG	• ↓ Oocyst shedding. • ↓ NO production. • ↓ MDA concentration.	[147]
	Partridge Shank broilers	50,000 *E. tenella*	125% and 150% (breeder recommendation)	No effect on BWG; ADFI 150% methionine reduced BWG	• ↓ VH and lesion score. • ↓ CD4+ and CD8+ cells. • ↓ IFN-γ.	[151]
		100 × Advent vaccine (*E. acervulina*, *E. maxima*, and *E. tenella*)	0.6, 0.8, 0.9, and 1.0% TSAA	Below the requirement reduced BWG, ADFI, and gain: feed	• ↓ TSAA reduced IgA production.	[151]
	Cobb 500	*E. acervulina* (62,500), *E. tenella* (12,500), and *E. maxima* (12,500)	16% CP + 0.75% Met	Negatively affected BWG, FI, and FCR	• ↑ Oocyst shedding. • ↓ Intestinal morphology.	[159]
Threonine	New Hampshire × Colombian	1500 *E. maxima*	125% of threonine (breeder recommendation)	Improved BWG, ADFI, and FCR	• ↓ IL-1β. • ↑ IL-12.	[160]
	Ross 708	Coccivac®-B	61.25% (NRC, 1994)	Exaggerated performance	• ↑ Oocyst shedding. • ↑ Gut permeability, IgA production. • ↓ Intestinal morphology. • ↓ Goblet cell counts. • ↓ Lymphocytes expressing CD4+, CD8+, CD3 cells.	[158]
	Cobb 500	Mixed *Eimeria* oocysts	112%, 124%, and 136% (breeder recommendation)	Improved BWG and FCR	• ↑ Intestinal morphology. • ↓ Oocyst shedding. • ↑ IgG and IgM production.	[157]
	Cobb 500	*E. acervulina* (62,500), *E. tenella* (12,500), and *E. maxima* (12,500)	16% CP and 0.75% threonine	Improvement in BWG, FI, and FCR	• ↓ Oocyst shedding. • ↑ Intestinal morphology. • ↓ Claudin-1 expression.	[159]
Glutamine	Cobb 500	20 × dose of Coccivac B	0.5% and 1% glutamine	No effect on performance	• ↓ Inflammatory and proinflammatory cytokines (IL-10, IFN-γ). • ↑ Expression of tight junction proteins. • ↓ Intestinal CD and ↑VH.	[154]

AA: amino acid; TSAA: total sulfur amino acid; CP: crude protein; BWG: body weight gain; ADFI: average daily feed intake; FCR: feed conversion ratio; NO: nitric oxide; IL-1β: Interleukin 1 β; MDA: malondialdehyde; GPx: glutathione peroxidase; SOD: superoxide dismutase; TLR-4: toll-like receptor-4; IFN-γ: Interferon γ; IL-2: Interleukin 2; IL-6: Interleukin 6; IL-10: Interleukin 10; IL-12: Interleukin 12; VH: villus height; CD: crypt depth; IgG: immunoglobin G; IgM: immunoglobin M.

7. Role of Phytogenic Feed Additives

Phytogenic feed additives (PFAs) are plant-derived natural bioactive compounds or products that, when fed to animals, have a beneficial effect on performance and health [161]. Phytogenic feed additives have a wide range of bioactive compounds with antimicrobial, antioxidant, or anti-inflammatory properties and are used in traditional human medicines [161,162]. Phytochemicals are generally recognized as safe in the United States, indicating their safety for consumption. This designation strengthens the potential utilization of phytochemicals in poultry production for coccidiosis control. In the poultry industry, PFAs are gaining considerable attention mainly because of improvements in the performance of birds by improving gastrointestinal health alongside antioxidative and immunomodulatory effects [161]. The efficacy of various phytogenic feed additives, including *Artemisia annua*, curcumin, oregano, thyme, and their essential oils, has been investigated in broiler chickens infected with coccidiosis, demonstrating some beneficial effects.

Dietary inclusion of curcumin powder (100–200 mg/kg) has been shown to reduce the lesion score and improve the oxidative status of broilers challenged with mixed *Eimeria* spp. [43]. Furthermore, essential oils of oregano have been shown to improve BWG and the FCR, reduce intestinal lesions and oxidative stress, and improve the gut morphology in broilers infected with coccidiosis [163,164]. The beneficial effects of these PFAs include supporting the host by immunomodulatory effects and providing protection against free radicals by scavenging reactive oxygen species and interfering directly with parasitic metabolism, reducing oocyst shedding and cecal short-chain fatty acid production [45,163,165]. Furthermore, Felici et al. (2023) [166] reported that bioactive compounds from a PFA can inhibit the intracellular replication of *Eimeria* and reduce schizont numbers.

Artemisinin, a bioactive flavonoid found in *Artemisia annua* leaves (AA), has been shown to inhibit the growth of several stages of *Plasmodium* spp. [167,168]. The use of AA and its extract artemisinin has been shown to demonstrate anticoccidial effects against *E. tenella* [169–173] when infected alone. The use of either dried AA leaves or artemisinin has been shown to improve lesion scores, reduce oocyst shedding and sporulation, and modulate the humoral and immune response in chickens [169,174–178]. The mechanism of action of artemisinin is promoting apoptosis of infected host cells, thus neutralizing parasites [172]. Furthermore, artemisinin from AA is able to alter the cell wall formation of the oocysts, leading to the death of developing oocysts and a reduced sporulation rate [170]. The use of phytogenic feed additives in laying hens infected with coccidiosis has not been studied. However, results from laying hen studies reported that PFA inclusion in diets improved the performance, immune response, and antioxidant status of laying hens [179–183]. Since coccidiosis mainly affects the performance of birds with immune suppression and increased oxidative stress, PFAs can be helpful in laying hens infected with coccidiosis, as in broilers.

8. Role of Prebiotics, Probiotics, and Symbiotics

8.1. Probiotics

Probiotics are selective nonpathogenic microorganisms that, when administered in adequate amounts, offer a beneficial advantage to the host and improve gut functions by altering the gut microflora and reducing pathogenic bacteria colonization in the gastrointestinal tract [184–187]. By selectively eliminating pathogenic microorganisms from the GI tract through competitive exclusion, probiotics promote the growth of beneficial bacteria. Additionally, their metabolites, including short-chain organic fatty acids and hydrogen peroxide, exhibit antimicrobial properties, effectively inhibiting the growth of pathogenic bacteria [188,189]. Probiotic microorganisms help develop immune components in the GI tract and innate or adaptive immune responses and modulate the phosphorylation of cytoskeletal and tight junction proteins, improving the intestinal barrier [188,190,191]. Some of the commonly used probiotics in the poultry industry include *Bacillus*, *Lactobacillus*, *Enterococcus*, *Bifidobacterium*, and *Lactococcus*, and yeasts such as *Aspergillus*, *Candida*, and *Saccharomyces* [188,192].

The inclusion of *Bacillus* spp. in diets has been shown to improve the performance, tight junction integrity of the intestine, and immune response as well as positively influenced the cecal microbiome of broilers challenged with coccidiosis [193–196]. Similar results were observed in birds fed lactobacillus-based probiotics, including reduced mortality and oocyst shedding, upregulation of intestinal integrity, increased intestinal intraepithelial lymphocytes expressing CD4+ and CD8+ cells, and antibody titers in broilers challenged with coccidiosis [197–199]. In the case of laying hens, the dietary inclusion of probiotics (*Saccharomyces*, *Pediococcus*, *Lactobacillus*, or *Bacillus*), either alone or in combination, has improved the performance, oxidative status, immune responses, intestinal morphometry, and microbial composition of ceca positively in a non-challenge model [197,200,201]. The effectiveness of probiotics (*Bacillus amyloliquefaciens*, *Bacillus licheniformis*, and *Bacillus pumilus*) against *Salmonella* Enteritidis in laying hens and observed a significant reduction in salmonella colonization in ceca of mature laying hens [202,203].

8.2. Prebiotics

Prebiotics are non-digestible or selectively fermentable feed ingredients that, when incorporated into diets, are utilized by the host intestinal microbiota, selectively promoting the growth or activity of specific bacterial populations, positively influencing the microbiome, and improving the gut health of the host [18,204,205]. While selecting prebiotics, it is essential to consider (i) digestibility (non-digestible by host enzymes), (ii) absorption (should not be absorbed directly by host cells), (iii) selective fermentation by intestinal microbiota, (iv) selective promotion of the growth of beneficial bacterial population, and (v) stimulation of the immune response of the host [188,204–206]. Some of the prebiotics commonly used in the poultry industry are oligosaccharides (mannan-, galacto-, and xylo-oligosaccharides), β-glucan, and fructans [188,206–208].

The efficacy of prebiotics in improving the gut health of poultry has been investigated using both challenged and unchallenged models [209–213]. In experimentally infected broilers with coccidiosis, the administration of chitosan oligosaccharide (1 g/kg) improved various parameters, including BWG, FCR, oocyst shedding, intestinal tight junctions, and morphometry and reduced intestinal inflammation [214]. Similarly, broilers fed mannooligosaccharides (0.8 g/kg), xylooligosaccharides (0.5 g/kg), galactoglucomannan oligosaccharide (4%), or yeast cell wall polysaccharides (0.5 g/kg) exhibited improved performance, intestinal tight junction integrity and nutrient digestibility and mitigation of hostile cecal microbial populations and fermentation induced by *Eimeria* infection [209–211,215].

Feeding prebiotics such as dry whey powder (6 g/kg), mannan oligosaccharides (0.25–2 kg/kg), or fructooligosaccharides (0.25–1 g/kg) to laying hens has shown positive effects on their intestinal microbial populations. Specifically, it promotes beneficial bacteria like *Lactobacillus* spp. and *Olensella* spp. while reducing the abundance of harmful bacteria such as *Clostridium perfringens*, *Escherichia coli*, and *Salmonella enteritidis* [216–218]. Additionally, oligosaccharide supplementation has been linked to improved performance, digestibility, and upregulation of toll-like receptor-4, interferon-γ, and antibody production in laying hens [217,218].

8.3. Synbiotics

Synbiotics refer to the combined application of both prebiotics and probiotics, which exhibit synergistic effects on the host by enhancing gut health, the immune response, and microbial balance [219]. The rationale behind synbiotic uses lies in the premise that prebiotics facilitate the survival and colonization of probiotics in the host's gastrointestinal tract, thereby positively influencing the health of the host [188,219]. In broiler chickens, both in ovo and in vivo studies have demonstrated the beneficial effects of synbiotics on performance, gut health, and the immune response, irrespective of whether the birds were challenged with coccidiosis or necrotic enteritis [220–225]. Similarly, in laying hens, supplementation with synbiotics containing both prebiotics and probiotics has been shown to enhance production performance, decrease levels of inflammatory cytokines, and increase populations of beneficial bacteria in the intestinal tract [226,227]. The efficacy of probiotics, prebiotics, and synbiotics in chickens infected with coccidiosis is summarized in Table 5. The effect of probiotics, prebiotics, and synbiotics in laying hens has been investigated in both normal and diseased (Salmonellosis) conditions and has shown positive impacts on the performance and wellness of hens. Although they have shown beneficial effects in broilers under coccidiosis, their effect in laying hens infected with *Eimeria* spp. hens has not been tested yet.

Table 5. Effects of prebiotics, probiotics, and synbiotics supplementation on performance and intestinal health of chickens infected with *Eimeria* species.

Nutritional Interventions	Breed	*Eimeria* Infection	Dosage	Impacts on Performance	Impact on Health	Reference
Galacto-glucomannan oligosaccharide-arabinoxylan	Ross × Ross	*E. acervulina* (1,000,000)	1, or 2, or 4%	Improved FI but not BW	• ↓ Propionate in ceca. • ↑ IFN-γ, IL-1β, IL-6, IL-12 expression. • ↑ *Bifidobacterium* spp., *Lactobacillus* spp. in ceca.	[209]
Mannan-oligosaccharides	Ross 308	*E. tenella* (20,000–30,000)	0.8 g/kg	Improved BWG, FI, and FCR	• ↓ Oocyst shedding. • ↓ Cecal lesions.	[215]
Chitosan-oligosaccharide	Cobb 500	15× Coccivac®-B-52 (*E. acervulina*, *E. maxima*, *E. maxima MF*, *E. mivati*, and *E. tenella*)	1 g/kg	Improved BW, BWG, FI, and FCR	• ↓ Oocyst shedding. • ↑ Intestinal morphology. • ↑ Expression of tight junction protein. • ↓ Expression of TNF-α, IFN-γ, and TLR-4. • ↑ Expression of IL-6, IL-10, and IL-1β.	[214]
Nucleotide-rich yeast extract	Ross 708	*E. acervulina* (25,000) and *E. maxima* (5000)	0.5 g/kg	Improved BW, BWG, and FCR	• ↑ Jejunal morphology. • ↓ Cecal SCFA concentration.	[210]
Xylo-oligosaccharides	Ross 308	12× Paracox 8	0.025%	Improved BWG and FI	• ↑ Propionic and butyric acid in ceca. • ↓ %G+C profile of cecal bacteria.	[228]
Xylo-oligosaccharides	Cobb 500	*E. acervulina* (62,500), *E. tenella* (12,500), and *E. maxima* (12,500)	0.5 or 1 g/kg	No effect on performance	• ↓ Duodenal lesions. • ↓ CLDN-1 overexpression. • ↓ Branched-chain fatty acids.	[211,229]
Yeast cell wall	Ross 308	*E. tenella* (5000)	0.1 or 0.2%	Improved BWG, FI, and FCR	• ↑ Jejunal morphology. • ↓ Bursa follicle length and area.	[230]
Lactobacillus spp.	Ross 308	*E. acervulina* (10,000)	1 g/kg	-	• ↓ Oocyst shedding. • ↑ Intraepithelial lymphocytes expressing CD4, CD8 cells.	[199]
Saccharomyces cerevisiae	Broilers	*E. tenella*, *E. maxima*, and *E. necatrix* (70,000 oocysts total)	0.1, or 1, or 10 g/kg	Improved BWG and FCR	• ↑ Cellular immune response. • ↑ IgM and IgG concentration.	[197]
Lactobacillus salivarius and *L. jhonsonii*			10^6, or 10^7, or 10^8 CFU/L drinking water		• ↓ Oocyst shedding and lesion score.	
Bacillus subtilis 747	Ross 708	*E. maxima* (10,000)	1.5 × 10^5 CFU/g feed	Improved BWG and FCR	• ↓ Oocyst shedding and lesion score. • ↓ IL-1β, IL-6, IL-2, and IFN-γ. • ↑ Expression of tight junction protein.	[195]
Bacillus licheniformis-A *Bacillus amyloliquefaciens*-B *Bacillus amyloliquefaciens*-D	Ross × Ross	*E. tenella* (5000), *E. maxima* (5000), and *E. acervulina* (5000)	1.5 × 10^5 CFU/g of feed	Improved performance	• ↓ Oocyst shedding and lesion score; • ↑ IL-6, IL-8, and IL-10. • ↑ JAM-2.	[196]
Bacillus subtilis (BS-9)	Ross 708	*E. maxima* (25,000) and *E. acervulina* (100,000)	10^8 CFU/bird/d in drinking water	No effect on BW, FI, and FCR	• ↑ Intestinal morphology. • ↓ Bursa weight.	[231]

Table 5. *Cont.*

Nutritional Interventions	Breed	*Eimeria* Infection	Dosage	Impacts on Performance	Impact on Health	Reference
Lactobacillus rhamnosus	Yellow broilers	*E. acervulina*, *E. maxima*, and *E. tenella* (100,000)	10^6, or 10^8 CFU in ovo	-	• ↓ Internal organ growth. • ↓ Intestinal morphology. • Exaggerated coccidia infection.	[232]
Lactobacillus plantalum + *Shallot extract*	Broilers	*E. tenella* (20,000)	10^6 cfu/mL of Lactobacillus plantalum + 2% Shallot extract in water	-	• ↓ Cecal lesion score.	[220]
Lactobacillus reuteri, *Enterococcus faecium*, *Bifidobacterium animalis*, *Pediococcus acidilactici*, and a fructooligosaccharide	Turkey poult	*E. adenoides* and *E. meleagrimitis*	-	Improved BWG	• ↓ Oocyst shedding and lesion score.	[221]
Postbiotics (Mantol)	Ross 708	*E. maxima* (10,000)	10 mg/kg	No effect on performance	• ↓ Oocyst shedding and lesion score. • ↓ IL-1β, IL-6, IL-17, IL-10, and IFN-γ.	[233]

BW: body weight; BWG: body weight gain; FI: feed intake; FCR: feed conversion ratio; NO: nitric oxide; IL-1β: Interleukin 1-β; IL-6: Interleukin 6; IL-10: Interleukin 10; IL-12: Interleukin 12; TLR-4: toll-like receptor-4; IFN-γ: Interferon γ; TNF-α: tumor necrosis factor-α; SCFA: short chain fatty acid; CLDN-1: Claudin -1; IL-2: Interleukin 2; IL-8: Interleukin 8; JAM-2: junction adhesion molecule-2; CFU: colony-forming unit.

8.4. Postbiotics

In recent years, it has been established that the beneficial effects of probiotics are not only limited to when the microorganisms are alive but also after their death, leading to the development of new antibiotic alternatives, named postbiotics [234,235]. Postbiotics are non-viable bacterial cells or cell walls or metabolites derived from probiotics that have biological activity and offer physiological benefits to the host [235]. Different bioactive metabolites from probiotics with biological activity may include a range of compounds such as short-chain fatty acids, peptides, enzymes, polysaccharides, vitamins, cell surface proteins, and organic acids [235]. Recent studies conducted in chickens have shown that postbiotics confer beneficial effects on the host by positively modulating the intestinal microbiome, immune responses, and oxidative status [233,236–238]. In laying hen pullets challenged with *Salmonella* spp., dietary inclusion of postbiotics from *Saccharomyces cerevisiae* (1–1.5 kg/MT) or oral inoculation of *Lactobacillus* spp. postbiotics reduced the intestinal colonization of *Salmonella* [239,240]. Furthermore, *Lactobacillus*-derived postbiotics have been shown to enhance the performance, oxidative status, and intestinal morphology of broilers under heat stress [236,241]. Additionally, in broilers with experimentally induced necrotic enteritis, postbiotics from *Enterococcus* spp. *Pediococcus* spp., *Lactobacillus* spp., *Enterococcus* spp., and *Lactiplantibacillus* spp. were able to improve performance, lesion scores, and oxidative status and reduce proinflammatory responses [237,238,242]. Moreover, the inclusion of maltol (a metabolite of Bacillus subtilis) in a broiler diet infected with *E. maxima* was able to reduce intestinal damage and inflammatory response [233]. Although the use of postbiotics in broilers and laying hens infected with *Eimeria* spp. has not been extensively explored, previous studies have demonstrated their efficacy in reducing the symptoms associated with coccidiosis in other stress conditions.

9. Conclusions

To conclude, this review summarizes the effect of coccidiosis on laying hen health and performance and its potential to cause a significant economic loss to the egg industry. Since the ban on antibiotics as growth promoters in animal production, there has been an increase in the incidences of economically important diseases in poultry, such as coccidiosis and necrotic enteritis. Several antibiotic alternatives (vitamins, functional amino acids, phytogenic feed additives, prebiotics, probiotics, synbiotics, and postbiotics) have been tested in broilers with positive effects on minimizing the effect of coccidiosis, but their beneficial effects have not been explored in laying hens. These nutritional strategies have been shown to have the potential to mitigate the negative effects of coccidiosis in laying hens as well. However, before applying these nutritional strategies either to boost immunity after vaccination or to mitigate the negative effects of coccidiosis, they need to be tested in laying hens. Additionally, while implementing these nutritional strategies to mitigate the adverse effects of coccidiosis, it is important to consider that these strategies are not capable of curing coccidiosis completely, and poultry producers must evaluate the severity of the conditions. Future studies should focus on evaluating nutritional strategies either alone or in combination, focusing on performance, gut health, and immune responses against coccidiosis in laying hens.

Author Contributions: Conceptualization M.K.S. and W.K.K.; writing—review and editing, M.K.S. and W.K.K.; supervision, W.K.K.; funding acquisition, W.K.K. All authors have read and agreed to the published version of the manuscript.

Funding: This research received no external funding.

Conflicts of Interest: The authors declare no conflicts of interest.

References

1. Blake, D.P.; Knox, J.; Dehaeck, B.; Huntington, B.; Rathinam, T.; Ravipati, V.; Ayoade, S.; Gilbert, W.; Adebambo, A.O.; Jatau, I.D.; et al. Re-Calculating the Cost of Coccidiosis in Chickens. *Vet. Res.* **2020**, *51*, 115. [CrossRef] [PubMed]
2. Conway, D.P.; Mckenzie, M.E.; Conway, D.P.; Mckenzie, M.E. *Poultry Coccidiosis*; Wiley-Blackwell: Oxford, UK, 2007; ISBN 9780813822020.
3. Su, S.; Miska, K.B.; Fetterer, R.H.; Jenkins, M.C.; Wong, E.A. Expression of Digestive Enzymes and Nutrient Transporters in *Eimeria acervulina*-Challenged Layers and Broilers. *Poult. Sci.* **2014**, *93*, 1217–1226. [CrossRef] [PubMed]
4. Abdisa, T.; Hasen, R.; Tagesu, T.; Regea, G.; Tadese, G. Poultry Coccidiosis and Its Prevention, Control. *J. Vet. Anim. Res.* **2019**, *2*, 1–6.
5. Swayne, D.E.; Glisson, J.R.; McDougald, L.R.; Nolan, L.K.; Suarez, D.L.; Nair, V. *Diseases of Poultry*; Blackwell Pub: Oxford, UK, 2019; ISBN 9781119371199.
6. Teng, P.Y.; Choi, J.; Tompkins, Y.; Lillehoj, H.; Kim, W. Impacts of Increasing Challenge with Eimeria Maxima on the Growth Performance and Gene Expression of Biomarkers Associated with Intestinal Integrity and Nutrient Transporters. *Vet. Res.* **2021**, *52*, 1–12. [CrossRef] [PubMed]
7. Teng, P.Y.; Yadav, S.; Castro, F.L.d.S.; Tompkins, Y.H.; Fuller, A.L.; Kim, W.K. Graded Eimeria Challenge Linearly Regulated Growth Performance, Dynamic Change of Gastrointestinal Permeability, Apparent Ileal Digestibility, Intestinal Morphology, and Tight Junctions of Broiler Chickens. *Poult. Sci.* **2020**, *99*, 4203–4216. [CrossRef] [PubMed]
8. Chapman, H.D. *Coccidiosis in Egg Laying Poultry*; Elsevier Inc.: Philadelphia, PA, USA, 2017; ISBN 9780128011515.
9. Chapman, H.D. Applied Strategies for the Control of Coccidiosis in Poultry. *CAB Rev. Perspect. Agric. Vet. Sci. Nutr. Nat. Resour.* **2018**, *13*, 1–11. [CrossRef]
10. Soares, R.; Cosstick, T.; Lee, E.H. Control of Coccidiosis in Caged Egg Layers: A Paper Plate Vaccination Method. *J. Appl. Poult. Res.* **2004**, *13*, 360–363. [CrossRef]
11. Rodenburg, T.B.; Tuyttens, F.A.M.; Sonck, B.; De Reu, K.; Herman, L.; Zoons, J. Welfare, Health, and Hygiene of Laying Hens Housed in Furnished Cages and in Alternative Housing Systems. *J. Appl. Anim. Welf. Sci.* **2005**, *8*, 211–226. [CrossRef] [PubMed]
12. Fossum, O.; Jansson, D.S.; Etterlin, P.E.; Vgsholm, I. Causes of Mortality in Laying Hens in Different Housing Systems in 2001 to 2004. *Acta Vet. Scand.* **2009**, *51*, 3. [CrossRef]
13. Kaufmann-Bat, M.; Hoop, R.K. Diseases in Chicks and Laying Hens during the First 12 Years after Battery Cages Were Banned in Switzerland. *Vet. Rec.* **2009**, *164*, 203–207. [CrossRef] [PubMed]
14. Lunden, A.; Thebo, P.; Gunnarsson, S.; Hooshmand-Rad, P.; Tauson, R.; Uggla, A. Eimeria Infections in Litter-Based, High Stocking Density Systems for Loose-Housed Laying Hens in Sweden. *Br. Poult. Sci.* **2000**, *41*, 440–447. [CrossRef] [PubMed]
15. Price, K.R. Use of Live Vaccines for Coccidiosis Control in Replacement Layer Pullets. *J. Appl. Poult. Res.* **2012**, *21*, 679–692. [CrossRef]
16. Blake, D.P.; Vrba, V.; Xia, D.; Jatau, I.D.; Spiro, S.; Nolan, M.J.; Underwood, G.; Tomley, F.M. Genetic and Biological Characterisation of Three Cryptic Eimeria Operational Taxonomic Units That Infect Chickens (*Gallus gallus domesticus*). *Int. J. Parasitol.* **2021**, *51*, 621–634. [CrossRef] [PubMed]
17. Shirley, M.W.; Smith, A.L.; Tomley, F.M. The Biology of Avian Eimeria with an Emphasis on Their Control by Vaccination. *Adv. Parasitol.* **2005**, *60*, 285–330. [PubMed]
18. Lee, Y.; Lu, M.; Lillehoj, H.S. Coccidiosis: Recent Progress in Host Immunity and Alternatives to Antibiotic Strategies. *Vaccines* **2022**, *10*, 215. [CrossRef] [PubMed]
19. Shivaramaiah, C.; Barta, J.R.; Hernandez-Velasco, X.; Téllez, G.; Hargis, B.M. Coccidiosis: Recent Advancements in the Immunobiology of Eimeria Species, Preventive Measures, and the Importance of Vaccination as a Control Tool against These Apicomplexan Parasites. *Vet. Med. Res. Rep.* **2014**, *5*, 23–34.
20. Lillehoj, H.S.; Lillehoj, E.P. Avian Coccidiosis. A Review of Acquired Intestinal Immunity and Vaccination Strategies. *Avian Dis.* **2000**, *44*, 408–425. [CrossRef] [PubMed]
21. Lillehoj, H.S.; Trout, J.M. Coccidia: A Review of Recent Advances on Immunity and Vaccine Development. *Avian Pathol.* **1993**, *22*, 3–31. [CrossRef] [PubMed]
22. Yun, C.H.; Lillehoj, H.S.; Lillehoj, E.P. Intestinal Immune Responses to Coccidiosis. *Dev. Comp. Immunol.* **2000**, *24*, 303–324. [CrossRef] [PubMed]
23. Wallach, M.; Smith, N.C.; Petracca, M.; Miller, C.M.D.; Eckert, J.; Braun, R. Eimeria Maxima Gametocyte Antigens: Potential Use in a Subunit Maternal Vaccine against Coccidiosis in Chickens. *Vaccine* **1995**, *13*, 347–354. [CrossRef] [PubMed]
24. Wallach, M.; Pillemer, G.; Yarus, S.; Halabi, A.; Pugatsch, T.; Mencher, D. Passive Immunization of Chickens against Eimeria Maxima Infection with a Monoclonal Antibody Developed against a Gametocyte Antigen. *Infect. Immun.* **1990**, *58*, 557–562. [CrossRef] [PubMed]
25. Lillehoj, H.S.; Trout, J.M. Avian Gut-Associated Lymphoid Tissues and Intestinal Immune Responses to Eimeria Parasites. *Clin. Microbiol. Rev.* **1996**, *9*, 349–360. [CrossRef] [PubMed]
26. Sharma, M.K.; Regmi, P.; Applegate, T.; Chai, L.; Kim, W.K.; Sharma, M.K.; Regmi, P.; Applegate, T.; Chai, L.; Kim, W.K. Osteoimmunology: A Link between Gastrointestinal Diseases and Skeletal Health in Chickens. *Animals* **2023**, *13*, 1816. [CrossRef] [PubMed]
27. Fathima, S.; Al Hakeem, W.G.; Shanmugasundaram, R.; Selvaraj, R.K. Necrotic Enteritis in Broiler Chickens: A Review on the Pathogen, Pathogenesis, and Prevention. *Microorganisms* **2022**, *10*, 1958. [CrossRef] [PubMed]

28. Lillehoj, H.S.; Choi, K.D. Recombinant Chicken Interferon-Gamma-Mediated Inhibition of *Eimeria tenella* Development in Vitro and Reduction of Oocyst Production and Body Weight Loss Following *Eimeria acervulina* Challenge Infection. *Avian Dis.* **1998**, *42*, 307–314. [CrossRef] [PubMed]
29. Rose, M.E.; Hesketh, P.; Rennie, M. Coccidiosis: Rapid Depletion of Circulating Lymphocytes after Challenge of Immune Chickens with Parasite Antigens. *Infect. Immun.* **1984**, *45*, 166–171. [CrossRef] [PubMed]
30. Lillehoj, H.S.; Min, W.; Dalloul, R.A. Recent Progress on the Cytokine Regulation of Intestinal Immune Responses to Eimeria. *Poult. Sci.* **2004**, *83*, 611–623. [CrossRef] [PubMed]
31. Liu, G.; Sharma, M.K.; Tompkins, Y.H.; Teng, P.Y.; Kim, W.K. Impacts of Varying Methionine to Cysteine Supplementation Ratios on Growth Performance, Oxidative Status, Intestinal Health, and Gene Expression of Immune Response and Methionine Metabolism in Broilers under Eimeria Spp. Challenge. *Poult. Sci.* **2024**, *103*, 103300. [CrossRef] [PubMed]
32. Fernando, M.A.; McCraw, B.M. Mucosal Morphology and Cellular Renewal in the Intestine of Chickens Following a Single Infection of *Eimeria acervulina*. *J. Parasitol.* **1973**, *59*, 493–501. [CrossRef] [PubMed]
33. Major, P.; Tóth, Š.; Goldová, M.; Révajová, V.; Kožárová, I.; Levkut, M.; Mojžišová, J.; Hisira, V.; Mihok, T. Dynamic of Apoptosis of Cells in Duodenal Villi Infected with *Eimeria acervulina* in Broiler Chickens. *Biologia* **2011**, *66*, 696–700. [CrossRef]
34. Sharma, M.K.; Liu, G.; White, D.L.; Kim, W.K. Graded Levels of Eimeria Infection Linearly Reduced the Growth Performance, Altered the Intestinal Health, and Delayed the Onset of Egg Production of Hy-Line W-36 Laying Hens When Infected at the Prelay Stage. *Poult. Sci.* **2024**, *103*, 103174. [CrossRef] [PubMed]
35. Sharma, M.K.; Liu, G.; White, D.L.; Tompkins, Y.H.; Kim, W.K. Effects of Mixed Eimeria Challenge on Performance, Body Composition, Intestinal Health, and Expression of Nutrient Transporter Genes of Hy-Line W-36 Pullets (0–6 Wks of Age). *Poult. Sci.* **2022**, *101*, 102083. [CrossRef] [PubMed]
36. Sharma, M.K.; Singh, A.K.; Goo, D.; Choppa, V.S.R.; Ko, H.; Shi, H.; Kim, W.K. Graded Levels of Eimeria Infection Modulated Gut Physiology and Temporarily Ceased the Egg Production of Laying Hens at Peak Production. *Poult. Sci.* **2024**, *103*, 103229. [CrossRef] [PubMed]
37. Choi, J.; Goo, D.; Sharma, M.K.; Ko, H.; Liu, G.; Paneru, D.; Choppa, V.S.R.; Lee, J.; Kim, W.K. Effects of Different Eimeria Inoculation Doses on Growth Performance, Daily Feed Intake, Gut Health, Gut Microbiota, Foot Pad Dermatitis, and Eimeria Gene Expression in Broilers Raised in Floor Pens for 35 Days. *Animals* **2023**, *13*, 2237. [CrossRef] [PubMed]
38. Hansen, V.L.; Kahl, S.; Proszkowiec-Weglarz, M.; Jiménez, S.C.; Vaessen, S.F.C.; Schreier, L.L.; Jenkins, M.C.; Russell, B.; Miska, K.B. The Effects of Tributyrin Supplementation on Weight Gain and Intestinal Gene Expression in Broiler Chickens during Eimeria Maxima-Induced Coccidiosis. *Poult. Sci.* **2021**, *100*, 100984. [CrossRef] [PubMed]
39. Morris, B.C.; Danforth, H.D.; Caldwell, D.J.; Pierson, F.W.; McElroy, A.P. Intestinal Mucosal Mast Cell Immune Response and Pathogenesis of Two *Eimeria acervulina* Isolates in Broiler Chickens. *Poult. Sci.* **2004**, *83*, 1667–1674. [CrossRef] [PubMed]
40. Belote, B.L.; Soares, I.; Tujimoto-Silva, A.; Sanches, A.W.D.; Kraieski, A.L.; Santin, E. Applying I See inside Histological Methodology to Evaluate Gut Health in Broilers Challenged with Eimeria. *Vet. Parasitol.* **2019**, *276*, 100004. [CrossRef] [PubMed]
41. Cliffe, L.J.; Humphreys, N.E.; Lane, T.E.; Potten, C.S.; Booth, C.; Grencis, R.K. Immunology—Accelerated Intestinal Epithelial Cell Turnover: A New Mechanism of Parasite Expulsion. *Science* **2005**, *308*, 1463–1465. [CrossRef] [PubMed]
42. Ulluwishewa, D.; Anderson, R.C.; McNabb, W.C.; Moughan, P.J.; Wells, J.M.; Roy, N.C. Regulation of Tight Junction Permeability by Intestinal Bacteria and Dietary Components. *J. Nutr.* **2011**, *141*, 769–776. [CrossRef]
43. Yadav, S.; Teng, P.Y.; Souza dos Santos, T.; Gould, R.L.; Craig, S.W.; Lorraine Fuller, A.; Pazdro, R.; Kim, W.K. The Effects of Different Doses of Curcumin Compound on Growth Performance, Antioxidant Status, and Gut Health of Broiler Chickens Challenged with Eimeria Species. *Poult. Sci.* **2020**, *99*, 5936–5945. [CrossRef] [PubMed]
44. Georgieva, N.V.; Gabrashanska, M.; Georgieva, N.V.; Gabrashanska, M.; Koinarski, V.; Ermidou-Pollet, S. Antioxidant Status in *Eimeria acervulina* Infected Chickens after Dietary Selenium Treatment. *Trace Elem. Electrolytes* **2011**, *28*, 42. [CrossRef]
45. Idris, M.; Abbas, R.Z.; Masood, S.; Rehman, T.; Farooq, U.; Babar, W.; Hussain, R.; Raza, A.; Riaz, U. World's Poultry Science Journal The Potential of Antioxidant Rich Essential Oils against Avian Coccidiosis. *Poult. Sci. J.* **2017**, *73*, 89–104.
46. Tompkins, Y.H.; Choi, J.; Teng, P.Y.; Yamada, M.; Sugiyama, T.; Kim, W.K. Reduced Bone Formation and Increased Bone Resorption Drive Bone Loss in Eimeria Infected Broilers. *Sci. Rep.* **2023**, *13*, 616. [CrossRef] [PubMed]
47. Teng, P.Y.; Liu, G.; Choi, J.; Yadav, S.; Wei, F.; Kim, W.K. Effects of Levels of Methionine Supplementations in Forms of L- or DL-Methionine on the Performance, Intestinal Development, Immune Response, and Antioxidant System in Broilers Challenged with *Eimeria* spp. *Poult. Sci.* **2023**, *102*, 102586. [CrossRef]
48. Tompkins, Y.H.; Teng, P.; Pazdro, R.; Kim, W.K. Long Bone Mineral Loss, Bone Microstructural Changes and Oxidative Stress after Eimeria Challenge in Broilers. *Front. Physiol.* **2022**, *13*, 945740. [CrossRef] [PubMed]
49. Galli, G.M.; Baldissera, M.D.; Griss, L.G.; Souza, C.F.; Fortuoso, B.F.; Boiago, M.M.; Gris, A.; Mendes, R.E.; Stefani, L.M.; da Silva, A.S. Intestinal Injury Caused by *Eimeria* spp. Impairs the Phosphotransfer Network and Gain Weight in Experimentally Infected Chicken Chicks. *Parasitol. Res.* **2019**, *118*, 1573–1579. [CrossRef] [PubMed]
50. Conway, D.P.; Sasai, K.; Gaafar, S.M.; Smothers, C.D. Effects of Different Levels of Oocyst Inocula of *Eimeria acervulina*, E. Tenella, and E. Maxima on Plasma Constituents, Packed Cell Volume, Lesion Scores, and Performance in Chickens. *Avian Dis.* **1993**, *37*, 118–123. [CrossRef] [PubMed]
51. Kipper, M.; Andretta, I.; Lehnen, C.R.; Lovatto, P.A.; Monteiro, S.G. Meta-Analysis of the Performance Variation in Broilers Experimentally Challenged by *Eimeria* spp. *Vet. Parasitol.* **2013**, *196*, 77–84. [CrossRef] [PubMed]

52. Bun, S.D.; Guo, Y.M.; Guo, F.C.; Ji, F.J.; Cao, H. Influence of Organic Zinc Supplementation on the Antioxidant Status and Immune Responses of Broilers Challenged with *Eimeria tenella*. *Poult. Sci.* **2011**, *90*, 1220–1226. [CrossRef] [PubMed]
53. Botsoglou, N. Effect of Supplementing Feed with Oregano and/or α-Tocopheryl Acetate on Growth of Broiler Chickens and Oxidative Stability of Meat. *Artic. J. Anim. Feed. Sci.* **2005**, *14*, 521–535.
54. Iuspa, M.A.M.; Soares, I.; Belote, B.L.; Kawazoe, U.; Santin, E. Comparing Performance and Resistance of Two Broilers Breeds Challenged by *Eimeria acervulina*. *Vet. Parasitol.* **2020**, *287*, 109235. [CrossRef] [PubMed]
55. Rochell, S.J.; Parsons, C.M.; Dilger, R.N. Effects of *Eimeria acervulina* Infection Severity on Growth Performance, Apparent Ileal Amino Acid Digestibility, and Plasma Concentrations of Amino Acids, Carotenoids, and A1-Acid Glycoprotein in Broilers. *Poult. Sci.* **2016**, *95*, 1573–1581. [CrossRef]
56. Choi, J.; Ko, H.; Tompkins, Y.H.; Teng, P.Y.; Lourenco, J.M.; Callaway, T.R.; Kim, W.K. Effects of *Eimeria tenella* Infection on Key Parameters for Feed Efficiency in Broiler Chickens. *Animals* **2021**, *11*, 3428. [CrossRef] [PubMed]
57. Hegde, K.S.; Reid, W.M. Effects of Six Single Species of Coccidia on Egg Production and Culling Rate of Susceptible Layers. *Poult. Sci.* **1969**, *48*, 928–932. [CrossRef] [PubMed]
58. Fitz-coy, A.S.H.; Edgar, S.A.; Edgar, S.A. Effects of *Eimeria mitis* on Egg Production of Single-Comb White Leghorn Hens. *Am. Assoc. Avian Pathol. Stable JSTOR* **2020**, *36*, 718–721. [CrossRef]
59. McDougald, L.R.; Fuller, A.L.; McMurray, B.L. An Outbreak of *Eimeria necatrix* Coccidiosis in Breeder Pullets: Analysis of Immediate and Possible Long-Term Effects on Performance. *Avian Dis.* **1990**, *34*, 485–487. [CrossRef] [PubMed]
60. Mayhew, R.L. Studies on Coccidiosis. *Poult. Sci.* **1934**, *13*, 148–154. [CrossRef]
61. Adams, C.; Vahl, H.A.; Veldman, A. Interaction between Nutrition and *Eimeria acervulina* Infection in Broiler Chickens: Diet Compositions That Improve Fat Digestion during *Eimeria acervulina* Infection. *Br. J. Nutr.* **1996**, *75*, 875–880. [CrossRef] [PubMed]
62. Liu, X.; Lin, X.; Mi, Y.; Li, J.; Zhang, C. Grape Seed Proanthocyanidin Extract Prevents Ovarian Aging by Inhibiting Oxidative Stress in the Hens. *Oxid. Med. Cell Longev.* **2018**, *2018*, 9390810. [CrossRef]
63. Zhu, M.; Miao, S.; Zhou, W.; Elnesr, S.S.; Dong, X.; Zou, X. MAPK, AKT/FoxO3a and MTOR Pathways Are Involved in Cadmium Regulating the Cell Cycle, Proliferation and Apoptosis of Chicken Follicular Granulosa Cells. *Ecotoxicol. Environ. Saf.* **2021**, *214*, 112091. [CrossRef] [PubMed]
64. Chapman, H.D. Milestones in Avian Coccidiosis Research: A Review. *Poult. Sci.* **2014**, *93*, 501–511. [CrossRef] [PubMed]
65. Hauck, R.; Macklin, K.S. Vaccination Against Poultry Parasites. *Avian Dis.* **2023**, *67*, 441–449. [CrossRef] [PubMed]
66. Adhikari, P.; Kiess, A.; Adhikari, R.; Jha, R. An Approach to Alternative Strategies to Control Avian Coccidiosis and Necrotic Enteritis. *J. Appl. Poult. Res.* **2020**, *29*, 515–534. [CrossRef]
67. Good, E.; Practice, C. Guidance for Industry Guidance for Industry. *Fed. Regist.* **2004**, *505*, 79.
68. Cervantes, H.M.; McDougald, L.R. Raising Broiler Chickens without Ionophore Anticoccidials. *J. Appl. Poult. Res.* **2023**, *32*, 100347. [CrossRef]
69. Martins, R.R.; Silva, L.J.G.; Pereira, A.M.P.T.; Esteves, A.; Duarte, S.C.; Pena, A. Coccidiostats and Poultry: A Comprehensive Review and Current Legislation. *Foods* **2022**, *11*, 2738. [CrossRef] [PubMed]
70. Smith, C.K.; Galloway, R.B.; White, S.L. Effect of Ionophores on Survival, Penetration, and Development of *Eimeria tenella* Sporozoites In Vitro. *J. Parasitol.* **1981**, *67*, 511–516. [CrossRef]
71. Chapman, H.D. Biochemical, Genetic and Applied Aspects of Drug Resistance in Eimeria Parasites of the Fowl. *Avian Pathol.* **1997**, *26*, 221–244. [CrossRef] [PubMed]
72. Beach, J.R.; Corl, J.C. Studies in the Control of Avian Coccidiosis. *Poult. Sci.* **1925**, *4*, 83–93. [CrossRef]
73. McDonald, V.; Shirley, M.W. Past and Future: Vaccination against Eimeria. *Parasitology* **2009**, *136*, 1477–1489. [CrossRef]
74. Chapman, H.D.; Blake, D.P. Genetic Selection of Eimeria Parasites in the Chicken for Improvement of Poultry Health: Implications for Drug Resistance and Live Vaccine Development. *Avian Pathol.* **2022**, *51*, 521–534. [CrossRef] [PubMed]
75. Dalloul, R.A.; Lillehoj, H.S. Poultry Coccidiosis: Recent Advancements in Control Measures and Vaccine Development. *Expert. Rev. Vaccines* **2006**, *5*, 143–163. [CrossRef] [PubMed]
76. Chapman, H.D.; Cherry, T.E.; Danforth, H.D.; Richards, G.; Shirley, M.W.; Williams, R.B. Sustainable Coccidiosis Control in Poultry Production: The Role of Live Vaccines. *Int. J. Parasitol.* **2002**, *32*, 617–629. [CrossRef]
77. Sharman, P.A.; Smith, N.C.; Wallach, M.G.; Katrib, M. Chasing the Golden Egg: Vaccination against Poultry Coccidiosis. *Parasite Immunol.* **2010**, *32*, 590–598. [CrossRef] [PubMed]
78. Dalloul, R.A.; Lillehoj, H.S. Recent Advances in Immunomodulation and Vaccination Strategies against Coccidiosis. *Avian Dis.* **2005**, *49*, 1–8. [CrossRef] [PubMed]
79. Kim, W.K.; Bloomfield, S.A.; Ricke, S.C. Effects of Age, Vitamin D3, and Fructooligosaccharides on Bone Growth and Skeletal Integrity of Broiler Chicks. *Poult. Sci.* **2011**, *90*, 2425–2432. [CrossRef] [PubMed]
80. White, D.; Chen, C.; Kim, W.K. Effect of the Combination of 25-Hydroxyvitamin D3 and Higher Level of Calcium and Phosphorus in the Diets on Bone 3D Structural Development in Pullets. *Front. Physiol.* **2023**, *14*, 1056481. [CrossRef] [PubMed]
81. Chen, C.; Turner, B.; Applegate, T.J.; Litta, G.; Kim, W.K. Role of Long-Term Supplementation of 25-Hydroxyvitamin D3 on Laying Hen Bone 3-Dimensional Structural Development. *Poult. Sci.* **2020**, *99*, 5771–5782. [CrossRef] [PubMed]
82. Fritts, C.A.; Waldroup, P.W. Effect of Source and Level of Vitamin D on Live Performance and Bone Development in Growing Broilers. *J. Appl. Poult. Res.* **2003**, *12*, 45–52. [CrossRef]

83. Oikeh, I.; Sakkas, P.; Blake, D.P.; Kyriazakis, I. Interactions between Dietary Calcium and Phosphorus Level, and Vitamin D Source on Bone Mineralization, Performance, and Intestinal Morphology of Coccidia-Infected Broilers. *Poult. Sci.* **2019**, *98*, 5679–5690. [CrossRef] [PubMed]
84. Sakkas, P.; Oikeh, I.; Blake, D.P.; Smith, S.; Kyriazakis, I. Dietary Vitamin D Improves Performance and Bone Mineralisation, but Increases Parasite Replication and Compromises Gut Health in Eimeria-Infected Broilers. *Br. J. Nutr.* **2019**, *122*, 676–688. [CrossRef] [PubMed]
85. Morris, A.; Shanmugasundaram, R.; McDonald, J.; Selvaraj, R.K. Effect of In Vitro and In Vivo 25-Hydroxyvitamin D Treatment on Macrophages, T Cells, and Layer Chickens during a Coccidia Challenge. *J. Anim. Sci.* **2015**, *93*, 2894–2903. [CrossRef] [PubMed]
86. Fakhoury, H.M.A.; Kvietys, P.R.; AlKattan, W.; Al Anouti, F.; Elahi, M.A.; Karras, S.N.; Grant, W.B. Vitamin D and Intestinal Homeostasis: Barrier, Microbiota, and Immune Modulation. *J. Steroid Biochem. Mol. Biol.* **2020**, *200*, 105663. [CrossRef] [PubMed]
87. Pike, J.W.; Christakos, S. Biology and Mechanisms of Action of the Vitamin D Hormone. *Endocrinol. Metab. Clin. N. Am.* **2017**, *46*, 815. [CrossRef] [PubMed]
88. Morris, A.; Selvaraj, R.K. In Vitro 25-Hydroxycholecalciferol Treatment of Lipopolysaccharide-Stimulated Chicken Macrophages Increases Nitric Oxide Production and MRNA of Interleukin-1beta and 10. *Vet. Immunol. Immunopathol.* **2014**, *161*, 265–270. [CrossRef]
89. Shanmugasundaram, R.; Morris, A.; Selvaraj, R.K. Effect of 25-Hydroxycholecalciferol Supplementation on Turkey Performance and Immune Cell Parameters in a Coccidial Infection Model. *Poult. Sci.* **2019**, *98*, 1127–1133. [CrossRef] [PubMed]
90. Upadhaya, S.D.; Cho, S.H.; Chung, T.K.; Kim, I.H. Anti-Coccidial Effect of Essential Oil Blends and Vitamin D on Broiler Chickens Vaccinated with Purified Mixture of Coccidian Oocyst from *Eimeria tenella* and *Eimeria maxima*. *Poult. Sci.* **2019**, *98*, 2919–2926. [CrossRef] [PubMed]
91. Leyva-Jimenez, H.; Gardner, K.; Al-Jumaa, Y.; Padgett, J.C.; Bailey, C.A. Partial Replacement of Dietary Cholecalciferol with 25-Hydroxycholecalciferol on Broiler Chickens Subjected to a Coccidiosis Vaccine Challenge. *J. Appl. Poult. Res.* **2019**, *28*, 743–754. [CrossRef]
92. Suarez, J.C.; Knape, K.; Ali, N.A.; Aljuboori, J.; Al-Alwani, T.; Gutierrez, O.; Carey, J.B. Dietary Supplementation of 25-Hydroxycholecalciferol and Its Impact on Performance, Intestinal Morphology, and Vitamin D Status in Broilers Subjected to Coccidiosis Vaccine. *J. Appl. Poult. Res.* **2023**, *32*, 100360. [CrossRef]
93. Lopes, T.S.B.; Shi, H.; White, D.; Araújo, I.C.S.; Kim, W.K. Effects of 25-Hydroxycholecalciferol on Performance, Gut Health, and Bone Quality of Broilers Fed with Reduced Calcium and Phosphorus Diet during Eimeria Challenge. *Poult. Sci.* **2024**, *103*, 103267. [CrossRef] [PubMed]
94. Grzesiak, M. Vitamin D3 Action within the Ovary—An Updated Review. *Physiol. Res.* **2020**, *69*, 371. [CrossRef] [PubMed]
95. Xu, F.; Wolf, S.; Green, O.; Xu, J. Vitamin D in Follicular Development and Oocyte Maturation. *Reproduction* **2021**, *161*, R129–R137. [CrossRef] [PubMed]
96. Ruschkowski, S.R.; Hart, L.E. Ionic and Endocrine Characteristics of Reproductive Failure in Calcium-Deficient and Vitamin D-Deficient Laying Hens. *Poult. Sci.* **1992**, *71*, 1722–1732. [CrossRef] [PubMed]
97. Wang, J.; Ding, X.; Zeng, Q.; Bai, S.; Zhang, K.; Mao, X.; Xu, S.; Zhuo, Y.; Xuan, Y.; Peng, H.; et al. Dietary 25-Hydroxyvitamin D Improves Productive Performance and Intestinal Health of Laying Hens under Escherichia Coli Lipopolysaccharide Challenge. *Poult. Sci.* **2023**, *102*, 102371. [CrossRef] [PubMed]
98. Geng, Y.; Ma, Q.; Wang, Z.; Guo, Y. Dietary Vitamin D3 Supplementation Protects Laying Hens against Lipopolysaccharide-Induced Immunological Stress. *Nutr. Metab.* **2018**, *15*, 58. [CrossRef] [PubMed]
99. Sharma, M.K.; Liu, G.; White, D.L.; Tompkins, Y.H.; Kim, W.K. Graded Levels of Eimeria Challenge Altered the Microstructural Architecture and Reduced the Cortical Bone Growth of Femur of Hy-Line W-36 Pullets at Early Stage of Growth (0–6 Wk of Age). *Poult. Sci.* **2023**, *102*, 102888. [CrossRef] [PubMed]
100. Konjufca, V.K.; Bottje, W.G.; Bersi, T.K.; Erf, G.F. Influence of Dietary Vitamin E on Phagocytic Functions of Macrophages in Broilers. *Poult. Sci.* **2004**, *83*, 1530–1534. [CrossRef] [PubMed]
101. Allen, P.C.; Fetterer, R.H. Interaction of Dietary Vitamin E with *Eimeria maxima* Infections in Chickens. *Poult. Sci.* **2002**, *81*, 41–48. [CrossRef]
102. Fu, Z.; Zhong, T.; Wan, X.; Xu, L.; Yang, H.; Han, H.; Wang, Z.; Fu, Z.; Zhong, T.; Wan, X.; et al. Effects of Dietary Vitamin E Supplementation on Reproductive Performance, Egg Characteristics, Antioxidant Capacity, and Immune Status in Breeding Geese during the Late Laying Period. *Antioxidants* **2022**, *11*, 2070. [CrossRef] [PubMed]
103. Tappel, A.L. Vitamin E and free radical peroxidation of lipids. *Ann. N. Y. Acad. Sci.* **1972**, *203*, 12–28. [CrossRef] [PubMed]
104. Gao, J.; Lin, H.; Wang, X.J.; Song, Z.G.; Jiao, H.C. Vitamin E Supplementation Alleviates the Oxidative Stress Induced by Dexamethasone Treatment and Improves Meat Quality in Broiler Chickens. *Poult. Sci.* **2010**, *89*, 318–327. [CrossRef] [PubMed]
105. Traber, M.G.; Atkinson, J. Vitamin E, Antioxidant and Nothing More. *Free Radic. Biol. Med.* **2007**, *43*, 4–15. [CrossRef] [PubMed]
106. Colnago, G.L.; Jensen, L.S.; Long, P.L. Effect of Selenium and Vitamin E on the Development of Immunity to Coccidiosis in Chickens. *Poult. Sci.* **1984**, *63*, 1136–1143. [CrossRef] [PubMed]
107. Al-Hamadani, A.A.; Al-Dhanki, Z.T.M. Effect of Locally Produced Dual Virulent Coccidian Vaccine and Supplementation of Vitamin A and E on Performance, Some Physiological Traits, Protein and Fat Digestibility of Broiler. *Al-Anbar J. Vet. Sci.* **2013**, *6*, 80–95.

108. Perez-Carbajal, C.; Caldwell, D.; Farnell, M.; Stringfellow, K.; Pohl, S.; Casco, G.; Pro-Martinez, A.; Ruiz-Feria, C.A. Immune Response of Broiler Chickens Fed Different Levels of Arginine and Vitamin E to a Coccidiosis Vaccine and Eimeria Challenge. *Poult. Sci.* **2010**, *89*, 1870–1877. [CrossRef] [PubMed]
109. Liu, Y.J.; Zhao, L.H.; Mosenthin, R.; Zhang, J.Y.; Ji, C.; Ma, Q.G. Protective Effect of Vitamin E on Laying Performance, Antioxidant Capacity, and Immunity in Laying Hens Challenged with Salmonella Enteritidis. *Poult. Sci.* **2019**, *98*, 5847–5854. [CrossRef] [PubMed]
110. da Silva, I.C.M.; Ribeiro, A.M.L.; Canal, C.W.; Vieira, M.M.; Pinheiro, C.C.; Gonçalves, T.; de Moraes, M.L.; Ledur, V.S. Effect of Vitamin E Levels on the Cell-Mediated Immunity of Broilers Vaccinated against Coccidiosis. *Braz. J. Poult. Sci.* **2011**, *13*, 53–56. [CrossRef]
111. Jafari, R.A. Effect of Dietary Vitamin E on *Eimeria tenella*-Induced Oxidative Stress in Broiler Chickens. *Afr. J. Biotechnol.* **2012**, *11*.
112. Dominguez, P.A.; Pro-Martinez, A.; Narciso-Gaytán, C.; Hernández-Cázares, A.; Sosa-Montes, E.; Perez-Hernandez, P.; Caldwell, D.; Ruiz-Feria, C.A. Supplémentation Simultanée En Arginine et En Vitamines E et c Antioxydantes Réduit Le Stress Oxydatif Chez Les Poulets à Griller Après Exposition à *Eimeria* spp. *Can. J. Anim. Sci.* **2015**, *95*, 143–153. [CrossRef]
113. Dalloul, R.A.; Lillehoj, H.S.; Shellem, T.A.; Doerr, J.A. Effect of Vitamin A Deficiency on Host Intestinal Immune Response to *Eimeria acervulina* in Broiler Chickens. *Poult. Sci.* **2002**, *81*, 1509–1515. [CrossRef] [PubMed]
114. Stephensen, C.B. Vitamin A, infection, and immune function*. *Annu. Rev. Nutr.* **2003**, *21*, 167–192. [CrossRef] [PubMed]
115. Green, A.S.; Fascetti, A.J. Meeting the Vitamin A Requirement: The Efficacy and Importance of β-Carotene in Animal Species. *Sci. World J.* **2016**, *2016*, 7393620. [CrossRef] [PubMed]
116. Taylor, M.W.; Russell, W.C. The Provitamin A Requirement of Growing Chickens. *Poult. Sci.* **1947**, *26*, 234–242. [CrossRef]
117. Erasmus, J.; Scott, M.L.; Levine, P.P. A Relationship Between Coccidiosis and Vitamin A Nutrition in Chickens. *Poult. Sci.* **1960**, *39*, 565–572. [CrossRef]
118. Panda, B.; Combs, G.F.; Devolt, H.M. Studies on Coccidiosis and Vitamin A Nutrition of Broilers 1. *Poult. Sci.* **1964**, *43*, 154–164. [CrossRef]
119. Li, P.; Liu, C.; Niu, J.; Zhang, Y.; Li, C.; Zhang, Z.; Guo, S.; Ding, B. Effects of Dietary Supplementation with Vitamin A on Antioxidant and Intestinal Barrier Function of Broilers Co-Infected with Coccidia and Clostridium Perfringens. *Animals* **2022**, *12*, 3431. [CrossRef] [PubMed]
120. Boyera, N.; Galey, I.; Bernard, B.A. Effect of Vitamin C and Its Derivatives on Collagen Synthesis and Cross-Linking by Normal Human Fibroblasts. *Int. J. Cosmet. Sci.* **1998**, *20*, 151–158. [CrossRef] [PubMed]
121. Carr, A.C.; Maggini, S. Vitamin C and Immune Function. *Nutrients* **2017**, *9*, 1211. [CrossRef] [PubMed]
122. McKee, J.S.; Harrison, P.C. Effects of Supplemental Ascorbic Acid on the Performance of Broiler Chickens Exposed to Multiple Concurrent Stressors. *Poult. Sci.* **1995**, *74*, 1772–1785. [CrossRef]
123. Little, P.L.; Edgar, S.A. The Effect of Vitamin C on Performance of Coccidia Infected Chickens Fed Omplete and Vitamin Deficient Semi-Purified Diets. *Poult. Sci.* **1971**, *50*, 26–34. [CrossRef] [PubMed]
124. Hutsko, S.; Wick, M.; Bielke, L.; Selvaraj, R. The Effects of an Anti-Coccidial Vaccination in Conjunction with Supplemental Protease, Vitamin C and Differing Levels of Dietary Protein on the Production and Gut. Ph.D. Thesis, The Ohio State University, Columbus, OH, USA, 2017.
125. Stephens, J.F.; Tugwell, R.L. Sources and Levels of Vitamin K in Relation to Cecal Coccidiosis. *Poult. Sci.* **1960**, *39*, 1183–1187. [CrossRef]
126. Liu, G.; Kim, W.K. The Functional Roles of Methionine and Arginine in Intestinal and Bone Health of Poultry: Review. *Animals* **2023**, *13*, 2949. [CrossRef] [PubMed]
127. de Souza Castro, F.L.; Kim, W.K. Secondary Functions of Arginine and Sulfur Amino Acids in Poultry Health: Review. *Animals* **2020**, *10*, 2106. [CrossRef] [PubMed]
128. Lee, J.T.; Rochell, S.J.; Kriseldi, R.; Kim, W.K.; Mitchell, R.D. Functional Properties of Amino Acids: Improve Health Status and Sustainability. *Poult. Sci.* **2023**, *102*, 102288. [CrossRef] [PubMed]
129. Rochell, S.J.; Helmbrecht, A.; Parsons, C.M.; Dilger, R.N. Interactive Effects of Dietary Arginine and *Eimeria acervulina* Infection on Broiler Growth Performance and Metabolism. *Poult. Sci.* **2017**, *96*, 659–666. [CrossRef] [PubMed]
130. Albina, J.E.; Abate, J.A.; Mastrofrancesco, B. Role of Ornithine as a Proline Precursor in Healing Wounds. *J. Surg. Res.* **1993**, *55*, 97–102. [CrossRef] [PubMed]
131. Stechmiller, J.K.; Childress, B.; Cowan, L. Arginine Supplementation and Wound Healing. *Nutr. Clin. Pract.* **2005**, *20*, 52–61. [CrossRef]
132. Allen, P.C.; Fetterer, R.H. Effect of *Eimeria acervulina* Infections on Plasma L-Arginine. *Poult. Sci.* **2000**, *79*, 1414–1417. [CrossRef] [PubMed]
133. Allen, P.C. Effects of Daily Oral Doses of L-Arginine on Coccidiosis Infections in Chickens. *Poult. Sci.* **1999**, *78*, 1506–1509. [CrossRef]
134. Tan, J.; Applegate, T.J.; Liu, S.; Guo, Y.; Eicher, S.D. Supplemental Dietary L-Arginine Attenuates Intestinal Mucosal Disruption during a Coccidial Vaccine Challenge in Broiler Chickens. *Br. J. Nutr.* **2014**, *112*, 1098–1109. [CrossRef] [PubMed]
135. Laika, M.; Jahanian, R. Increase in Dietary Arginine Level Could Ameliorate Detrimental Impacts of Coccidial Infection in Broiler Chickens. *Livest. Sci.* **2016**, *195*, 38–44. [CrossRef]

136. Castro, F.L.S.; Teng, P.Y.; Yadav, S.; Gould, R.L.; Craig, S.; Pazdro, R.; Kim, W.K. The Effects of L-Arginine Supplementation on Growth Performance and Intestinal Health of Broiler Chickens Challenged with *Eimeria* spp. *Poult. Sci.* **2020**, *99*, 5844–5857. [CrossRef] [PubMed]
137. Liu, G.; Ajao, A.M.; Shanmugasundaram, R.; Taylor, J.; Ball, E.; Applegate, T.J.; Selvaraj, R.; Kyriazakis, I.; Olukosi, O.A.; Kim, W.K. The Effects of Arginine and Branched-Chain Amino Acid Supplementation to Reduced-Protein Diet on Intestinal Health, Cecal Short-Chain Fatty Acid Profiles, and Immune Response in Broiler Chickens Challenged with *Eimeria* spp. *Poult. Sci.* **2023**, *102*, 102773. [CrossRef]
138. Yazdanabadi, F.I.; Moghaddam, G.H.; Nematollahi, A.; Daghighkia, H.; Sarir, H. Effect of Arginine Supplementation on Growth Performance, Lipid Profile, and Inflammatory Responses of Broiler Chicks Challenged with Coccidiosis. *Prev. Vet. Med.* **2020**, *180*, 105031. [CrossRef] [PubMed]
139. Yazdanabadi, F.I.; Mohebalian, H.; Moghaddam, G.; Abbasabadi, M.; Sarir, H.; Vashan, S.J.H.; Haghparast, A. Influence of Eimeria Spp. Infection and Dietary Inclusion of Arginine on Intestine Histological Parameters, Serum Amino Acid Profile and Ileal Amino Acids Digestibility in Broiler Chicks. *Vet. Parasitol.* **2020**, *286*, 109241. [CrossRef]
140. Izadi, F.; Moghaddam, G.; Nematollahi, A.; Khodadmehr, M.; Abbasabadi, M. The Effects of Different Levels of Arginine on Cecum Microbial Population and Serum Antioxidant Properties of Healthy and Coccidia-Challenged Broiler Chicks. *Vet. Clin. Pathol.* **2020**, *14*, 127–145.
141. Tan, J.; Liu, S.; Guo, Y.; Applegate, T.J.; Eicher, S.D. Dietary L-Arginine Supplementation Attenuates Lipopolysaccharide-Induced Inflammatory Response in Broiler Chickens. *Br. J. Nutr.* **2014**, *111*, 1394–1404. [CrossRef]
142. Manwar, S.J.; Moudgal, R.P.; Sastry, K.V.H.; Tyagi, I.S.; Saxena, Y.K. Nitric Oxide Regulating Immune Functions in Egg Type Japanese Quail. *Indian J. Poult. Sci.* **2003**, *38*, 133–136.
143. Basiouni, G.F. The Effect of Feeding an Extra Amounts of Arginine to Local Saudi Hens on Luteinizing Hormone Secretion. *J. Biol. Sci.* **2009**, *9*, 617–620. [CrossRef]
144. Uyanga, V.A.; Xin, Q.; Sun, M.; Zhao, J.; Wang, X.; Jiao, H.; Onagbesan, O.M.; Lin, H. Research Note: Effects of Dietary L-Arginine on the Production Performance and Gene Expression of Reproductive Hormones in Laying Hens Fed Low Crude Protein Diets. *Poult. Sci.* **2022**, *101*, 101816. [CrossRef] [PubMed]
145. Sasaki, J.; Nakashima, N. Methionine-Independent Initiation of Translation in the Capsid Protein of an Insect RNA Virus. *Proc. Natl. Acad. Sci. USA* **2000**, *97*, 1512–1515. [CrossRef] [PubMed]
146. Swennen, Q.; Geraert, P.A.; Mercier, Y.; Everaert, N.; Stinckens, A.; Willemsen, H.; Li, Y.; Decuypere, E.; Buyse, J. Effects of Dietary Protein Content and 2-Hydroxy-4-Methylthiobutanoic Acid or Dl-Methionine Supplementation on Performance and Oxidative Status of Broiler Chickens. *Br. J. Nutr.* **2011**, *106*, 1845–1854. [CrossRef] [PubMed]
147. Pourali, M.; Kermanshahi, H.; Golian, A.; Razmi, G.R.; Soukhtanloo, M. Antioxidant and Anticoccidial Effects of Garlic Powder and Sulfur Amino Acids on Eimeria-Infected and Uninfected Broiler Chickens. *J. Biol. Sci.* **2014**, *15*, 227–232.
148. Lai, A.; Dong, G.; Song, D.; Yang, T.; Zhang, X. Responses to Dietary Levels of Methionine in Broilers Medicated or Vaccinated against Coccidia under *Eimeria tenella*-Challenged Condition. *BMC Vet. Res.* **2018**, *14*, 140. [CrossRef] [PubMed]
149. Lai, A.; Yuan, Z.; Wang, Z.; Chen, B.; Zhi, L.; Huang, Z.; Zhang, Y.; Lai, A.; Yuan, Z.; Wang, Z.; et al. Dietary Methionine Increased the Growth Performances and Immune Function of Partridge Shank Broilers after Challenged with Coccidia. *Animals* **2023**, *13*, 613. [CrossRef] [PubMed]
150. Castro, F.L.S.; Tompkins, Y.H.; Pazdro, R.; Kim, W.K. The Effects of Total Sulfur Amino Acids on the Intestinal Health Status of Broilers Challenged with *Eimeria* spp. *Poult. Sci.* **2020**, *99*, 5027–5036. [CrossRef] [PubMed]
151. Ren, Z.; Bütz, D.E.; Whelan, R.; Naranjo, V.; Arendt, M.K.; Ramuta, M.D.; Yang, X.; Crenshaw, T.D.; Cook, M.E. Effects of Dietary Methionine plus Cysteine Levels on Growth Performance and Intestinal Antibody Production in Broilers during Eimeria Challenge. *Poult. Sci.* **2020**, *99*, 374–384. [CrossRef]
152. Castro, F.L.S.; Kim, Y.; Xu, H.; Kim, W.K. The Effect of Total Sulfur Amino Acid Levels on Growth Performance and Bone Metabolism in Pullets under Heat Stress. *Poult. Sci.* **2020**, *99*, 5783–5791. [CrossRef] [PubMed]
153. Yu, J.C.; Jiang, Z.M.; Li, D.M. Glutamine: A Precursor of Glutathione and Its Effect on Liver. *World J. Gastroenterol.* **1999**, *5*, 143. [CrossRef] [PubMed]
154. Oxford, J.H.; Selvaraj, R.K. Effects of Glutamine Supplementation on Broiler Performance and Intestinal Immune Parameters During an Experimental Coccidiosis Infection. *J. Appl. Poult. Res.* **2019**, *28*, 1279–1287. [CrossRef]
155. Bortoluzzi, C.; Rochell, S.J.; Applegate, T.J. Threonine, Arginine, and Glutamine: Influences on Intestinal Physiology, Immunology, and Microbiology in Broilers. *Poult. Sci.* **2018**, *97*, 937–945. [CrossRef] [PubMed]
156. Adedokun, S.A.; Ajuwon, K.M.; Romero, L.F.; Adeola, O. Ileal Endogenous Amino Acid Losses: Response of Broiler Chickens to Fiber and Mild Coccidial Vaccine Challenge. *Poult. Sci.* **2012**, *91*, 899–907. [CrossRef] [PubMed]
157. Mehdipour, Z.; Golian, A.; Nassiri-Moghadam, H.; Javadmanesh, A. Effect of Threonine Supplementation on Growth Performance, Metabolizable Energy, Morphological Changes and Immune Response in Broiler Chickens Challenged with Coccidia. *Poult. Sci. J.* **2020**, *8*, 95–107.
158. Zhang, Q.; Chen, X.; Eicher, S.D.; Ajuwon, K.M.; Applegate, T.J. Effect of Threonine Deficiency on Intestinal Integrity and Immune Response to Feed Withdrawal Combined with Coccidial Vaccine Challenge in Broiler Chicks. *Br. J. Nutr.* **2016**, *116*, 2030–2043. [CrossRef] [PubMed]

159. Teng, P.Y.; Choi, J.; Yadav, S.; Tompkins, Y.H.; Kim, W.K. Effects of Low-Crude Protein Diets Supplemented with Arginine, Glutamine, Threonine, and Methionine on Regulating Nutrient Absorption, Intestinal Health, and Growth Performance of Eimeria-Infected Chickens. *Poult. Sci.* **2021**, *100*, 101427. [CrossRef] [PubMed]
160. Wils-Plotz, E.L.; Jenkins, M.C.; Dilger, R.N. Modulation of the Intestinal Environment, Innate Immune Response, and Barrier Function by Dietary Threonine and Purified Fiber during a Coccidiosis Challenge in Broiler Chicks. *Poult. Sci.* **2013**, *92*, 735–745. [CrossRef] [PubMed]
161. Applegate, T.J.; Klose, V.; Steiner, T.; Ganner, A.; Schatzmayr, G. Probiotics and Phytogenics for Poultry: Myth or Reality? *J. Appl. Poult. Res.* **2010**, *19*, 194–210. [CrossRef]
162. Mohammadi Gheisar, M.; Kim, I.H. Phytobiotics in Poultry and Swine Nutrition–a Review. *Ital. J. Anim. Sci.* **2018**, *17*, 92–99. [CrossRef]
163. Bozkurt, M.; Ege, G.; Aysul, N.; Akşit, H.; Tüzün, A.E.; Küçükyllmaz, K.; Borum, A.E.; Uygun, M.; Akşit, D.; Aypak, S.; et al. Effect of Anticoccidial Monensin with Oregano Essential Oil on Broilers Experimentally Challenged with Mixed *Eimeria* spp. *Poult. Sci.* **2016**, *95*, 1858–1868. [CrossRef]
164. Mohiti-Asli, M.; Ghanaatparast-Rashti, M. Dietary Oregano Essential Oil Alleviates Experimentally Induced Coccidiosis in Broilers. *Prev. Vet. Med.* **2015**, *120*, 195–202. [CrossRef] [PubMed]
165. Gordillo Jaramillo, F.X.; Kim, D.H.; Lee, S.H.; Kwon, S.K.; Jha, R.; Lee, K.W. Role of Oregano and Citrus Species-Based Essential Oil Preparation for the Control of Coccidiosis in Broiler Chickens. *J. Anim. Sci. Biotechnol.* **2021**, *12*, 47. [CrossRef] [PubMed]
166. Felici, M.; Tugnoli, B.; De Hoest-Thompson, C.; Piva, A.; Grilli, E.; Marugan-Hernandez, V. Thyme, Oregano, and Garlic Essential Oils and Their Main Active Compounds Influence *Eimeria tenella* Intracellular Development. *Animals* **2023**, *14*, 77. [CrossRef]
167. Golenser, J.; Waknine, J.H.; Krugliak, M.; Hunt, N.H.; Grau, G.E. Current Perspectives on the Mechanism of Action of Artemisinins. *Int. J. Parasitol.* **2006**, *36*, 1427–1441. [CrossRef]
168. Meshnick, S.R. Artemisinin: Mechanisms of Action, Resistance and Toxicity. *Int. J. Parasitol.* **2002**, *32*, 1655–1660. [CrossRef] [PubMed]
169. Allen, P.C.; Lydon, J.; Danforth, H.D. Effects of Components of *Artemisia annua* on Coccidia Infections in Chickens. *Poult. Sci.* **1997**, *76*, 1156–1163. [CrossRef] [PubMed]
170. Del Cacho, E.; Gallego, M.; Francesch, M.; Quílez, J.; Sánchez-Acedo, C. Effect of Artemisinin on Oocyst Wall Formation and Sporulation during *Eimeria tenella* Infection. *Parasitol. Int.* **2010**, *59*, 506–511. [CrossRef] [PubMed]
171. Drăgan, L.; Györke, A.; Ferreira, J.F.S.; Pop, I.A.; Dunca, I.; Drăgan, M.; Mircean, V.; Dan, I.; Cozma, V. Effects of *Artemisia annua* and *Foeniculum vulgare* on Chickens Highly Infected with *Eimeria tenella* (Phylum Apicomplexa). *Acta Vet. Scand.* **2014**, *56*, 22. [CrossRef]
172. Jiao, J.Y.; Yang, Y.Q.; Liu, M.J.; Li, J.G.; Cui, Y.; Yin, S.J.; Tao, J.P. Artemisinin and Artemisia Annua Leaves Alleviate *Eimeria tenella* Infection by Facilitating Apoptosis of Host Cells and Suppressing Inflammatory Response. *Vet. Parasitol.* **2018**, *254*, 172–177. [CrossRef] [PubMed]
173. Hady, M.M.; Zaki, M.M. Efficacy of Some Herbal Feed Additives on Performance and Control of Cecal Coccidiosis in Broilers. *APCBEE Procedia* **2012**, *4*, 163–168. [CrossRef]
174. Kaboutari, J.; Arab, H.A.; Ebrahimi, K.; Rahbari, S. Prophylactic and Therapeutic Effects of a Novel Granulated Formulation of Artemisia Extract on Broiler Coccidiosis. *Trop. Anim. Health Prod.* **2014**, *46*, 43–48. [CrossRef] [PubMed]
175. Arab, H.A.; Katadj, J.K.; Rahbari, S.; Nabian, S.; Mohammadi, A.R.S.; Pirali-Kheirabadi, K. Comparison between Anticoccidial Effect of Granulated Artemisia Siberi Extract and Pure Artemisinin in Affected Broilers. *J. Vet. Res.* **2012**, *67*, 119–125.
176. Song, Z.; Cheng, K.; Zhang, L.; Wang, T. Dietary Supplementation of Enzymatically Treated Artemisia Annua Could Alleviate the Intestinal Inflammatory Response in Heat-Stressed Broilers. *J. Therm. Biol.* **2017**, *69*, 184–190. [CrossRef]
177. Yang, S.; Zhang, J.; Jiang, Y.; Xu, Y.Q.; Jin, X.; Yan, S.M.; Shi, B.L. Effects of *Artemisia argyi* Flavonoids on Growth Performance and Immune Function in Broilers Challenged with Lipopolysaccharide. *Anim. Biosci.* **2021**, *34*, 1169–1180. [CrossRef] [PubMed]
178. Wan, X.L.; Niu, Y.; Zheng, X.C.; Huang, Q.; Su, W.P.; Zhang, J.F.; Zhang, L.L.; Wang, T. Antioxidant Capacities of *Artemisia annua* L. Leaves and Enzymatically Treated *Artemisia annua* L. in Vitro and in Broilers. *Anim. Feed Sci. Technol.* **2016**, *221*, 27–34. [CrossRef]
179. Abdelli, N.; Solà-Oriol, D.; Pérez, J.F. Phytogenic Feed Additives in Poultry: Achievements, Prospective and Challenges. *Animals* **2021**, *11*, 3471. [CrossRef] [PubMed]
180. Aktaran Bala, D.; Matur, E.; Ergul Ekiz, E.; Akyazi, I.; Eraslan, E.; Ozcan, M.; Ergen, E.; Erek, M.; Gursel, F.E.; Eseceli, H.; et al. Effects of dietary thyme on immune cells, the antioxidant defence system, cytokine cascade, productive performance and egg quality in laying hens. *J. Anim. Plant Sci.* **2020**, *31*, 394–402.
181. Mao, X.; Dou, Y.; Fan, X.; Yu, B.; He, J.; Zheng, P.; Yu, J.; Luo, J.; Luo, Y.; Yan, H.; et al. The Effect of Dietary *Yucca schidigera* Extract Supplementation on Productive Performance, Egg Quality, and Gut Health in Laying Hens with Clostridium Perfringens and Coccidia Challenge. *Poult. Sci.* **2023**, *102*, 102822. [CrossRef] [PubMed]
182. Laptev, G.Y.; Yildirim, E.A.; Ilina, L.A.; Filippova, V.A.; Kochish, I.I.; Gorfunkel, E.P.; Dubrovin, A.V.; Brazhnik, E.A.; Narushin, V.G.; Novikova, N.I.; et al. Effects of Essential Oils-Based Supplement and Salmonella Infection on Gene Expression, Blood Parameters, Cecal Microbiome, and Egg Production in Laying Hens. *Animals* **2021**, *11*, 360. [CrossRef] [PubMed]
183. Sharma, M.K.; Dinh, T.; Adhikari, P.A. Production Performance, Egg Quality, and Small Intestine Histomorphology of the Laying Hens Supplemented with Phytogenic Feed Additive. *J. Appl. Poult. Res.* **2020**, *29*, 362–371. [CrossRef]

184. Sharma, M.K.; White, D.L.; Singh, A.K.; Liu, H.; Tan, Z.; Peng, X.; Kim, W.K. Effect of Dietary Supplementation of Probiotic Aspergillus Niger on Performance and Cecal Microbiota in Hy-Line W-36 Laying Hens. *Animals* **2022**, *12*, 2406. [CrossRef] [PubMed]
185. Castañeda, C.D.; Dittoe, D.K.; Wamsley, K.G.S.; McDaniel, C.D.; Blanch, A.; Sandvang, D.; Kiess, A.S. In Ovo Inoculation of an Enterococcus Faecium—Based Product to Enhance Broiler Hatchability, Live Performance, and Intestinal Morphology. *Poult. Sci.* **2020**, *99*, 6163–6172. [CrossRef] [PubMed]
186. Adhikari, P.; Lee, C.H.; Cosby, D.E.; Cox, N.A.; Kim, W.K. Effect of Probiotics on Fecal Excretion, Colonization in Internal Organs and Immune Gene Expression in the Ileum of Laying Hens Challenged with *Salmonella enteritidis*. *Poult. Sci.* **2019**, *98*, 1235–1242. [CrossRef] [PubMed]
187. Adhikari, P.A.; Cosby, D.E.; Cox, N.A.; Lee, J.H.; Kim, W.K. Effect of Dietary Bacteriophage Supplementation on Internal Organs, Fecal Excretion, and Ileal Immune Response in Laying Hens Challenged by *Salmonella enteritidis*. *Poult. Sci.* **2017**, *96*, 3264–3271. [CrossRef]
188. Markowiak, P.; Ślizewska, K. The Role of Probiotics, Prebiotics and Synbiotics in Animal Nutrition. *Gut Pathog.* **2018**, *10*, 21. [CrossRef] [PubMed]
189. Mountzouris, K.C.; Tsitrsikos, P.; Palamidi, I.; Arvaniti, A.; Mohnl, M.; Schatzmayr, G.; Fegeros, K. Effects of Probiotic Inclusion Levels in Broiler Nutrition on Growth Performance, Nutrient Digestibility, Plasma Immunoglobulins, and Cecal Microflora Composition. *Poult. Sci.* **2010**, *89*, 58–67. [CrossRef]
190. Kogut, M.H.; Lee, A.; Santin, E. Microbiome and Pathogen Interaction with the Immune System. *Poult. Sci.* **2020**, *99*, 1906–1913. [CrossRef] [PubMed]
191. Broom, L.J.; Kogut, M.H. The Role of the Gut Microbiome in Shaping the Immune System of Chickens. *Vet. Immunol. Immunopathol.* **2018**, *204*, 44–51. [CrossRef] [PubMed]
192. Jha, R.; Das, R.; Oak, S.; Mishra, P. Probiotics (Direct-Fed Microbials) in Poultry Nutrition and Their Effects on Nutrient Utilization, Growth and Laying Performance, and Gut Health: A Systematic Review. *Animals* **2020**, *10*, 1863. [CrossRef] [PubMed]
193. Memon, F.U.; Yang, Y.; Zhang, G.; Leghari, I.H.; Lv, F.; Wang, Y.; Laghari, F.; Khushk, F.A.; Si, H. Chicken Gut Microbiota Responses to Dietary *Bacillus subtilis* Probiotic in the Presence and Absence of Eimeria Infection. *Microorganisms* **2022**, *10*, 1548. [CrossRef]
194. Lee, K.W.; Lillehoj, H.S.; Jang, S.I.; Li, G.; Lee, S.H.; Lillehoj, E.P.; Siragusa, G.R. Effect of Bacillus-Based Direct-Fed Microbials on *Eimeria maxima* Infection in Broiler Chickens. *Comp. Immunol. Microbiol. Infect. Dis.* **2010**, 33, e105–e110. [CrossRef] [PubMed]
195. Park, I.; Lee, Y.; Goo, D.; Zimmerman, N.P.; Smith, A.H.; Rehberger, T.; Lillehoj, H.S. The Effects of Dietary *Bacillus subtilis* Supplementation, as an Alternative to Antibiotics, on Growth Performance, Intestinal Immunity, and Epithelial Barrier Integrity in Broiler Chickens Infected with *Eimeria maxima*. *Poult. Sci.* **2020**, *99*, 725–733. [CrossRef] [PubMed]
196. Chaudhari, A.A.; Lee, Y.; Lillehoj, H.S. Beneficial Effects of Dietary Supplementation of Bacillus Strains on Growth Performance and Gut Health in Chickens with Mixed Coccidiosis Infection. *Vet. Parasitol.* **2020**, *277*, 109009. [CrossRef] [PubMed]
197. Awais, M.M.; Jamal, M.A.; Akhtar, M.; Hameed, M.R.; Anwar, M.I.; Ullah, M.I. Immunomodulatory and Ameliorative Effects of *Lactobacillus* and *Saccharomyces* Based Probiotics on Pathological Effects of Eimeriasis in Broilers. *Microb. Pathog.* **2019**, *126*, 101–108. [CrossRef] [PubMed]
198. Li, P.; Zheng, L.; Qi, Y.; Liu, Z.; Du, E.; Wei, J.; Zhang, Z.; Guo, S.; Ding, B. Dietary *Lactobacillus fermentum* and *Lactobacillus paracasei* Improve the Intestinal Health of Broilers Challenged with *Coccidia* and *Clostridium perfringens*. *Front. Vet. Sci.* **2022**, *9*, 1025677. [CrossRef] [PubMed]
199. Dalloul, R.A.; Lillehoj, H.S.; Shellem, T.A.; Doerr, J.A. Enhanced Mucosal Immunity against *Eimeria acervulina* in Broilers Fed a Lactobacillus-Based Probiotic. *Poult. Sci.* **2003**, *82*, 62–66. [CrossRef] [PubMed]
200. Zhang, J.L.; Xie, Q.M.; Ji, J.; Yang, W.H.; Wu, Y.B.; Li, C.; Ma, J.Y.; Bi, Y.Z. Different Combinations of Probiotics Improve the Production Performance, Egg Quality, and Immune Response of Layer Hens. *Poult. Sci.* **2012**, *91*, 2755–2760. [CrossRef] [PubMed]
201. Xiang, Q.; Wang, C.; Zhang, H.; Lai, W.; Wei, H.; Peng, J. Effects of Different Probiotics on Laying Performance, Egg Quality, Oxidative Status, and Gut Health in Laying Hens. *Animals* **2019**, *9*, 1110. [CrossRef]
202. Connor Padgett, J.; Allen Byrd, J.; Bailey, C.; Thomas Price, P.; Anthony Bailey, C. Salmonella Enteritidis Control in Mature Laying Hens Through Dry Fed Parietal Yeast Fraction or Bacillus Blend Probiotic. *Int. J. Anim. Sci. Technol.* **2021**, *5*, 1–6. [CrossRef]
203. Price, P.T.; Gaydos, T.A.; Berghaus, R.D.; Baxter, V.; Hofacre, C.L.; Sims, M.D. Salmonella Enteritidis Reduction in Layer Ceca with a Bacillus Probiotic. *Vet. World* **2020**, *13*, 184. [CrossRef] [PubMed]
204. Patterson, J.A.; Burkholder, K.M. Application of Prebiotics and Probiotics in Poultry Production. *Poult. Sci.* **2003**, *82*, 627–631. [CrossRef]
205. Adhikari, P.A.; Kim, W.K. Overview Of prebiotics and probiotics: Focus On performance, Gut Health and Immunity—A Review. *Anim. Sci* **2017**, *17*, 949–966. [CrossRef]
206. Teng, P.Y.; Kim, W.K. Review: Roles of Prebiotics in Intestinal Ecosystem of Broilers. *Front. Vet. Sci.* **2018**, *5*, 398676. [CrossRef]
207. Teng, P.Y.; Adhikari, R.; Llamas-Moya, S.; Kim, W.K. Effects of Combination of Mannan-Oligosaccharides and β-Glucan on Growth Performance, Intestinal Morphology, and Immune Gene Expression in Broiler Chickens. *Poult. Sci.* **2021**, *100*, 101483. [CrossRef] [PubMed]

208. Yadav, S.; Singh, A.K.; Selvaraj, R.K.; Applegate, T.J.; Bhattacharya, P.; Shinall, S.B.; Fenn, L.S.; Shanmugasundaram, R.; Kim, W.K. Research Note: Effect of Dietary Xylo-Oligosaccharide on Growth Performance, Intestinal Histomorphology, and Specific Cecal Bacteria in Broiler Chickens. *Poult. Sci.* **2024**, *103*, 103189. [CrossRef]
209. Faber, T.A.; Dilger, R.N.; Hopkins, A.C.; Price, N.P.; Fahey, J.C. The Effects of a Galactoglucomannan Oligosaccharide-Arabinoxylan (GGMO-AX) Complex in Broiler Chicks Challenged with *Eimeria acervulina*. *Poult. Sci.* **2012**, *91*, 1089–1096. [CrossRef] [PubMed]
210. Leung, H.; Yitbarek, A.; Snyder, R.; Patterson, R.; Barta, J.R.; Karrow, N.; Kiarie, E. Responses of Broiler Chickens to Eimeria Challenge When Fed a Nucleotide-Rich Yeast Extract. *Poult. Sci.* **2019**, *98*, 1622–1633. [CrossRef] [PubMed]
211. Lin, Y.; Teng, P.Y.; Olukosi, O.A. The Effects of Xylo-Oligosaccharides on Regulating Growth Performance, Nutrient Utilization, Gene Expression of Tight Junctions, Nutrient Transporters, and Cecal Short Chain Fatty Acids Profile in Eimeria-Challenged Broiler Chickens. *Poult. Sci.* **2022**, *101*, 102125. [CrossRef] [PubMed]
212. Baurhoo, B.; Ferket, P.R.; Zhao, X. Effects of Diets Containing Different Concentrations of Mannanoligosaccharide or Antibiotics on Growth Performance, Intestinal Development, Cecal and Litter Microbial Populations, and Carcass Parameters of Broilers. *Poult. Sci.* **2009**, *88*, 2262–2272. [CrossRef] [PubMed]
213. Fernandez, F.; Hinton, M.; Van Gils, B. Dietary Mannan-Oligosaccharides and Their Effect on Chicken Caecal Microflora in Relation to *Salmonella* Enteritidis Colonization. *Avian Pathol.* **2002**, *31*, 49–58. [CrossRef] [PubMed]
214. Osho, S.O.; Adeola, O. Impact of Dietary Chitosan Oligosaccharide and Its Effects on Coccidia Challenge in Broiler Chickens. *Br. Poult. Sci.* **2019**, *60*, 766–776. [CrossRef] [PubMed]
215. Chand, N.; Faheem, H.; Khan, R.U.; Qureshi, M.S.; Alhidary, I.A.; Abudabos, A.M. Anticoccidial Effect of Mananoligosacharide against Experimentally Induced Coccidiosis in Broiler. *Environ. Sci. Pollut. Res.* **2016**, *23*, 14414–14421. [CrossRef] [PubMed]
216. Pineda-Quiroga, C.; Borda-Molina, D.; Chaves-Moreno, D.; Ruiz, R.; Atxaerandio, R.; Camarinha-Silva, A.; García-Rodríguez, A. Microbial and Functional Profile of the Ceca from Laying Hens Affected by Feeding Prebiotics, Probiotics, and Synbiotics. *Microorganisms* **2019**, *7*, 123. [CrossRef]
217. Ghasemian, M.; Jahanian, R. Dietary Mannan-Oligosaccharides Supplementation Could Affect Performance, Immunocompetence, Serum Lipid Metabolites, Intestinal Bacterial Populations, and Ileal Nutrient Digestibility in Aged Laying Hens. *Anim. Feed Sci. Technol.* **2016**, *213*, 81–89. [CrossRef]
218. Adhikari, P.; Cosby, D.E.; Cox, N.A.; Franca, M.S.; Williams, S.M.; Gogal, R.M.; Ritz, C.W.; Kim, W.K. Effect of Dietary Fructooligosaccharide Supplementation on Internal Organs Salmonella Colonization, Immune Response, Ileal Morphology, and Ileal Immunohistochemistry in Laying Hens Challenged with Salmonella Enteritidis. *Poult. Sci.* **2018**, *97*, 2525–2533. [CrossRef] [PubMed]
219. Gibson, G.R.; Roberfroid, M.B. Dietary Modulation of the Human Colonic Microbiota: Introducing the Concept of Prebiotics. *J. Nutr.* **1995**, *125*, 1401–1412. [CrossRef]
220. Chalorsuntisakul, S.; Sirithunyalug, J.; Chaiyasuta, C.; Aengwanich, W.; Pewnim, T. Effect of Synbiotics on Caecal Morphology and Lesion Score in Broilers Infected with *Eimeria tenella*. *Avian Biol. Res.* **2010**, *3*, 187–190. [CrossRef]
221. Duff, A.F.; Briggs, W.N.; Chasser, K.M.; Lilburn, M.S.; Syed, B.; Ramirez, S.; Murugesan, R.; Pender, C.; Bielke, L.R. Effect of Dietary Synbiotic Supplementation on Performance Parameters in Turkey Poults Administered a Mixed Eimeria Species Inoculation I. *Poult. Sci.* **2020**, *99*, 4235–4241. [CrossRef] [PubMed]
222. Shanmugasundaram, R.; Markazi, A.; Mortada, M.; Ng, T.T.; Applegate, T.J.; Bielke, L.R.; Syed, B.; Pender, C.M.; Curry, S.; Murugesan, G.R.; et al. Research Note: Effect of Synbiotic Supplementation on Caecal *Clostridium perfringens* Load in Broiler Chickens with Different Necrotic Enteritis Challenge Models. *Poult. Sci.* **2020**, *99*, 2452–2458. [CrossRef]
223. White, M.B. In Ovo and Feed Application of Probiotics or Synbiotics and Response of Broiler Chicks to Post-Hatch Necrotic Enteritis Academic Abstract. Ph.D. Thesis, Virginia Polytechnic Institute and State University, Blacksburg, VA, USA, 2021.
224. Ghasemi, H.A.; Shivazad, M.; Esmaeilnia, K.; Kohram, H.; Karimi, M.A. The Effects of a Synbiotic Containing Enterococcus Faecium and Inulin on Growth Performance and Resistance to Coccidiosis in Broiler Chickens. *J. Poult. Sci.* **2010**, *47*, 149–155. [CrossRef]
225. Shah, B.R.; Hakeem, W.A.; Shanmugasundaram, R.; Selvaraj, R.K. Effect of Synbiotic Supplementation on Production Performance and Severity of Necrotic Enteritis in Broilers during an Experimental Necrotic Enteritis Challenge. *Poult. Sci.* **2023**, *102*, 102959. [CrossRef]
226. Song, D.; Wang, W.; Chen, B.; Li, A.; Song, G.; Cheng, J.; Qiao, L.; Zhu, R.; Min, Y. Dietary Supplemental Synbiotic—Yucca Extract Compound Preparation Modulates Production Performance, Immune Status and Faecal Microflora Diversity in Laying Hens. *Food Agric. Immunol.* **2022**, *33*, 360–376. [CrossRef]
227. Murate, L.S.; Paião, F.G.; de Almeida, A.M.; Berchieri, A.; Shimokomaki, M. Efficacy of Prebiotics, Probiotics, and Synbiotics on Laying Hens and Broilers Challenged with Salmonella Enteritidis. *J. Poult. Sci.* **2015**, *52*, 52–56. [CrossRef]
228. Craig ID, A.D.; Khattak, F.; Hastie, P.; Bedford, M.R.; Olukosi, O.A. The Similarity of the Effect of Carbohydrase or Prebiotic Supplementation in Broilers Aged 21 Days, Fed Mixed Cereal Diets and Challenged with Coccidiosis Infection. *PLoS ONE* **2020**, *15*, e0229281. [CrossRef]
229. Lin, Y.; Lourenco, J.M.; Olukosi, O.A. The Effects of Protease, Xylanase, and Xylo-Oligosaccharides on Growth Performance, Nutrient Utilization, Short-Chain Fatty Acids, and Microbiota in Eimeria-Challenged Broiler Chickens Fed Low-Protein Diet. *Poult. Sci.* **2023**, *102*, 102789. [CrossRef]

230. M'Sadeq, S.A. Ameliorative Effect of Yeast Cell Walls on Broiler Chickens' Performance and Gut Health under Coccidiosis Challenge. *Czech J. Anim. Sci.* **2023**, *68*, 346–355.
231. Gelinas, A.; Sudan, S.; Patterson, R.; Li, J.; Huyben, D.; Barta, J.R.; Kiarie, E.G. Growth Performance, Organs Weight, Intestinal Histomorphology, and Oocyst Shedding in Broiler Chickens Offered Novel Single Strain Bacillus Subtilis Isolated from Camel Dung and Challenged with Eimeria. *Poult. Sci.* **2024**, *103*, 103519. [CrossRef] [PubMed]
232. Huang, N.; Ma, Y.; Chai, J.; Li, Z.; You, X.; Wang, X.; Huang, Y.; Shi, H. In Ovo Injection Dosage of Lactobacillus Rhamnosus on Intestinal Health and Microbial Composition of Yellow Broilers with or without Eimeria Challenge. *J. Appl. Poult. Res.* **2024**, *33*, 100411. [CrossRef]
233. Park, I.; Goo, D.; Nam, H.; Wickramasuriya, S.S.; Lee, K.; Zimmerman, N.P.; Smith, A.H.; Rehberger, T.G.; Lillehoj, H.S. Effects of Dietary Maltol on Innate Immunity, Gut Health, and Growth Performance of Broiler Chickens Challenged with *Eimeria maxima*. *Front. Vet. Sci.* **2021**, *8*, 667425. [CrossRef] [PubMed]
234. Reuben, R.C.; Sarkar, S.L.; Roy, P.C.; Anwar, A.; Hossain, M.A.; Jahid, I.K. Prebiotics, Probiotics and Postbiotics for Sustainable Poultry Production. *Worlds Poult. Sci. J.* **2021**, *77*, 825–882. [CrossRef]
235. Aguilar-Toalá, J.E.; Garcia-Varela, R.; Garcia, H.S.; Mata-Haro, V.; González-Córdova, A.F.; Vallejo-Cordoba, B.; Hernández-Mendoza, A. Postbiotics: An Evolving Term within the Functional Foods Field. *Trends Food Sci. Technol.* **2018**, *75*, 105–114. [CrossRef]
236. Humam, A.M.; Loh, T.C.; Foo, H.L.; Izuddin, W.I.; Awad, E.A.; Idrus, Z.; Samsudin, A.A.; Mustapha, N.M. Dietary Supplementation of Postbiotics Mitigates Adverse Impacts of Heat Stress on Antioxidant Enzyme Activity, Total Antioxidant, Lipid Peroxidation, Physiological Stress Indicators, Lipid Profile and Meat Quality in Broilers. *Animals* **2020**, *10*, 982. [CrossRef] [PubMed]
237. Chang, H.M.; Loh, T.C.; Foo, H.L.; Lim, E.T.C. Lactiplantibacillus Plantarum Postbiotics: Alternative of Antibiotic Growth Promoter to Ameliorate Gut Health in Broiler Chickens. *Front. Vet. Sci.* **2022**, *9*, 883324. [CrossRef] [PubMed]
238. Abd El-Ghany, W.A.; Abdel-Latif, M.A.; Hosny, F.; Alatfeehy, N.M.; Noreldin, A.E.; Quesnell, R.R.; Chapman, R.; Sakai, L.; Elbestawy, A.R. Comparative Efficacy of Postbiotic, Probiotic, and Antibiotic against Necrotic Enteritis in Broiler Chickens. *Poult. Sci.* **2022**, *101*, 101988. [CrossRef] [PubMed]
239. Gingerich, E.; Frana, T.; Logue, C.M.; Smith, D.P.; Pavlidis, H.O.; Chaney, W.E. Effect of Feeding a Postbiotic Derived from Saccharomyces Cerevisiae Fermentation as a Preharvest Food Safety Hurdle for Reducing Salmonella Enteritidis in the Ceca of Layer Pullets. *J. Food Prot.* **2021**, *84*, 275–280. [CrossRef] [PubMed]
240. Ribeiro, G.C.; Mogollón-García, H.D.; de Moraes, A.C.I.; Dias, G.S.; de Brito Viana, G.; Milbradt, E.L.; Andreatti-Filho, R.L.; Okamoto, A.S. Research Note: The Effects of a Lactobacillus Helveticus ATCC 15009-Derived Postbiotic Mitigating Salmonella Gallinarum Colonization in Commercial Layer Chicks. *Poult. Sci.* **2023**, *102*, 103095. [CrossRef] [PubMed]
241. Humam, A.M.; Loh, T.C.; Foo, H.L.; Samsudin, A.A.; Mustapha, N.M.; Zulkifli, I.; Izuddin, W.I. Effects of Feeding Different Postbiotics Produced by Lactobacillus Plantarum on Growth Performance, Carcass Yield, Intestinal Morphology, Gut Microbiota Composition, Immune Status, and Growth Gene Expression in Broilers under Heat Stress. *Animals* **2019**, *9*, 644. [CrossRef] [PubMed]
242. Johnson, C.N.; Kogut, M.H.; Genovese, K.; He, H.; Kazemi, S.; Arsenault, R.J. Administration of a Postbiotic Causes Immunomodulatory Responses in Broiler Gut and Reduces Disease Pathogenesis Following Challenge. *Microorganisms* **2019**, *7*, 268. [CrossRef] [PubMed]

animals

Article

Effects of Starter Diet Energy Concentration on Nutrient Digestibility and Subsequent Growth Performance and Meat Yields of Broilers under Two Coccidiosis Control Programs

Alyson G. Myers [1] and Samuel J. Rochell [1,2,*]

1 Department of Poultry Science, University of Arkansas, Fayetteville, AR 72701, USA
2 Department of Poultry Science, Auburn University, Auburn, AL 36849, USA
* Correspondence: rochesj@auburn.edu

Simple Summary: Simple Summary: Coccidiosis vaccination is a widely used and effective tool for protecting poultry against clinical coccidiosis outbreaks. However, live *Eimeria* vaccines induce mild, transient epithelial damage and inflammation as immunity develops during oocyst cycling. This process can impair nutrient utilization, particularly for dietary lipids. Therefore, this experiment studied the influence of dietary energy level (standard, moderate, or high), adjusted by varying soybean oil inclusion, during the starter phase (0 to 18 d post-hatch) on nutrient and energy digestibility and subsequent performance and carcass characteristics of broilers either vaccinated against coccidiosis on day of hatch or not vaccinated and provided an in-feed anticoccidial drug. Vaccination impaired nutrient and energy digestibility, weight gain up to 31 d, feed conversion, and carcass yield of broilers compared with those fed an in-feed anticoccidial. However, the impacts of vaccination on feed conversion ratio were dependent upon starter diet energy/lipid concentration, even during feeding phases after the starter period, with greater impacts observed at higher energy/lipid levels. Thus, compensating for impaired lipid and energy utilization by coccidiosis-vaccinated broilers through increasing dietary lipid content is an ineffective, and perhaps detrimental, strategy for vaccinated broilers.

Citation: Myers, A.G.; Rochell, S.J. Effects of Starter Diet Energy Concentration on Nutrient Digestibility and Subsequent Growth Performance and Meat Yields of Broilers under Two Coccidiosis Control Programs. *Animals* **2024**, *14*, 1524. https://doi.org/10.3390/ani14111524

Academic Editor: Kenneth Bafundo

Received: 22 March 2024
Revised: 8 May 2024
Accepted: 10 May 2024
Published: 22 May 2024

Abstract: An experiment was conducted to evaluate the live performance, processing characteristics, and apparent ileal digestibility (AID) of nutrients and energy (IDE) in broilers under two coccidiosis control programs (CCP) and fed three starter diet energy levels. Treatments were a factorial arrangement of CCP [in-feed diclazuril (ACD) or vaccinated after hatch (VAC)] and three starter diet energy levels [3008 (standard), 3058 (moderate), and 3108 (high) kcal/kg apparent ME_n] achieved with different soybean oil concentrations. Birds were reared in floor pens (12 per pen) and received experimental starter diets from 0 to 18 d and common grower and finisher diets to 43 d. At d 11, VAC birds had higher ($p < 0.05$) excreta oocyst counts and lower ($p < 0.05$) plasma carotenoids, nutrient AID, and IDE than ACD birds. From 0 to 18 and 0 to 31 d, VAC decreased ($p < 0.05$) body weight gain and increased (energy × CCP, $p < 0.05$) feed conversion ratio of birds fed the moderate and high-energy diets but not for those fed the standard energy diet. From 0 to 43 d, VAC only increased the feed conversion ratio of birds fed the moderate-energy starter diet (energy × CCP, $p < 0.05$). Carcass yields were lower ($p < 0.05$) for VAC birds than for ACD birds, and interactive effects ($p < 0.05$) were observed for wing yield. In summary, increasing dietary lipid concentration to account for *Eimeria*-induced reductions in lipid digestibility during the starter period of coccidiosis-vaccinated broilers may exacerbate, rather than ameliorate, these impacts on bird performance.

Keywords: coccidiosis; broiler; nutrition; energy; lipid; digestibility

1. Introduction

Consumer demand for poultry products originating from birds reared without antibiotics has increased the use of vaccines to replace ionophore anticoccidial drugs for the control of coccidiosis. As such, vaccines are increasingly used in rotational or "bioshuttle" programs with chemical anticoccidial drugs, which are often permitted in these systems but are susceptible to resistance. Although live oocyst vaccination is effective in promoting immunity and protection against clinical coccidiosis, it can induce a mild transient form of coccidiosis, referred to as "coccidiasis", that usually occurs between 14 and 28 d post-hatch impairs broiler performance [1]. Impaired feed efficiency during vaccine cycling is presumably due in part to nutrient malabsorption associated with intestinal damage and inflammation attributed to the sub-clinical, vaccine-induced infection [1–3].

Previous work conducted in our laboratory longitudinally characterized nutrient digestibility in floor-reared broilers given a coccidiosis vaccine on the day of hatch and showed that nutrient digestibility was impaired the most at 12 d post-hatch, with the greatest duration and magnitude of this impact occurring for lipid digestibility [4]. Impaired lipid digestibility also reduces the absorption of fat-soluble nutrients, including vitamins and carotenoids, which may lead to indirect consequences on bone health and pigmentation [5]. Intestinal damage associated with an *Eimeria* infection also increases epithelial cell turnover and induces an inflammatory immune response, both of which are energetically costly and occur at the expense of broiler performance [6,7]. Indeed, the energetic costs of impaired digestibility and metabolic changes during a coccidial infection are consequential [8].

Providing an increased concentration or a more available source of nutrients for which digestibility is impaired is a potential strategy to support the performance of broilers during coccidial vaccine cycling. For example, previous research reported that increased dietary digestible amino acid content improved the performance of broilers challenged with a coccidiosis vaccine [3,9,10]. Korver et al. [11] reported that increasing dietary metabolizable energy from 2714 to 3303 kcal/kg by increasing corn oil supplementation improved the performance of lipopolysaccharide-challenged broilers, whereas increased tallow supplementation did not, indicating that both dietary lipid type and level may influence inflammatory responses. However, to our knowledge, responses of coccidiosis-vaccinated broilers to increased dietary energy resulting from higher lipid inclusion have not been investigated. Therefore, the objective of this experiment was to test the hypothesis that live performance and processing characteristics (i.e., deboned parts yield) of coccidiosis-vaccinated broilers can be improved by increasing dietary soybean oil inclusion during the starter phase (0 to 18 d) to account for the expected vaccine-induced reductions in energy digestibility during this period.

2. Materials and Methods

2.1. General Bird Husbandry and Dietary Treatments

A total of 1368 male broiler by-product chicks from a Cobb 500 female line were obtained from a commercial hatchery on the day of hatch. Upon arrival, one-half (684) of the chicks was orally gavaged with the manufacturer's recommended dose of a live oocyst vaccine (Coccivac®-B52; Merck Animal Health, Intervet Inc., Millsboro, DE, USA) (VAC), whereas the other half was not vaccinated and received in-feed diclazuril (Clinacox, Huvepharma, Peachtree City, GA, USA), a chemical anticoccidial drug (ACD), throughout the entire experiment. The vaccine was delivered by oral gavage (0.25 mL/bird) using a stainless-steel gavage needle to provide uniform administration. All chicks were group-weighed and distributed to 114 floor pens on unused pine shavings. Throughout the trial, litter was sprayed daily with water to increase litter moisture and promote oocyst sporulation. Each floor pen (0.91 × 1.22 m) was equipped with a hanging feeder and a nipple drinker line and contained 12 birds (0.09 m^2 per bird). Birds were provided access to feed and water ad libitum throughout the 43 d experiment, and the lighting schedule and temperature targets were adjusted according to published management guidelines [12].

From 0 to 18 d of age, ACD and VAC birds were provided one of three isonitrogenous experimental starter diets that contained 2.10, 2.67, or 3.24% soybean oil (ME value of soybean oil used in dietary formulation was 8800 kcal/kg) included at the expense of cellulose, with calculated AME_n contents of 3008 (standard), 3058, (moderate) and 3108 (high) kcal/kg, respectively (Table 1). Throughout the remainder of the trial, ACD and VAC birds were fed common grower (18 to 31 d) and finisher (31 to 43 d) diets. Corn, soybean meal, and distiller's dried grains with solubles were used as the primary ingredients, and all diets in Table 1 were formulated to meet or exceed published nutrient recommendations [13].

Table 1. Composition of diets fed to broilers from 0 to 43 d post-hatch [1].

Ingredient, % as-Fed	Experimental (0–18 d)			Common Grower (18–31 d)	Common Finisher (31–43 d)
	Standard	Moderate	High		
Corn	53.94	53.94	53.94	58.46	60.45
Soybean meal (45%)	34.81	34.81	34.81	30.12	27.84
DDGS	4.00	4.00	4.00	6.00	6.00
Soybean oil	2.10	2.67	3.24	2.29	2.97
Limestone	1.09	1.09	1.09	1.09	1.03
Dicalcium phosphate	1.03	1.03	1.03	0.83	0.63
Salt	0.34	0.34	0.34	0.33	0.31
DL-methionine	0.31	0.31	0.31	0.25	0.21
L-lysine HCl	0.23	0.23	0.23	0.18	0.10
L-threonine	0.09	0.09	0.09	0.06	0.08
Trace mineral premix [2]	0.10	0.10	0.10	0.10	0.10
Vitamin premix [3]	0.10	0.10	0.10	0.10	0.10
Se premix [4] (0.06%)	0.02	0.02	0.02	0.02	0.02
Choline chloride (60%)	0.10	0.10	0.10	0.05	0.05
Santoquin	0.02	0.02	0.02	0.02	0.02
Phytase [5]	0.01	0.01	0.01	0.01	0.01
Xylanase [6]	0.03	0.03	0.03	0.03	0.03
Titanium dioxide	0.50	0.50	0.50	-	-
Inert filler [7]	1.19	0.62	0.05	0.05	0.05
Calculated composition, % unless noted otherwise					
AME_n, kcal/kg	3008	3058	3108	3108	3175
CP	22.00	22.00	22.00	20.00	19.00
Digestible lysine	1.18	1.18	1.18	1.05	0.95
Digestible TSAA	0.89	0.89	0.89	0.80	0.74
Digestible threonine	0.77	0.77	0.77	0.69	0.65
Calcium	0.90	0.90	0.90	0.84	0.76
Available *p*	0.45	0.45	0.45	0.42	0.38
Analyzed composition, % unless noted otherwise					
Gross energy, kcal/kg	4027	4082	4062	4038	4097
CP	23.15	23.65	22.95	21.55	20.55
Ether extract	5.00	5.41	6.00	4.71	5.35

[1] DDGS—distiller's dried grains with solubles; AME_n—nitrogen-corrected apparent metabolizable energy. [2] Supplied the following per kg of diet: manganese, 100 mg; zinc, 100 mg; copper, 10.0 mg; iodine, 1.0 mg; iron, 50 mg; magnesium, 27 mg. [3] Supplied the following per kg of diet: vitamin A, 30,863 IU; vitamin D3, 22,045 ICU; vitamin E, 220 IU; vitamin B12, 0.05 mg; menadione, 6.0 mg; riboflavin, 26 mg; d-pantothenic acid, 40 mg; thiamine, 6.2 mg; niacin, 154 mg; pyridoxine, 11 mg; folic acid, 3.5 mg; biotin, 0.33 mg. [4] Supplied 0.12 mg of selenium per kg of diet. [5] Optiphos® (Huvepharma Inc., Peachtree City, GA, USA) provided 250 FTU/kg of diet. [6] Hostazym® X 250 (Huvepharma Inc., Peachtree City, GA, USA) provided 250 FTU/kg of diet. [7] Solka-Floc (Cellulose International Fiber Corporation, North Tonawanda, NY, USA) was used as the inert filler to provide space for Clinacox® (Huvepharma Inc., Peachtree City, GA, USA) and the addition of soybean oil.

2.2. Determination of Growth Performance, Plasma Carotenoids, and Excreta Oocyst Shedding

Birds and feeders were weighed at 0, 11, 18, 32, and 43 d post-hatch for calculation of body weight gain (BWG), feed intake (FI), and feed conversion ratio (FCR) to assess growth performance. All dead and culled birds were weighed individually, and FCR calculations were adjusted to include the weight gain of dead birds. From 0 to 11 d post-hatch, performance parameters were represented by 19 replicate pens for each treatment.

At 11 d post-hatch, after all pens had been weighed, all birds in 7 replicate pens were euthanized for the collection of blood and ileal digesta, whereas the remaining 12 replicate pens were used for performance measurements throughout the remainder of the trial. All birds from each treatment's seven replicate sampling pens were euthanized by CO_2 inhalation. Blood was collected from two randomly selected birds per pen via cardiac puncture and placed into tubes containing EDTA. After collection, tubes were placed on ice and subsequently centrifuged for 15 min at 1300× *g* and 4 °C to separate plasma. Plasma from birds within a pen were pooled, aliquoted, and stored at −80 °C until further analysis. All blood processing and carotenoid analysis procedures were conducted under yellow light and were determined by spectrophotometry as previously described by Allen [14]. Duplicate aliquots of 0.1 mL of plasma from each bird were diluted in 0.9 mL of denatured ethanol, centrifuged at 3000 rpm for 10 min to sediment protein, and 200 µL of supernatant from each sample was transferred into duplicate wells of a microtiter plate. Absorbance at 530 nm was subtracted from absorbance at 474 nm. The difference in these values was multiplied by 81 as the extinction coefficient of lutein to calculate lutein equivalents.

Oocyst shedding, as indicated by the number of oocysts per gram of pooled excreta samples collected from each pen, was determined before bird placement and at 11 d post-vaccination to confirm vaccine cycling. Samples of unused litter shavings were taken before placement, whereas fresh excreta samples at 11 d were collected from each pen using wax paper placed on the litter 12 h prior to collection. Each sample consisted of multiple droppings pooled within pen. All samples were placed in airtight conical tubes and kept refrigerated until processing. Samples were soaked in water overnight and homogenized by vigorous stirring. Following homogenization, 1 mL of sample was further diluted with 9 mL of saturated salt solution and pipetted into the chamber of a McMaster counting slide. Duplicate counts were made for each sample using the following equation:

$$\text{Oocysts per gram of excreta} = (\text{Oocyst count} \times \text{dilution} \times \text{volume})/(\text{volume of counting chamber} \times \text{weight of sample})$$

where the dilution was 10, and the volume of the counting chamber was 0.15 mL.

2.3. Determination of Apparent Ileal Digestibility and Ileal Digestible Energy

Digesta contents of the distal half of the ileum from all birds in the 7 replicate pens used for sampling were collected by gently flushing with deionized water. Digesta samples within each pen were pooled and frozen (−20 °C) until analysis. Frozen digesta samples were lyophilized and finely ground using an electric coffee grinder. Diet and digesta samples were analyzed for dry matter, gross energy, nitrogen, and ether extract content. Gross energy was determined with a bomb calorimeter (Parr 6200 bomb calorimeter, Parr Instruments Co., Moline, IL, USA). Nitrogen was determined using the combustion method (Fisions NA-2000, CE Elantech, Lakewood, NJ, USA) standardized with EDTA (method 990.03, AOAC International [15]) and ether extract was determined according to AOAC [15] method 920.39. Titanium dioxide was included in the feed at 0.5% as an indigestible marker, and diet and digesta TiO_2 concentrations were determined in duplicate following the procedures of Short et al. [16]. Apparent ileal digestibility (AID) of dry matter, gross energy, ether extract, and nitrogen were calculated using the following equation:

$$\text{AID, \%} = \{[(X/TiO_2)_{diet} - (X/TiO_2)_{digesta}]/(X/TiO_2)_{diet}\} \times 100,$$

where (X/TiO_2) = ratio of nutrient concentration (%) to TiO_2 (%) in the diet or ileal digesta. Energy digestibility (%) values obtained from the above equation were multiplied by the gross energy content of the feed to calculate apparent ileal digestible energy (IDE) in units of kcal/kg.

2.4. Processing Characteristics

At 43 d post-hatch, all birds from eight replicate pens were selected for processing on d 44 after 12 h of feed withdrawal. Birds were transported in coops to the University of Arkansas Pilot Processing Plant. Individual live bird weights were recorded immediately before live-hanging. Birds were humanely euthanized, processed, and eviscerated. Hot carcass and abdominal fat pad weights were collected, and carcasses were chilled for at least 2 h before deboning. All carcasses from each pen were deboned for collection of parts weights and yields, with yield calculated as a percentage of pre-slaughter live body weight. Processing outcomes included weights and yields of the following: hot and chilled carcass, hot abdominal fat pad, pectoralis major (breast), pectoralis minor (tenders), leg quarters (thigh and drum), and wings.

2.5. Statistical Analysis

Treatments were comprised of a factorial arrangement of coccidiosis control program (CCP; ACD or VAC) × 3 starter dietary energy levels in a completely randomized block design. The pen was considered the experiment unit for all measurements, and pen location was a random blocking factor. Oocyst count data were arcsin square root transformed to achieve normality prior to statistical analysis. Data within each time point were subjected to a two-way ANOVA using the MIXED procedure of SAS 9.4 to assess the main effects of dietary energy level, vaccination status, and their interaction. Statistically different main effect means were separated using a Tukey's multiple comparison test. When interactions were present, single degree of freedom orthogonal contrasts were used to compare ACD and VAC birds within a dietary energy level, with differences based on this comparison denoted with an asterisk. Statistical significance was considered at $p < 0.05$ in all cases.

3. Results

3.1. Oocyst Shedding and Plasma Carotenoids

No oocysts were detected in the shavings before bird placement. At 11 d post-hatch, VAC birds had higher ($p < 0.05$) excreta oocyst output and lower ($p < 0.05$) plasma carotenoid concentrations than ACD birds, with no main or interactive effects ($p > 0.05$) of energy level (Table 2).

Table 2. Effects of coccidiosis vaccination and starter diet energy concentrations on oocyst per gram of excreta sample (OPG) and plasma carotenoid concentration (µg/mL) at 11 d post-hatch [1,2].

Item	Standard		Moderate		High		SEM	*p*-Values [3]		
	ACD	VAC	ACD	VAC	ACD	VAC		Energy	CCP	Interaction
OPG [4]	0.016 (268)	0.086 (8113)	0.026 (1416)	0.091 (9826)	0.021 (567)	0.090 (14,746)	0.0074	0.543	<0.001	0.908
Plasma carotenoids	1.96	1.17	1.69	1.05	1.86	1.05	0.191	0.557	0.001	0.876

[1] Values are LSMeans of 18 or 19 replicate pens for OPG and 7 replicate pens for plasma carotenoids. [2] Abbreviations: ACD—birds unvaccinated and given an in-feed anticoccidial drug; VAC—vaccinated, birds were given a commercial dose of vaccine on the day of hatch. CCP—coccidiosis control program. Standard = 3008 kcal/kg; moderate = 3058 kcal/kg; high = 3108 kcal/kg of nitrogen-corrected apparent metabolizable energy. [3] Overall ANOVA *p*-values for the effects of energy, vaccination, and their interaction. [4] Statistical analysis was performed on arcsin square root transformed data for OPG. Raw means are presented in parentheses.

3.2. Growth Performance and Apparent Ileal Digestibility of Nutrients and Ileal Digestible Energy at 11 d Post-Hatch

Growth performance from 0 to 11 d post-hatch and nutrient and energy digestibility data are presented in Table 3. At 11 d, VAC birds had lower ($p < 0.05$) BWG and higher ($p < 0.05$) FCR than ACD birds, with no differences ($p > 0.05$) in FI. Main effects of CCP ($p < 0.05$) and energy level ($p < 0.05$) were observed for AID of nutrients and IDE at 11 d post-hatch. Vaccinated birds had a lower ($p < 0.05$) AID of dry matter, nitrogen, ether extract, and energy and IDE compared with ACD birds. The AID of nitrogen was influenced by energy level ($p < 0.05$) and was highest for the high-energy diet (81%), intermediate for the moderate-energy diet (80%), and lowest for the standard energy diet (78%). Energy level also influenced the AID of ether extract ($p < 0.05$) and was highest for the standard energy diet (91%), intermediate for the moderate-energy diet (88%), and lowest for the high-energy diet (86%). Furthermore, energy level also influenced IDE ($p < 0.05$) and was highest for the high-energy diet (3426 kcal/kg), intermediate for the moderate-energy diet (3384 kcal/kg), and lowest for the standard energy diet (3279 kcal/kg). Dietary energy level did not influence ($p > 0.05$) AID of dry matter or energy.

Table 3. Effects of coccidiosis vaccination and starter diet energy concentrations on broiler growth performance and apparent ileal digestibility of nutrients and energy in broilers at 11 d post-hatch [1,2].

Item	Standard		Moderate		High		SEM	*p*-Values [3]		
	ACD	VAC	ACD	VAC	ACD	VAC		Energy	CCP	Interaction [4]
Performance from 0 to 11 d post-hatch										
Body weight gain, kg/bird	0.288	0.281	0.295	0.277	0.283	0.276	0.004	0.277	0.002	0.316
Feed intake, kg/bird	0.349	0.340	0.353	0.339	0.334	0.341	0.005	0.127	0.182	0.061
FCR	1.219	1.232	1.200	1.247	1.189	1.244	0.014	0.817	0.001	0.270
Apparent ileal digestibility at 11 d post-hatch										
Dry matter, %	69.1	63.4	70.4	64.7	71.4	65.8	1.10	0.077	0.001	0.999
Nitrogen, %	80.3	76.2	82.0	77.2	83.4	79.1	0.88	0.003	0.001	0.767
Ether extract, %	92.3	89.2	90.6	84.7	90.9	81.3	1.86	0.024	0.001	0.142
Energy, %	72.3	66.6	73.5	67.5	74.8	68.4	1.14	0.116	0.001	0.955
IDE [5], kcal/kg	3432	3125	3530	3239	3581	3271	54	0.014	0.001	0.980

[1] Values are LSMeans of 19 replicate pens for broiler performance and 6 or 7 replicate pens for nutrient digestibility. [2] Abbreviations: ACD—birds unvaccinated and given an in-feed anticoccidial drug; VAC—vaccinated, birds were given a commercial dose of vaccine on the day of hatch. CCP—coccidiosis control program. Standard = 3008 kcal/kg; moderate = 3058 kcal/kg; high = 3108 kcal/kg of nitrogen-corrected apparent metabolizable energy. [3] Overall ANOVA *p*-values for the effects of energy, vaccination, and their interaction. [4] In the case of an interaction, an asterisk (*) denotes statistical significance ($p < 0.05$) due to CCP within a dietary energy level based on a single degree of freedom orthogonal contrasts. [5] IDE—ileal digestible energy.

3.3. Growth Performance to 43 d Post-Hatch and Processing Characteristics

At 18 d post-hatch, VAC birds had lower ($p < 0.05$) BWG and FI than ACD birds, with no main or interactive effects of energy level (Table 4). An energy level by CCP interaction was observed for FCR, where FCR was increased ($p < 0.05$) by VAC in birds fed the moderate and high-energy diets but was not influenced by VAC ($p > 0.05$) in birds fed the standard energy diet. After 18 d post-hatch, birds were fed a common grower diet until 31 d post-hatch and a common finisher diet was fed from 31 to 43 d post-hatch. At 31 d post-hatch, BWG ($p < 0.05$) was lower for VAC birds than for ACD birds, with no dietary effects. There were no treatment effects on FI ($p > 0.05$), but similar to the starter period, an energy level by CCP interaction was observed for FCR ($p < 0.05$) whereby VAC increased FCR in birds fed the moderate and high-energy diets, but not those ($p > 0.05$) fed the standard energy diet. At 43 d post-hatch, neither CCP nor energy level impacted broiler BWG or FI ($p > 0.05$). However, an energy level by CCP interaction for FCR persisted whereby FCR was increased ($p < 0.05$) by VAC in birds fed the moderate-energy diet but was not influenced by VAC ($p > 0.05$) in birds fed the standard or high-energy diets.

At 44 d post-hatch, hot and chilled carcass and wing yields were lower ($p < 0.05$) for VAC birds than for ACD birds, with no effects of dietary energy (Table 5). A diet by CCP interaction for wing weight was observed, whereby VAC reduced ($p < 0.05$) the wing weight in birds fed the moderate and high-energy diets but not those ($p > 0.05$) fed the standard energy diet. No other processing measurement was influenced by energy level or CCP ($p > 0.05$).

Table 4. Effects of coccidiosis vaccination and starter (0 to 18 d) diet energy concentrations on broiler growth performance [1,2].

Item	Standard		Moderate		High		SEM	*p*-Values [3]		
	ACD	VAC	ACD	VAC	ACD	VAC		Energy	CCP	Interaction [4]
0 to 18 d post-hatch										
Body weight gain, kg/bird	0.700	0.681	0.711	0.659	0.703	0.667	0.009	0.740	0.001	0.206
Feed intake, kg/bird	0.899	0.879	0.899	0.859	0.881	0.881	0.012	0.671	0.044	0.233
FCR	1.290	1.316	1.279	1.328 *	1.264	1.340 *	0.008	0.981	0.001	0.006
0 to 31 d post-hatch										
Body weight gain, kg/bird	2.096	2.094	2.115	2.046	2.112	2.049	0.021	0.703	0.007	0.178
Feed intake, kg/bird	2.972	2.956	2.963	2.919	2.975	2.965	0.029	0.579	0.330	0.814
FCR	1.445	1.443	1.415	1.479 *	1.416	1.459 *	0.009	0.507	0.001	0.001
0 to 43 d post-hatch										
Body weight gain, kg/bird	3.419	3.439	3.424	3.384	3.429	3.392	0.028	0.628	0.377	0.447
Feed intake, kg/bird	5.392	5.406	5.363	5.350	5.409	5.395	0.434	0.505	0.892	0.936
FCR	1.603	1.595	1.579	1.627 *	1.584	1.602	0.008	0.381	0.001	0.002

[1] Values are LSMeans of 12 replicate pens. [2] Abbreviations: ACD—birds unvaccinated and given an in-feed anticoccidial drug; VAC—vaccinated, birds were given a commercial dose of vaccine on the day of hatch. CCP—coccidiosis control program. Standard = 3008 kcal/kg; moderate = 3058 kcal/kg; high = 3108 kcal/kg of nitrogen-corrected apparent metabolizable energy. [3] Overall ANOVA *p*-values for the effects of energy, vaccination, and their interaction. [4] In the case of an interaction, an asterisk (*) denotes statistical significance ($p < 0.05$) due to CCP within a dietary energy level based on single degree of freedom orthogonal contrasts.

Table 5. Effects of coccidiosis vaccination and starter diet energy concentrations on broilers processed at 44 d post-hatch [1,2].

Item	Standard		Moderate		High		SEM	*p*-Values [3]		
	ACD	VAC	ACD	VAC	ACD	VAC		Energy	CCP	Interaction [4]
Hot Carcass										
Weight, kg	2.541	2.530	2.533	2.506	2.557	2.520	0.029	0.777	0.301	0.900
Yield, %	75.12	74.73	75.26	75.05	75.40	74.87	0.175	0.351	0.012	0.672
Fat Pad										
Weight, kg	0.034	0.032	0.034	0.035	0.037	0.034	0.002	0.203	0.404	0.360
Yield, %	1.00	0.95	0.99	1.05	1.08	1.02	0.043	0.160	0.533	0.210
Chilled Carcass										
Weight, kg	2.575	2.565	2.569	2.538	2.592	2.555	0.029	0.766	0.284	0.895
Yield, %	76.13	75.75	76.34	76.01	76.43	75.94	0.176	0.312	0.008	0.894
Breast										
Weight, kg	0.683	0.680	0.684	0.673	0.688	0.680	0.010	0.757	0.202	0.965
Yield, %	20.46	20.04	20.31	20.12	20.23	20.15	0.150	0.925	0.063	0.510
Tender										
Weight, kg	0.132	0.134	0.130	0.130	0.133	0.132	0.002	0.196	0.937	0.624
Yield, %	3.91	3.91	3.86	3.89	3.90	3.92	0.030	0.455	0.495	0.848
Leg Quarters										
Weight, kg	0.785	0.788	0.779	0.781	0.789	0.792	0.010	0.517	0.740	0.999
Yield, %	23.14	23.26	23.15	23.41	23.30	23.54	0.133	0.248	0.053	0.829
Wings										
Weight, kg	0.269	0.270	0.273	0.263 *	0.277	0.265 *	0.002	0.489	0.001	0.023
Yield, %	7.96	7.97	8.12	7.89	8.08	7.89	0.617	0.797	0.008	0.094

[1] Values are LSMeans of eight replicate pens. [2] Abbreviations: ACD—birds unvaccinated and given an in-feed anticoccidial drug; VAC—vaccinated, birds were given a commercial dose of vaccine on the day of hatch. CCP—coccidiosis control program. Standard = 3008 kcal/kg; moderate = 3058 kcal/kg; high = 3108 kcal/kg of nitrogen-corrected apparent metabolizable energy. [3] Overall ANOVA *p*-values for the effects of energy, vaccination, and their interaction. [4] In the case of an interaction, an asterisk (*) denotes statistical significance ($p < 0.05$) due to CCP within a dietary energy level based on single degree of freedom orthogonal contrasts.

4. Discussion

Coccidiosis vaccination with live *Eimeria* induces immunity in chickens after several rounds of oocyst cycling [17]. In the current experiment, increased excreta oocyst output confirmed vaccinal oocyst cycling and indicated a successful model for comparing the responses of birds to dietary energy under two coccidiosis control programs. Plasma carotenoids, a sensitive indicator of coccidial-induced intestinal damage in birds [18], reflected malabsorption in VAC birds compared with those given an ACD. Coccidiosis vaccination on the day of hatch also reduced BWG and impaired FCR of broilers at 11 d post-hatch, with no effects on FI. This impaired FCR was likely due in part to nutrient malabsorption, as vaccination decreased IDE by 303 kcal/kg and AID of dry matter, nitrogen, and ether extract by 5.7, 4.5, and 6.2 percentage units, respectively, when compared with ACD birds at 11 d post-hatch. This vaccine-induced decrease in nutrient digestibility, particularly for lipids (ether extract), was expected based on previous reports [19].

The magnitude of reductions in IDE associated with coccidiosis vaccination in previous experiments in our lab [4,20] informed the dietary energy concentrations evaluated in the current study. As the energy density of starter diets increased with soybean oil supplementation, the AID of nitrogen and IDE also increased. Likewise, previous research has shown that supplemental lipids can reduce the digesta passage rate through the intestinal tract, allowing more time for the digestion and absorption of all nutrients present in the diet, including protein and starch [21,22]. The reduction in AID of ether extract that occurred as soybean oil inclusion increased, regardless of vaccination status, may reflect a limited ability of broilers at this age to efficiently digest or absorb lipids beyond a certain threshold [23]. Furthermore, despite being fat-soluble components, plasma carotenoids were not influenced by soybean oil content, indicating that the amount of added lipid in the standard energy diet was likely not limiting carotenoid absorption for birds in either coccidiosis control program.

The reduction in BWG for VAC birds observed at 18 and 31 d diminished by 43 d post-hatch. Previous literature has also reported broilers administered an *Eimeria* vaccine at day of hatch had a 3 to 4% reduction in BWG at 17 and 21 d post-hatch, with no differences in BWG observed at 28 d post-hatch when compared with non-vaccinated broilers [4,24,25]. Therefore, as intended, coccidiosis vaccination provided early *Eimeria* exposure, which allowed birds to compensate for the minor reduction in weight before the end of the grow-out period. Similar to the final live weights, vaccination did not impact hot or chilled carcass weights or yields at 44 d post-hatch in the current experiment. The reduction in wing weight for VAC birds fed moderate and high-energy diets may be due to differences in the protein and amino acid composition of the wing compared with other processing parts, which were not impacted by energy level or vaccination status [9]. Starter diet energy levels did not influence any processing characteristics. Similarly, Birk et al. [26] reported no differences in processing yields at 50 d of age when broilers were fed pre-starter diets that contained increased energy densities achieved by increased lipid inclusions at 1.3, 6, or 8% of the diet.

Despite the lack of difference in final BW and processing yields, increasing the dietary energy density by soybean oil supplementation in the starter period negatively impacted the FCR of VAC birds, not only during the starter period but throughout the grower and finisher periods, which was contrary to our hypothesis. Vaccinated birds fed the moderate and high-energy diets during the starter period had impaired FCR at 18 and 31 d post-hatch, and while no differences in FCR were observed between ACD and VAC birds fed the high-energy diet at 43 d post-hatch, the FCR of VAC birds fed the moderate-energy diet remained impaired. A lack of interaction between starter energy level and CCP on 11 d nutrient digestibility indicates that causes other than altered digestibility were likely responsible for the persistent energy level by vaccine interaction for FCR.

The unexpected, relatively greater impact of coccidiosis vaccination on FCR of birds fed the moderate and high-energy levels than for those fed the standard energy diet may have resulted from a proportionately greater need by vaccinated birds for glucose than for lipids as a fuel source. Benson et al. [27] reported that BWG and FCR of lipopolysaccharide-injected birds responded positively when dietary energy was increased from 2800 t 3200 kcal ME/kg using cornstarch, whereas the same increase in energy achieved wi corn oil exacerbated the growth depression caused by lipopolysaccharide injection. Oth experiments using lipopolysaccharide as an inflammatory agent have demonstrated creased demand for glucose by peripheral tissues and enhanced rates of gluconeogene from various precursors [28].

Enteric pathogenic infection likely also increases the demand for glucose per se as fuel source for enterocytes. Indeed, the demand for glutamine, a glucogenic amino acid, increases during coccidiosis vaccination, potentially for its role as an energy source for immune cells and enterocytes [29]. Additionally, serum lipid levels have been reported to increase following infection or inflammatory response, perhaps due to increased lipolysis and decreased lipoprotein lipase activity, which alters the uptake and utilization of fatty acids during an inflammatory response [30–32]. In humans, the consumption of high-fat diets has been shown to increase intestinal permeability via a reduction in mucin production and downregulation of tight junction proteins [33,34]. Therefore, the intestinal damage and inflammation associated with the sub-clinical, vaccine-induced infection in the current experiment may have been exacerbated when birds were fed increased amounts of soybean oil. As such, an increase in energy density by carbohydrate or amino acid supplementation, and not lipid, may be a more effective strategy for supporting the performance of coccidiosis-vaccinated broilers.

5. Conclusions

In conclusion, coccidiosis vaccination did not compromise final BWG, FI, or most processed parts weights of 43 d old broilers, and a lack of response in these measurements to dietary energy during the starter period (0 to 18 d) indicates that this was not limiting dietary component for these outcomes. However, contrary to our original hypothesis, feeding increased energy density diets through greater soybean oil inclusion during the starter period caused detrimental effects on the feed efficiency of vaccinated broilers that persisted beyond the period of vaccine cycling. As such, further research is warranted to understand the mechanisms of *Eimeria*-induced lipid malabsorption and its impacts on broiler gastrointestinal health beyond reduced energy availability and lipid-soluble vitamin absorption.

Author Contributions: A.G.M. and S.J.R. both contributed to conceptualization, research execution, data summary and analysis, and manuscript preparation and revision. All authors have read and agreed to the published version of the manuscript.

Funding: This research was primarily funded by the U. S. Poultry and Egg Association, Grant #700: Optimizing Amino Acid Digestibility and Energy Values Used in Feed Formulations for Broilers Vaccinated for the Control of Coccidiosis.

Institutional Review Board Statement: All animal care and experimental procedures were approved by the University of Arkansas Institutional Animal Care and Use Committee before initiation of the experiment (protocol #17028), ethics approval date: 3 November 2019.

Informed Consent Statement: Not applicable.

Data Availability Statement: Data are available by contacting the corresponding author.

Acknowledgments: The authors express appreciation to Merck Animal Health for donating the coccidiosis vaccine used in this experiment.

Conflicts of Interest: The authors declare no conflicts of interest.

References

1. Lehman, R.; Moran, E.T.; Hess, J.B. Response of Coccidiostat- versus Vaccination-Protected Broilers to Gelatin Inclusion in High and Low Crude Protein Diets. *Poult. Sci.* **2009**, *88*, 984–993. [CrossRef]

2. Lee, J.T.; Eckert, N.H.; Ameiss, K.A.; Stevens, S.M.; Anderson, P.N.; Anderson, S.M.; Barri, A.; McElroy, A.P.; Danforth, H.D.; Caldwell, D.J. The Effect of Dietary Protein Level on Performance Characteristics of Coccidiosis Vaccinated and Nonvaccinated Broilers Following Mixed-Species Eimeria Challenge. *Poult. Sci.* **2011**, *90*, 1916–1925. [CrossRef]
3. Adedokun, S.A.; Helmbrecht, A.; Applegate, T.J. Investigation of the Effect of Coccidial Vaccine Challenge on Apparent and Standardized Ileal Amino Acid Digestibility in Grower and Finisher Broilers and Its Evaluation in 21-Day-Old Broilers. *Poult. Sci.* **2016**, *95*, 1825–1835. [CrossRef]
4. Gautier, A.E.; Latorre, J.D.; Matsler, P.L.; Rochell, S.J. Longitudinal Characterization of Coccidiosis Control Methods on Live Performance and Nutrient Utilization in Broilers. *Front. Vet. Sci.* **2020**, *6*, 468. [CrossRef]
5. Oikeh, I.; Sakkas, P.; Blake, D.P.; Kyriazakis, I. Interactions between Dietary Calcium and Phosphorus Level, and Vitamin D Source on Bone Mineralization, Performance, and Intestinal Morphology of Coccidia-Infected Broilers. *Poult. Sci.* **2019**, *98*, 5679–5690. [CrossRef]
6. Choct, M. Managing Gut Health through Nutrition. *Br. Poult. Sci.* **2009**, *50*, 9–15. [CrossRef]
7. Persia, M.E.; Young, E.L.; Utterback, P.L.; Parsons, C.M. Effects of Dietary Ingredients and Eimeria Acervulina Infection on Chick Performance, Apparent Metabolizable Energy, and Amino Acid Digestibility. *Poult. Sci.* **2006**, *85*, 48–55. [CrossRef]
8. Teeter, R.; Beker, A.; Brown, C. Calorific Cost of Immunity Development to Coccidiosis. In Proceedings of the XXXIII World's Poultry Congress, Brisbane, Austrailia, 30 June 30–4 July 2008.
9. Cloft, S.E.; Rochell, S.J.; Macklin, K.S.; Dozier, W.A. Effects of Dietary Amino Acid Density and Feed Allocation during the Starter Period on 41 Days of Age Growth Performance and Processing Characteristics of Broiler Chickens given Coccidiosis Vaccination at Hatch. *Poult. Sci.* **2019**, *98*, 5005–5016. [CrossRef]
10. Cloft, S.E.; Rochell, S.J.; Macklin, K.S.; Dozier, W.A. Effects of Pre-Starter and Starter Diets Varying in Amino Acid Density given to Broiler Chickens That Received Coccidiosis Vaccination at Hatch. *Poult. Sci.* **2019**, *98*, 4878–4888. [CrossRef]
11. Korver, D.R.; Roura, E.; Klasing, K.C. Effect of Dietary Energy Level and Oil Source on Broiler Performance and Response to an Inflammatory Challenge. *Poult. Sci.* **1998**, *77*, 1217–1227. [CrossRef]
12. Cobb-Vantress. *Cobb Broiler Management Guide*; Cobb-Vantress: Siloam Springs, AR, USA, 2015.
13. Cobb-Vantress. *Cobb 500 Broiler Performance and Nutrition Supplement*; Cobb-Vantress: Siloam Springs, AR, USA, 2015.
14. Allen, P. Physiological Responses of Chicken Gut Tissue to Coccidial Infection: Comparative Effects of Eimeria Acervulina and Eimeria Mitis on Mucosal Mass, Carotenoid Content, and Brush Border Enzyme Activity. *Poult. Sci.* **1987**, *66*, 1306–1315. [CrossRef]
15. AOAC International. *Official Methods of Analysis of AOAC INTERNATIONAL*, 18th ed.; Horwitz, W., Latimer, G.W., Eds.; AOAC International: Gaithersburg, MD, USA, 2005.
16. Short, F.J.; Gorton, P.; Wiseman, J.; Boorman, K.N. Determination of Titanium Dioxide Added as an Inert Marker in Chicken Digestibility Studies. *Anim. Feed. Sci. Technol.* **1996**, *59*, 215–221. [CrossRef]
17. Chapman, H.D.; Cherry, T.E.; Danforth, H.D.; Richards, G.; Shirley, M.W.; Williams, R.B. Sustainable Coccidiosis Control in Poultry Production: The Role of Live Vaccines. *Int. J. Parasitol.* **2002**, *32*, 617–629. [CrossRef]
18. Hernández-Velasco, X.; Chapman, H.D.; Owens, C.M.; Kuttappan, V.A.; Fuente-Martínez, B.; Menconi, A.; Latorre, J.D.; Kallapura, G.; Bielke, L.R.; Rathinam, T.; et al. Absorption and Deposition of Xanthophylls in Broilers Challenged with Three Dosages of Eimeria Acervulina Oocysts. *Br. Poult. Sci.* **2014**, *55*, 167–173. [CrossRef]
19. Amerah, A.M.; Ravindran, V. Effect of Coccidia Challenge and Natural Betaine Supplementation on Performance, Nutrient Utilization, and Intestinal Lesion Scores of Broiler Chickens Fed Suboptimal Level of Dietary Methionine. *Poult. Sci.* **2015**, *94*, 673–680. [CrossRef]
20. Gautier, A.E.; Rochell, S.J. Influence of Coccidiosis Vaccination on Nutrient Utilization of Corn, Soybean Meal, and Distillers Dried Grains with Solubles in Broilers. *Poult. Sci.* **2020**, *99*, 3540–3549. [CrossRef]
21. Mateos, G.G.; Sell, J.L. Influence of Graded Levels of Fat on Utilization of Pure Carbohydrate by the Laying Hen. *J. Nutr.* **1980**, *110*, 1894–1903. [CrossRef]
22. Latshaw, J.D. Daily Energy Intake of Broiler Chickens Is Altered by Proximate Nutrient Content and Form of the Diet. *Poult. Sci.* **2008**, *87*, 89–95. [CrossRef]
23. Tancharoenrat, P.; Ravindran, V.; Zaefarian, F.; Ravindran, G. Influence of Age on the Apparent Metabolisable Energy and Total Tract Apparent Fat Digestibility of Different Fat Sources for Broiler Chickens. *Anim. Feed. Sci. Technol.* **2013**, *186*, 186–192. [CrossRef]
24. Parker, J.; Oviedo-Rondon, E.O.; Clack, B.A.; Clemente-Hernandez, S.; Osborne, J.; Remus, J.C.; Kettunen, H.; Makivuokko, H.; Pierson, E.M. Enzymes as Feed Additive to Aid in Responses against Eimeria Species in Coccidia-Vaccinated Broilers Fed Corn-Soybean Meal Diets. *Poult. Sci.* **2007**, *86*, 643–653. [CrossRef]
25. da Silva, I.C.M.; Ribeiro, A.M.L.; Canal, C.W.; Pinheiro, C.C.; de Moraes Vieira, M.; Gonçalves, T.A.; Pereira, R.A.; Lacerda, L. Broiler Chicken Responses to Immunological Stimuli as Mediated by Different Levels of Vitamin E in the Diet. *J. Appl. Poult. Res.* **2009**, *18*, 752–760. [CrossRef]
26. Birk, A.R.; Johnson, C.A.; Firman, J.D. Effects of High Fat Broiler Pre-Starter Rations on Performance and Cost. *Int. J. Poult. Sci.* **2016**, *15*, 467–474. [CrossRef]
27. Benson, B.N.; Calvert, C.C.; Roura, E.; Klasing, K.C. Dietary Energy Source and Density Modulate the Expression of Immunologic Stress in Chicks. *J. Nutr.* **1993**, *123*, 1714–1723. [CrossRef]

28. Spitzer, J.A.; Spitzer, J.J. Effect of LPS on Carbohydrate and Lipid Metabolism. In *Beneficial Effects of Endotoxins*; Springer: Boston, MA, USA, 1983; pp. 57–74.
29. Mussini, F.J.; Goodgame, S.D.; Lu, C.; Bradley, C.D.; Fiscus, S.M.; Waldroup, P.W. A Nutritional Approach to the Use of Anticoccidial Vaccines in Broilers: Glutamine Utilization in Critical Stages of Immunity Acquisition. *Int. J. Poult. Sci.* **2012**, *11*, 243–246. [CrossRef]
30. Blackburn, G.L. Lipid Metabolism in Infection. *Am. J. Clin. Nutr.* **1977**, *30*, 132–133. [CrossRef]
31. Grunfeld, C.; Feingold, K.R. Tumor Necrosis Factor, Interleukin, and Interferon Induced Changes in Lipid Metabolism as Part of Host Defense (43424). *Proc. Soc. Exp. Biol. Med.* **1992**, *200*, 224–227. [CrossRef]
32. Van Heugten, E.; Coffey, M.T.; Spears, J.W. Effects of Immune Challenge, Dietary Energy Density, and Source of Energy on Performance and Immunity in Weanling Pigs. *J. Anim. Sci.* **1996**, *74*, 2431–2440. [CrossRef]
33. Duan, Y.; Zeng, L.; Zheng, C.; Song, B.; Li, F.; Kong, X.; Xu, K. Inflammatory Links Between High Fat Diets and Diseases. *Front. Immunol.* **2018**, *9*, 2649. [CrossRef]
34. Rohr, M.W.; Narasimhulu, C.A.; Rudeski-Rohr, T.A.; Parthasarathy, S. Negative Effects of a High-Fat Diet on Intestinal Permeability: A Review. *Adv. Nutr.* **2020**, *11*, 77–91. [CrossRef]

animals

MDPI

Article

Thyme, Oregano, and Garlic Essential Oils and Their Main Active Compounds Influence *Eimeria tenella* Intracellular Development

Martina Felici [1,*], Benedetta Tugnoli [2], Callum De Hoest-Thompson [3], Andrea Piva [1,2], Ester Grilli [1,4,†] and Virginia Marugan-Hernandez [3,†]

1 DIMEVET, Dipartimento di Scienze Mediche Veterinarie, Università di Bologna, Via Tolara di Sopra 50, Ozzano dell'Emilia, 40064 Bologna, Italy; andrea.piva@unibo.it (A.P.); ester.grilli@unibo.it (E.G.)
2 Vetagro S.p.A., Via Porro 2, 42124 Reggio Emilia, Italy; benedetta.tugnoli@vetagro.com
3 Department of Pathobiology and Population Sciences, Royal Veterinary College, University of London, Hatfield AL9 7TA, UK; cdehoestthomp22@rvc.ac.uk (C.D.H.-T.); vhernandez@rvc.ac.uk (V.M.-H.)
4 Vetagro Inc., 17 East Monroe Street Suite #179, Chicago, IL 60603, USA
* Correspondence: martina.felici3@unibo.it
† These authors contributed equally to this work.

Simple Summary: Anticoccidial agents, crucial for controlling parasitic infections in poultry, should ideally interfere with multiple stages of the parasite's lifecycle. This study aimed to evaluate the influence of anticoccidial drugs, natural essential oils, and their pure bioactive compounds on the first round of schizogony of *Eimeria tenella* in vitro. The results revealed that both essential oils and conventional anticoccidial compounds were equally adept at preventing schizont formation. However, the pure bioactive compounds displayed only a slightly reduced level of development, indicating a potential decrease in pathogenicity. This investigation sheds light on the capacity of natural substances to disrupt the intracellular development of the *E. tenella* parasite, providing valuable insights into their mechanisms of action and their potential for safer alternatives in anticoccidial development.

Abstract: Coccidiosis poses a significant challenge in poultry production and is typically managed with ionophores and chemical anticoccidials. However, the emergence of drug resistance and limitations on their use have encouraged the exploration of alternative solutions, including botanical compounds and improvements in in vitro screening methods. Prior research focused only on the impact of these alternatives on *Eimeria* invasion, with intracellular development in cell cultures receiving limited attention. This study assessed the impact of thyme (*Thymus vulgaris*), oregano (*Origanum vulgare*), and garlic (*Allium sativum*) essential oils, as well as their bioactive compounds, on the initial phase of schizogony in Madin–Darby bovine kidney cells, comparing their effectiveness to two commercially used anticoccidial drugs. Using image analysis and quantitative PCR, the study confirmed the efficacy of commercial anticoccidials in reducing invasion and schizont formation, and it found that essential oils were equally effective. Notably, thymol and carvacrol exhibited mild inhibition of intracellular replication of the parasite but significantly reduced schizont numbers, implying a potential reduction in pathogenicity. In conclusion, this research highlights the promise of essential oils and their bioactive components as viable alternatives to traditional anticoccidial drugs for mitigating coccidiosis in poultry, particularly by disrupting the intracellular development of the parasites.

Keywords: avian coccidiosis; *Eimeria tenella*; in vitro; schizogony; botanicals; anticoccidials

Citation: Felici, M.; Tugnoli, B.; De Hoest-Thompson, C.; Piva, A.; Grilli, E.; Marugan-Hernandez, V. Thyme, Oregano, and Garlic Essential Oils and Their Main Active Compounds Influence *Eimeria tenella* Intracellular Development. *Animals* **2024**, *14*, 77. https://doi.org/10.3390/ani14010077

Academic Editor: Kenneth Bafundo

Received: 27 November 2023
Revised: 22 December 2023
Accepted: 23 December 2023
Published: 25 December 2023

1. Introduction

Avian coccidiosis is one of the most concerning parasitic diseases as it poses a significant threat to the health of birds and the poultry industry, which is estimated to be up to

USD 13 billion per year [1]. *Eimeria* spp. are recognized as the causative agent of the disease, and they are protozoan parasites that replicate inside the host's enterocytes, causing inflammation, ulceration, and bleeding. This results in poor nutrient absorption and growth. High mortality rates are also common in severe cases, while, in sub-clinical cases, an infection with *Eimeria* spp. can predispose to secondary opportunistic pathogen infections, such as *Clostridium perfringens*, *Campylobacter jejuni*, and *Salmonella infections* [2–4].

The treatment and prevention of coccidiosis relies on the use of ionophores and chemical anticoccidials; however, spreading of resistance due to their extensive use has raised concerns in the poultry industry [5]. Significant research efforts have been directed towards exploring alternatives to anticoccidials. Particularly, attention has been focused on botanical compounds and natural remedies, which allegedly exhibit the ability to disrupt multiple stages of parasite development, offering promising avenues for combating this parasitic disease [6].

Among the various compounds that have been tested, essential oils (EOs) derived from Lamiaceae and Amaryllidaceae (i.e., *Oregano* and *Allium* spp.) have shown positive outcomes in alleviating the clinical signs of coccidiosis in vivo and in reducing oocyst shedding after infection. In addition, in some cases, an improvement in the *Eimeria*-induced lesions has been assessed [7–9]. The positive effects of these substances are believed to be due to the bioactive molecules they contain. Thymol and carvacrol are two of the main monoterpene phenols of Lamiaceae plants, like thyme and oregano. They have well-documented antibacterial and immunomodulatory proprieties, which make them good candidates to treat enteric infections [10,11]. Also, recent studies have claimed them as anti-protozoal agents against *Lehismania* spp. and *Cryptosporidium parvum* [12,13]. *Allium* spp. contain a wide pool of biologically active organosulfur compounds, including allin, ajoene, allicin, diallyl sulphide, and S-allylcysteine, showcasing antibacterial, anti-inflammatory, antiseptic, antiparasitic, and immunomodulatory properties [14,15].

However, due to their high variability, these botanical mixtures are far from being adequately characterized, especially on their anticoccidial proprieties and modes of action. To do so, an improvement of the available screening methodologies for novel anticoccidial candidates is needed.

The challenges associated with obtaining reliable and standardized in vitro models have led to a predominant reliance on in vivo research methods for studying these alternatives to ionophores and chemical anticoccidials. However, conducting research in vivo presents certain limitations, including the intricate and variable nature of these models and ethical concerns associated with animal experimentation, as well as the high cost and restricted availability of necessary resources [16].

Among the various species causing chicken coccidiosis, *E. tenella* holds particular significance due to its high pathogenicity. For this reason, it has been the most studied avian *Eimeria* species. Consequently, it serves as a valuable model for studying and conducting research on these parasites in vitro [16]. In addition, unlike other species, *E. tenella* is able to undergo the first round of schizogony in immortalized mammalian cell lines, like Madin–Darby bovine kidney (MDBK) cells [17]. This property has been used in previous studies to establish a robust in vitro assay for evaluating the first round of schizogony [18]. Moreover, researchers utilized a transgenic fluorescent strain of *E. tenella*, which expresses yellow fluorescent protein (YFP), to visually examine the developmental timeline and the impact of certain anticoccidial drugs [18].

In the present study, this methodology has been used to examine the anticoccidial proprieties of three essential oils derived from thyme (*Thymus vulgaris*), oregano (*Oregano vulgaris*), and garlic (*Allium sativum*), and their main bioactive compounds ('nature-identical compounds', NIC), on the development of *E. tenella*. Also, the effects of these natural compounds have been compared to two common anticoccidial drugs (AC): salinomycin, an ionophore drug that alters the membrane gradient, and robenidine, a guanidine derivative whose main action is the inhibition of maturation of 1st generation schizonts [19].

2. Material and Methods

2.1. Parasites and Birds

The strain *E. tenella* dYFP, derived from the Wisconsin strain [20], was propagated in 3-week-old White Leghorn chickens reared under specific pathogen-free conditions [21]. Oocyst harvest, excystation, and sporozoite purification were conducted following the methods described in a previous study [22]. Freshly purified sporozoites (0.1×10^6 per replicate) were used to infect a cell monolayer in a 96-well plate format.

2.2. Cell Culture

Madin–Darby bovine kidney cells (MDBK-NBL-1, ECACC-Sigma-Aldrich, Salisbury, UK) were maintained in cell culture flasks in Advanced DMEM (Cat. #11520436—Thermo Fisher Scientific, Waltham, MA, USA) supplemented with 5% fetal bovine serum (Cat. #11573397—Thermo Fisher Scientific, Waltham, MA, USA), 1× penicillin/streptomycin solution (Cat. #11548876—Thermo Fisher Scientific, Waltham, MA, USA), and 1× Glutamax (Cat. #11574466—Thermo Fisher Scientific, Waltham, MA, USA). Four hours prior to invasion, the cells were seeded in 96-well plates (Cat. #10687551—Thermo Fisher Scientific, Waltham, MA, USA) at a concentration of 0.05×10^6 cells/well.

2.3. Chemicals and Reagents

Salinomycin from *Streptomyces albus* (Cat. # S4526, Sigma-Aldrich, St. Louis, MO, USA) and robenidine hydrochloride (Cat. # 33979, Sigma-Aldrich, St. Louis, MO, USA) were resuspended, respectively, with ethanol and dimethyl-sulfoxide (Cat. #D2650, Sigma-Aldrich, St. Louis, MO, USA) and used at a final concentration of 1 and 5 ppm, respectively.

All EOs in this study were analyzed at the same concentration (40 ppm), and the NIC were tested at 20 ppm. The choice of a single concentration was based on a previous study [23].

Oregano essential oil (OEO) was provided by Galen-N (Galen-N Ltd., Sofia, Bulgaria). Garlic oil (GAR) was purchased from Lluch Essence (Lluch Essence S.L.U, Barcelona, Spain) and thyme essential oil (TEO) from Grupo Indukern (Grupo Indukern, Madrid, Spain). All the stock solutions were prepared in ethanol and added to cells, with a final concentration of 40 ppm in supplemented Advanced DMEM. In all cases, the final concentration of ethanol was less than 0.5% (v/v).

Carvacrol (CAR—Cat. #W224511, analytical grade 99%, Sigma-Aldrich, St. Louis, MO, USA), diallyl-disulfide (DDS—Cat. #SMB00378, analytical grade $\geq$ 98%, Sigma-Aldrich, St. Louis, MO, USA), and thymol (THY—Cat. #T0501, analytical grade $\geq$ 98.5%, Sigma-Aldrich, St. Louis, MO, USA) were all diluted in ethanol to prepare stock solutions and used at a final concentration of 20 ppm in supplemented Advanced DMEM.

2.4. Development Assay

After purification, the sporozoites were added to the cells in the absence or presence of the described treatments. The cells were incubated at 41 °C with 5% CO_2. After 4 h of incubation, the non-invading sporozoites were removed from the wells, and the medium with the treatments was replenished. After 48 h, the cells were provided with fresh medium along with the respective treatments.

After 24, 48, 72, and 96 h post-infection (hpi), the cells were either fixed in 4% paraformaldehyde or harvested with RTL buffer (Qiagen, Hilden, Germany), as described by the manufacturer's protocol, for subsequent nucleic acid extraction. Each experiment was repeated at least twice.

2.5. Isolation of Nucleic Acids

Genomic DNA (gDNA) was isolated from the samples stored in RTL buffer. The AllPrep DNA/RNA 96 Kit (Cat. #80311—Qiagen, Hilden, Germany) was used for the isolation process, following the manufacturer's instructions. A vacuum manifold designed for processing 96-well plates (Cat. #9014579—Qiagen, Hilden, Germany) was employed for this purpose.

2.6. Real Time Quantitative PCR (qPCR)

To perform qPCR, the CFX96 Touch R Real-Time PCR Detection System (C1000—Bio-Rad, Hercules, CA, USA) was used. The procedures were carried out according to previously described methods, utilizing the DNA-binding dye SsoFastTM EvaGreen Supermix (Cat. #172-5203—Bio-Rad, Hercules, CA, USA) [24].

For parasite quantification, the number of haploid genomes (equivalent to single sporozoites) per well (four wells per sample as technical replicates) was determined. This quantification was achieved using gDNA-specific primers targeting *E. tenella* 5S rDNA (Fw_5S: TCATCACCCAAAGGGATT, Rv_5S: TTCATACTGCGTCTAATGCAC) [20]. A standard curve of sporozoite gDNA, extracted using the same methods, was generated [25]. This standard curve encompassed gDNA equivalent to 10^7 genomes, followed by serial dilution until 10^2 genomes.

The inhibition of invasion at 24 hpi was calculated with the following equation developed by Thabet et al. [26]:

$$\%\,of\ inhibition = 100 \times \left(1 - \frac{number\ of\ E.\ tenella\ gene\ copies\ in\ treated\ sample}{number\ of\ E.\ tenella\ gene\ copies\ in\ non-treated\ control}\right)$$

To compare the replication rates after treatments, excluding the effects of invasion, slopes of the replication lines were calculated between 24 and 72 hpi. The slopes were calculated with the formula below, as previously described by Arias-Maroto et al. [25]:

$$m\ (slope) = \frac{y\ (72\ \text{hpi}) - y\ (24\ \text{hpi})}{48\ (time\ lag)}$$

2.7. Image Analysis

The paraformaldehyde-fixed samples were observed using a fluorescence microscope (Nikon Eclipse Ti2—Nikon, Tokyo, Japan) at 20×, using Plan fluor Nikon lenses at a wavelength of 490 nm. Six random fields were captured and analyzed with a semi-automated procedure relying on a custom script designed on ImageJ. Each picture was converted into an 8-bit image before applying threshold masks separately and converting the picture into a binary image. On the binary images, two possible values (black or white) were given to each pixel, according to the intensity of the fluorescence. The final particle analysis involved identifying the edges of the black and white structures; this allowed the recording of particle numbers and sizes, both on average and individually. These data were collected and processed for statistical analysis. To assess general parasite growth, the average area, including any sort of stage (sporozoites and schizonts), was calculated for every field. Also, for specific cases where a certain level of growth was observed, the schizont area was calculated separately, assuming as schizonts the particles with an area higher than 50 μm^2. The rate of schizont growth was calculated as the slope of the regression line between 24 and 72 hpi, as indicated by the formula reported in the previous section.

2.8. Statistical Analysis

GraphPad Prism 9.4.1 was utilized for statistical analysis. For the qPCR data, the number of genomes was normalized based on the mean value of the challenged control at 24 hpi of each independent experiment. For all data, the descriptive analysis was conducted, and normality was assessed with the Shapiro–Wilk test ($p > 0.05$). Normally distributed data were analyzed using either a one-way ANOVA (for schizont size) or a two-way ANOVA (for repeated measurements). Non-normally distributed data were analyzed using the Kruskal-Wallis test (for schizont number) or mixed-effects analysis (for repeated measurements). To define significant differences among treatments, post hoc multiple comparison analyses were performed: the mean of each treatment was compared with the mean of every other treatment using Tukey's test for normally distributed data and mixed-effects analysis, while Dunn's test was adopted for non-normally distributed data. Differences were considered significant when the p value was ≤ 0.05 and were denoted as letters.

3. Results

3.1. Characterization of *E. tenella* Intracellular Growth in MDBK Cells

Infected MDBK cells were examined using a fluorescence microscope to evaluate the morphological characteristics of the different developmental stages obtained in culture on a 96-h time course. Figure 1 shows representative images of the development at each time point.

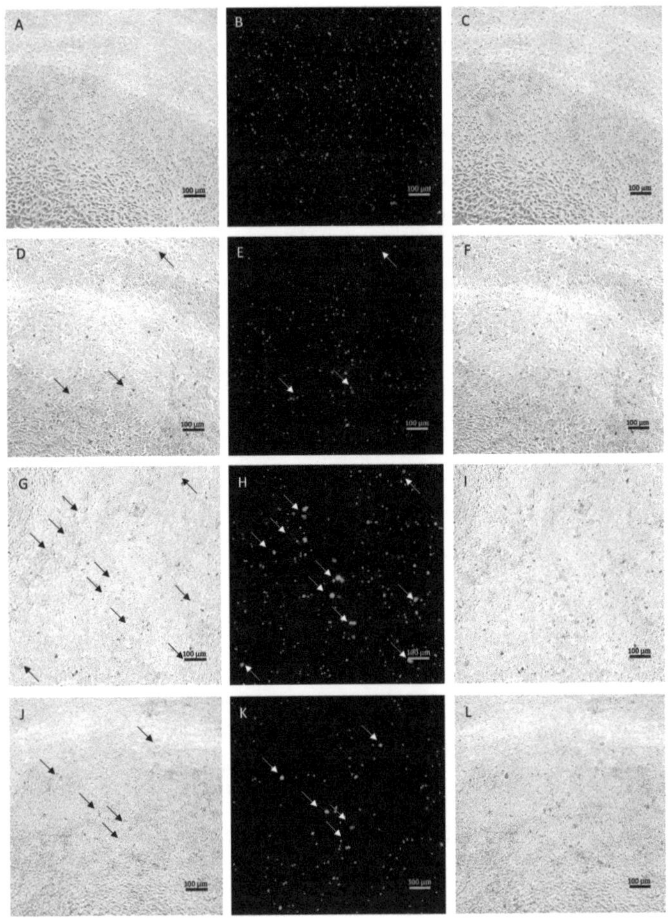

Figure 1. Intracellular development time course (20× objective). Images (**A–C**) indicate the intracellular sporozoites at 24 h post infection (hpi) ((**A**) = merge, (**B**) = green fluorescence, (**C**) = brightfield). Images (**D–F**) indicate intracellular sporozoites at 48 hpi and early stages of development, indicated by black arrows in merge pictures and white arrows in green fluorescence pictures ((**D**) = merge, (**E**) = green fluorescence, (**F**) = brightfield). Images (**G–I**) indicate intracellular parasite at 72 hpi and late stages of development, indicated by white arrows ((**G**) = merge, (**H**) = green fluorescence, (**I**) = brightfield). Images **J–L** indicate intracellular parasites at 96 hpi. Residual developmental stages are indicated by white arrows ((**J**) = merge, (**K**) = green fluorescence, (**L**) = brightfield). Size bar ~100 μm.

During the first phases of infection, occurring within 24 hpi, sporozoites that successfully invaded the cells were observed. At 48 hpi, it was already possible to observe some early stages of development, corresponding to early schizonts. Those appeared as round objects, larger in size compared to intracellular sporozoites. At 72 hpi, larger schizonts were clearly observed in the infected monolayer. Compared to early schizonts, which are difficult to distinguish using bright field, late ones appeared more condensed and, in most cases, were clearly discernable in the monolayer. After 96 h, the number and size of schizonts dropped visibly due to schizont breakage.

Figure 2 shows the average area increase of intracellular parasites during the time course, considering sporozoites at 24 hpi and only schizonts at later timepoints. Starting from an area of around 30.0 μm^2, the intracellular sporozoites started their maturation in schizonts. At 48 hpi, the average area was significantly bigger (181.0 μm^2), but the highest increase was visible at 72 hpi, as the schizonts' area reached an average of 377.0 μm^2. At 96 hpi, the area was slightly smaller than the previous timepoint (240.5 μm^2).

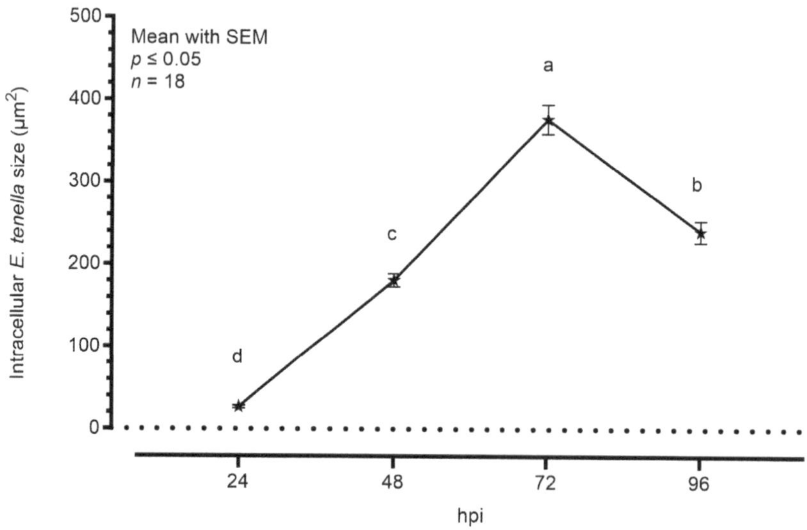

Figure 2. Average size of *E. tenella* intracellular stages measured by semi-automated imaging software. Data are represented as mean ± SEM (n = 18). The plotted values are 26.70 ± 1.72 at 24 hpi, 181.00 ± 7.94 at 48 hpi, 377.00 ± 17.91 at 72 hpi, and 240.50 ± 13.01 at 96 hpi. Different letters indicate significant differences among timepoints ($p \leq 0.05$).

3.2. Effects of Natural Compounds on Intracellular Growth

The effect of the different tested compounds on the average size of intracellular *E. tenella* stages is shown in Figure 3 and Table 1. Concerning AC and all the EO, no intracellular development was visible; in the cell monolayer only internalized sporozoites, characterized by an average size of 30 μm^2, were observed. In the case of NIC, THY 20 ppm allowed a low degree of intracellular development at 48 and 72 hpi, while CAR and DDS did not cause any significant reduction of the average *E. tenella* size compared to the challenged control.

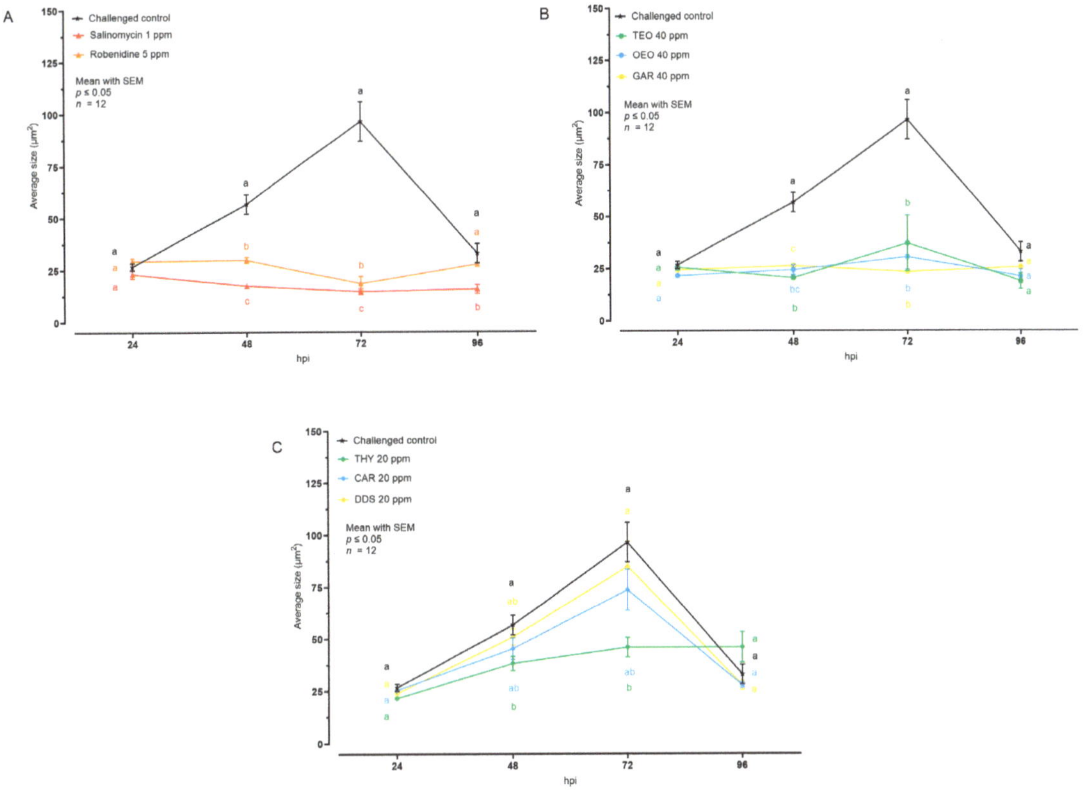

Figure 3. Average size of intracellular *E. tenella* after treatment with anticoccidial drugs (**A**), essential oils (**B**), or nature-identical compounds (**C**). Data are represented as mean with SEM (n = 12), and the values are listed in Table 1. Different letters indicate significant differences among groups within the same time point and the same graph ($p \leq 0.05$). Thyme essential oil (TEO); oregano essential oil (OEO); garlic oil (GAR); thymol (THY); carvacrol CAR); diallyl-disulfide (DDS).

Table 1. Average size of intracellular *E. tenella* after treatment. Data are represented as mean with SEM (n = 12). Values are represented as graph in Figure 3.

	Average Size (µm^2)			
	24 hpi	**48 hpi**	**72 hpi**	**96 hpi**
Challenged control	26.70 ± 1.72	56.65 ± 4.70	96.30 ± 9.49	32.8 ± 4.65
Robenidine 5 ppm	29.49 ± 1.32	29.90 ± 1.44	18.47 ± 3.38	27.60 ± 1.14
Salinomycin 1 ppm	23.21 ± 2.07	17.50 ± 1.13	14.54 ± 1.40	15.66 ± 2.12
TEO 40 ppm	25.62 ± 1.40	20.46 ± 1.14	37.01 ± 13.38	18.76 ± 3.81
OEO 40 ppm	21.58 ± 1.22	24.30 ± 2.41	30.41 ± 5.89	21.08 ± 3.39
GAR 40 ppm	24.54 ± 1.16	26.35 ± 0.60	23.38 ± 1.21	25.67 ± 0.97
THY 20 ppm	21.59 ± 0.92	38.26 ± 3.39	45.88 ± 4.71	45.85 ± 7.33
CAR 20 ppm	25.45 ± 1.04	45.31 ± 5.16	73.44 ± 9.94	27.71 ± 1.26
DDS 20 ppm	24.05 ± 1.60	51.02 ± 3.91	84.870 ± 11.98	27.76 ± 2.23

Thyme essential oil (TEO); oregano essential oil (OEO); garlic oil (GAR); thymol (THY); carvacrol CAR); diallyl-disulfide (DDS).

To better understand if the growth velocity was affected by the treatments, the rate of growth was calculated, considering average area values between 24 and 72 hpi (Table 2). In the challenged control group, the intracellular parasites grew by 1.21 µm^2/h. All AC and

EO managed to stop growth, keeping the growth rate close to 0. Among NIC, only THY was significantly lower than the challenged control.

Table 2. Rate of growth of intracellular *E. tenella* after treatment (between 24 and 72 hpi). Data are reported as mean with SEM (n = 12).

	Rate (μm^2/h)
Challenged control	1.21 ± 0.08 [a]
Robenidine 5 ppm	-0.23 ± 0.09 [d]
Salinomycin 1 ppm	-0.17 ± 0.04 [d]
TEO 40 ppm	-0.04 ± 0.03 [cd]
OEO 40 ppm	0.07 ± 0.06 [cd]
GAR 40 ppm	-0.03 ± 0.03 [cd]
THY 20 ppm	0.51 ± 0.10 [bc]
CAR 20 ppm	0.87 ± 0.24 [ab]
DDS 20 ppm	0.97 ± 0.27 [ab]

[a,b,c,d] Significant differences among groups ($p \leq 0.05$). Thyme essential oil (TEO); oregano essential oil (OEO); garlic oil (GAR); thymol (THY); carvacrol CAR); diallyl-disulfide (DDS).

The sizes and numbers of *E. tenella* schizonts were also assessed after treatment in cases where a certain degree of intracellular growth was observed. This was done to examine the impact of these substances specifically on the schizonts at the peak of development, observed at 72 hpi (Figure 4).

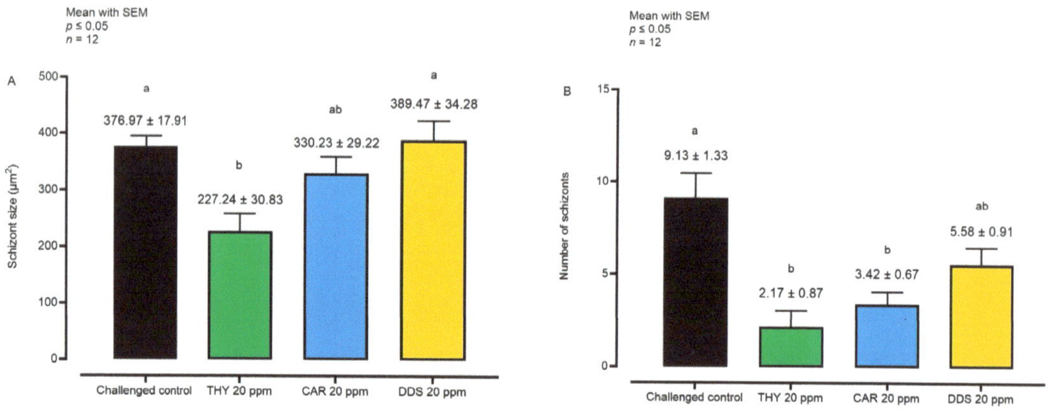

Figure 4. Analysis of schizont size (**A**) and number (**B**) at 72 hpi. Data are represented as mean with SEM (n = 12). Different letters indicate significant differences among groups ($p \leq 0.05$). Thymol (THY); carvacrol CAR); diallyl-disulfide (DDS).

Considering the schizont size at 72 hpi, only THY reduced the parameter compared to the challenged control. However, a reduction in the number of sporozoites per field was observable with both THY and CAR, while with DDS, only a non-significant numerical reduction was visible.

3.3. *Effects of Natural Compounds on Parasite Invasion and Replication*

Almost all the compounds significantly reduced the internalization of sporozoites at 24 hpi (Table 3). AC and EO lead to the most significant decrease of invasion, with values above 50% for both groups. Among NIC, only CAR significantly inhibited invasion regarding the control, with a level of 43%.

Table 3. Level of inhibition of invasion 24 hpi. Data are reported as mean with SEM (n = 8).

	Level of Inhibition (%)
Challenged control	0.00 ± 9.29 [bc]
Robenidine 5 ppm	51.65 ± 5.44 [a]
Salinomycin 1 ppm	61.40 ± 3.04 [a]
TEO 40 ppm	67.76 ± 5.88 [a]
OEO 40 ppm	55.86 ± 10.63 [a]
GAR 40 ppm	44.70± 10.81 [a]
THY 20 ppm	−26.08 ± 19.74 [c]
CAR 20 ppm	43.05 ± 3.62 [a]
DDS 20 ppm	38.96± 10.47 [ab]

[a,b,c] Significant differences among groups ($p \leq 0.05$). Thyme essential oil (TEO); oregano essential oil (OEO); garlic oil (GAR); thymol (THY); carvacrol CAR); diallyl-disulfide (DDS).

Figure 5 shows the quantification of *E. tenella* intracellular genomes as a percentage in relation to the challenged control at 24 hpi. Parasite replication was evident in the challenged control, as the number of genomes nearly tripled every 24 h, peaking at 72 hpi. However, at 96 hpi, the number of intracellular genomes dropped significantly (probably caused by schizont breakage, with release of merozoites into the washed supernatants).

Table 4. Quantification of intracellular *E. tenella* DNA by qPCR after treatment. Data are represented as mean with SEM (n = 8). Values are represented as a graph in Figure 5.

	Intracellular *E. tenella* DNA (%)			
	24 hpi	**48 hpi**	**72 hpi**	**96 hpi**
Challenged control	100.00 ± 9.29	291.99 ± 82.71	676.43 ± 162.14	311.58 ± 27.35
Robenidine 5 ppm	48.35 ± 5.44	21.93 ± 4.14	8.06 ± 2.36	12.87 ± 4.75
Salinomycin 1 ppm	38.61 ± 3.04	16.28 ± 2.62	13.34 ± 3.97	12.80 ± 3.81
TEO 40 ppm	32.24 ± 5.88	13.73 ± 2.88	13.34 ± 2.14	14.25 ± 3.14
OEO 40 ppm	44.14 ± 10.63	8.54 ± 2.18	8.98 ± 1.59	9.43 ± 2.77
GAR 40 ppm	64.15 ± 12.88	32.12 ± 3.72	16.99 ± 3.78	23.31 ± 7.41
THY 20 ppm	126.08 ± 19.74	94.65 ± 30.03	280.62 ± 103.94	250.85 ± 95.30
CAR 20 ppm	71.43 ± 14.82	121.33 ± 37.52	166.03 ± 51.33	203.15 ± 71.00
DDS 20 ppm	61.04 ± 10.47	208.01 ± 65.58	414.91 ± 33.20	310.68 ± 34.73

Thyme essential oil (TEO); oregano essential oil (OEO); garlic oil (GAR); thymol (THY); carvacrol CAR); diallyl-disulfide (DDS).

The ACs and the EOs successfully reduced the number of intracellular parasite genomes starting from the moment of infection (24 hpi). Subsequently, there was no further detected nuclear replication in any of these groups during the subsequent time points (Figure 5). Concerning NIC, a partial inhibition of replication was noticed in the groups THY and CAR. These groups did not exhibit a distinct peak of growth at 72 hpi, suggesting a certain level of efficacy in inhibiting parasite replication. The DDS group did not show any significant difference, but their numbers at the peak (72 hpi) were lower than in the challenged control, potentially influenced by the lower (but not significant) number of invading sporozoites (Figure 5 and Table 4).

The rate of replication was calculated to investigate the intracellular development velocity of between 24 and 72 hpi (Table 5). In the challenged control group, the replication rate was 6.44 genomes/h. Most of the compounds demonstrated a capacity to reduce this rate. Specifically, AC and EO significantly reduced it, reaching values below 0 (usually caused by non-developing sporozoites observed to leave host cells over time). Meanwhile, THY and CAR managed to decrease the rate to almost half of the challenged control, but not significantly.

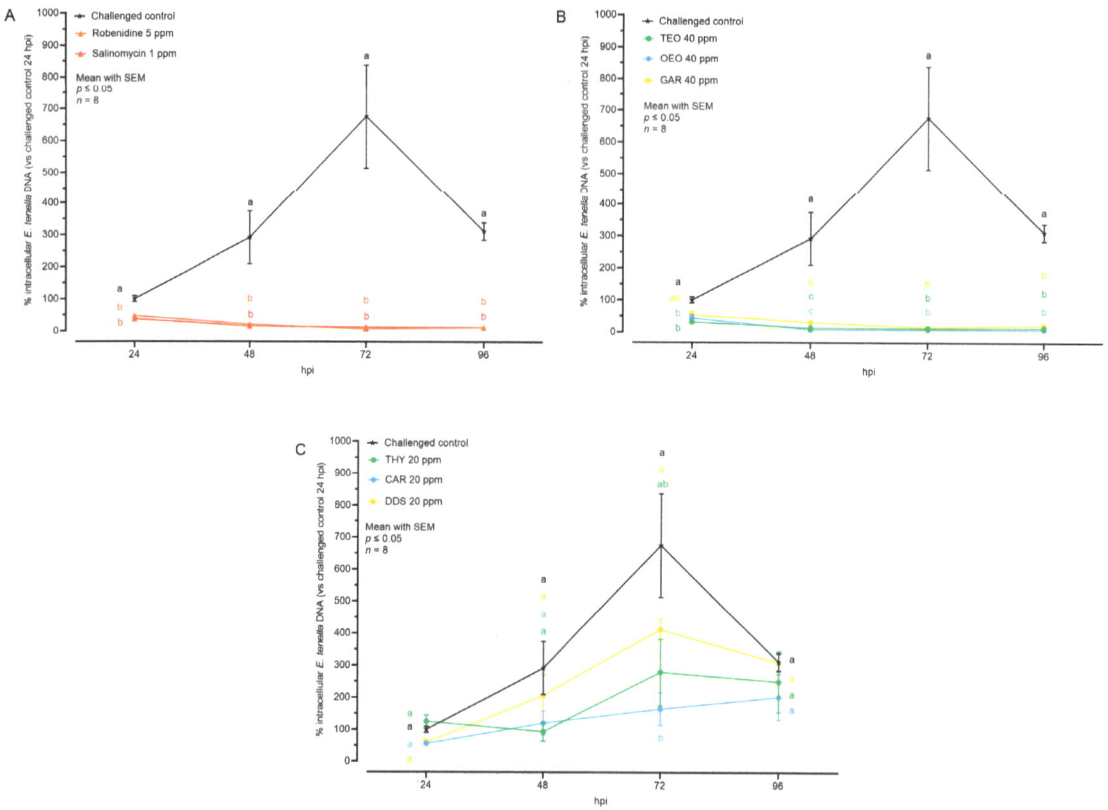

Figure 5. Quantification of intracellular *E. tenella* DNA by qPCR after treatment with anticoccidial drugs (**A**), essential oils (**B**), or nature-identical compounds (**C**). Values are expressed as percentage over the control at 24 hpi. Data are represented as mean with SEM (n = 8) and the values are listed in Table 4. Significant differences among groups within the same time point and the same graph are indicated with different letters ($p \leq 0.05$). Thyme essential oil (TEO); oregano essential oil (OEO); garlic oil (GAR); thymol (THY); carvacrol CAR); diallyl-disulfide (DDS).

Table 5. Rate of *E. tenella* DNA replication (between 24 and 72 hpi). Data are reported as mean with SEM (n = 8).

	Rate (genomes/h)
Challenged control	6.44 ± 0.94^{ab}
Robenidine 5 ppm	-0.84 ± 0.14^{c}
Salinomycin 1 ppm	-0.53 ± 0.09^{c}
TEO 40 ppm	-0.39 ± 0.09^{c}
OEO 40 ppm	-0.73 ± 0.23^{c}
GAR 40 ppm	-0.78 ± 0.30^{c}
THY 20 ppm	3.55 ± 2.45^{abc}
CAR 20 ppm	2.63 ± 1.16^{ac}
DDS 20 ppm	7.37 ± 0.82^{b}

^{a,b,c} Significant differences among groups ($p \leq 0.05$). Thyme essential oil (TEO); oregano essential oil (OEO); garlic oil (GAR); thymol (THY); carvacrol CAR); diallyl-disulfide (DDS).

4. Discussion

Anticoccidials of any nature should be safe for the host, easy to produce and apply, and, preferentially, they should interfere with more than one stage of the parasite's lifecycle, with a deep and comprehensive characterization of their mode of action [27]. For this purpose, in vitro systems that employ invasion and development assays on cell cultures have been increasingly used, as they allow a closer look at the undergoing processes and higher throughput testing compared to in vivo systems [27]. *Eimeria* spp., in particular *E. tenella*, are able to invade different cell lines and to complete the first round of schizogony in MDBK cells. Moreover, transgenic parasites expressing the YFP fluorescent marker have become a powerful tool for studies on *Eimeria*, especially to study intracellular development in cell cultures, as they allow the direct visualization of intracellular stages (i.e., schizonts) of development [18,27].

Previous research has explored the anticoccidial effects of thyme, oregano, and garlic essential oils, as well as their main bioactive molecules, both in vivo and in vitro, focusing primarily on the inhibition of sporozoite invasion [23,28]. Considering that parasite pathogenesis is related to intracellular stages, more extensive evaluation is required to comprehensively assess the impact of these novel compounds. [27].

4.1. Anticoccidial Drugs

In the present study, the efficacy of salinomycin and robenidine (commercial anticoccidials) has been confirmed, as both were able to decrease the number of sporozoites invading the cell monolayer and also to prevent the formation of schizonts. As previously reported, a pre-treatment with these molecules, before infection, is already enough to cause sporozoites death, thus stopping any further maturation [18]. In a similar study, the impact of 1 h of pre-incubation with salinomycin and robenidine on intracellular development was investigated by qPCR [25]. Pre-treatment with salinomycin 1 ppm lowered the level of invasion by 31%. Additionally, both robenidine and salinomycin pre-treatment managed to decrease the rate of replication between 24 and 44 hpi [25]. In the present study, the treatment with anticoccidials occurred along invasion and continued for the whole time course. This resulted in higher inhibition of invasion at 24 hpi and in no observed replication, suggesting that the anticoccidial power could be time-dependent.

4.2. Essential Oils

Similar to anticoccidials, thyme, oregano, and garlic essential oils were very effective in preventing the formation of schizonts. Inhibition of invasion was observed at 24 hpi with essential oils, suggesting that the action on sporozoites is so detrimental that internalized sporozoites struggle to undergo further development after meeting these compounds. The effect of similar essential oils on invasion has been explored previously by Sidiropoulou and colleagues [9,28]. *E. tenella* sporozoites were pre-treated for 1 h with different concentrations of garlic, oregano, and thyme essential oils from Greece. They recorded effective inhibition of invasion for oregano and garlic essential oils, in particular, at higher concentrations (100 ppm), whereas thyme had more consistent effects at lower concentrations [9,28]. Also, a previous study conducted on primary chicken enterocytes showed that 40 ppm of garlic, oregano, and thyme essential oils effectively reduced invasion at 24 hpi by around 45% [23]. In the first case, differences can be attributed to a different composition of the essential oils. In the second, where the source of the essential oils was the same as that in the current study, the levels of inhibition were slightly lower in the chicken enterocytes compared to MDBK cells. This could be due to differences in the protocols of invasion, but also, since a better predisposition of *E. tenella* to chicken enterocytes has been observed, differences caused by treatments could be dampened [23]. In addition to the cell model used, the effect of these essential oils on invasion and development was clear in the present study. To our knowledge, this is the first study to explore the impact of essential oils and bioactive molecules on *Eimeria* endogenous development in vitro, extending beyond invasion and reaffirming their potential as anticoccidial agents.

4.3. Nature-Identical Compounds

The power of essential oils lies in their diverse composition, which encompasses a wide range of bioactive molecules like terpenes, phenols, aldehydes, thiol compounds, and others. Often, these derivatives alone can drive the main effects of the original oil, as they maintain a more uniform composition than essential oils, which depends on specific extraction protocols; therefore, they have become a target of study for their therapeutic effects. In this study, thymol, carvacrol, and diallyl-disulfide, major components of thyme, oregano, and garlic essential oils, respectively, have been investigated to understand if these compounds solely drive the observed effects.

Diallyl-disulfide had no significant impact on development, in contrast to the relative essential oil (garlic), suggesting that the inhibitory effect is likely to be mediated by other molecules within the mixture. Garlic oil is rich in sulfur species that are able to oxidize thiols in protein residues, leading to loss of function. Other studies on sporozoite invasion in vitro evidenced that allicin, found in garlic species too, is a very strong anticoccidial. It shares the main chemical structure with diallyl-disulfide; however, allicin has a thiosulfinate group that makes the molecule very unstable and reactive [29]. A poorer anticoccidial performance of diallyl-disulfide compared to garlic oil was also observed in previous research on *E. tenella* invasion [23], concluding that thiosulfinates could be key effectors to drive the anticoccidial effects.

Thymol and carvacrol slightly reduced the intracellular number of *E. tenella* genomes; in particular, carvacrol reduced by 43% the level of invasion at 24 hpi. A similar effect on inhibition of invasion was observed in a previous study, where a mild level of inhibition was observed on chicken enterocytes, especially for carvacrol [23]. However, in the present study, thymol and carvacrol did not completely stop replication. Nevertheless, they proved to be effective in reducing both the number and size of developed schizonts at 72 hpi, factors associated with decreased pathogenicity, as seen in attenuated *Eimeria* strains [30]. Hence, these treatments may not completely halt development, but they can still result in a decreased severity of infection. When compared to thyme and oregano essential oils, thymol and carvacrol exhibit a milder impact on development. This suggests that a portion of sporozoites treated with these compounds managed to survive and develop within cells, whereas sporozoites treated with the essential oils ultimately succumbed.

Natural alternatives operate through a combination of synergistic effects among compounds found in complex mixtures, such as essential oils, or rely on the concentrated nature of single bioactive compounds. In this study, the pure bioactives proved to be less effective than the complex mixture, underscoring the importance of synergistic interaction between natural compounds in enhancing anticoccidial effects. Therefore, formulations combining bioactive compounds might offer a better alternative; however, this will need to be tested in vitro, and if high efficacy is shown, in vivo models will be necessary for efficacy validation.

5. Conclusions

In conclusion, this study has shed light on the promising potential of essential oils and their bioactive compounds as anticoccidial candidates. These findings highlight that these natural substances not only inhibit sporozoite invasion but also interfere with the formation of schizonts, a critical phase in the parasite's lifecycle. Further research on the mechanisms and applications of these natural compounds is necessary, holding promise for safer and more comprehensive anticoccidial solutions in the future.

Author Contributions: Conceptualization, M.F. and V.M.-H.; investigation, M.F. and C.D.H.-T.; writing—original draft preparation, M.F.; writing—review and editing, V.M.-H., B.T. and E.G.; supervision, V.M.-H., A.P. and E.G. All authors have read and agreed to the published version of the manuscript.

Funding: This research received no external funding.

Institutional Review Board Statement: Procedures used in this study for the generation of *E. tenella* oocysts were carried out in accordance with the Animals (Scientific Procedures) Act 1986 (United Kingdom Parliament Act). All animal studies and protocols have been approved by the Royal Veterinary College Animal Welfare and Ethical Review Board (AWERB) and performed under the United Kingdom Government Home Office project license PDAAD5C9D.

Informed Consent Statement: Not applicable.

Data Availability Statement: Data are contained within the article.

Conflicts of Interest: Andrea Piva serves as a professor at the University of Bologna and is a member of the board of directors of Vetagro S.p.A. (Reggio Emilia, Italy). Ester Grilli serves as an assistant professor at the University of Bologna and is a member of the board of directors of Vetagro, Inc. The remaining authors declare that research was conducted in the absence of any commercial or financial relationships that could be construed as a potential conflict of interest.

References

1. Blake, D.P.; Knox, J.; Dehaeck, B.; Huntington, B.; Rathinam, T.; Ravipati, V.; Ayoade, S.; Gilbert, W.; Adebambo, A.O.; Jatau, I.D.; et al. Re-calculating the cost of coccidiosis in chickens. *Vet. Res.* **2020**, *51*, 115. [CrossRef] [PubMed]
2. Collier, C.T.; Hofacre, C.L.; Payne, A.M.; Anderson, D.B.; Kaiser, P.; Mackie, R.I.; Gaskins, H.R. Coccidia-induced mucogenesis promotes the onset of necrotic enteritis by supporting Clostridium perfringens growth. *Vet. Immunol. Immunopathol.* **2008**, *122*, 104–115. [CrossRef] [PubMed]
3. Macdonald, S.E.; Van Diemen, P.M.; Martineau, H.; Stevens, M.P.; Tomley, F.M.; Stabler, R.A.; Blake, P. Impact of *Eimeria tenella* Coinfection on Campylobacter jejuni Colonization of the Chicken. *J. Hepatol.* **2019**, *87*, e00772-18. [CrossRef] [PubMed]
4. Rimet, C.S.; Maurer, J.J.; Berghaus, R.D.; Jordan, B.J.; Antoniassi Da Silva, L.H.; Stabler, L.J.; Johnson, K.K.; Tensa, L.R.; Segovia, K.M.; França, M.S. The Contribution of Eimeria Coinfection and Intestinal Inflammation to Cecal Colonization and Systemic Spread of Salmonella Typhimurium Deficient in Tetrathionate Reductase or Type III Secretion Systems Salmonella Pathogenicity Island 1 or 2. *Avian Dis.* **2019**, *63*, 559–567. [CrossRef] [PubMed]
5. Noack, S.; Chapman, H.D.; Selzer, P.M. Anticoccidial drugs of the livestock industry. *Parasitol. Res.* **2019**, *118*, 2009–2026. [CrossRef] [PubMed]
6. Muthamilselvan, T.; Kuo, T.F.; Wu, Y.C.; Yang, W.C. Herbal remedies for coccidiosis control: A review of plants, compounds, and anticoccidial actions. *Evid.-Based Complement. Altern. Med.* **2016**, *2016*, 2657981. [CrossRef] [PubMed]
7. Gordillo Jaramillo, F.X.; Kim, D.H.; Lee, S.H.; Kwon, S.K.; Jha, R.; Lee, K.W. Role of oregano and Citrus species-based essential oil preparation for the control of coccidiosis in broiler chickens. *J. Anim. Sci. Biotechnol.* **2021**, *12*, 47. [CrossRef]
8. Chang, L.Y.; Di, K.Q.; Xu, J.; Chen, Y.F.; Xi, J.Z.; Wang, D.H.; Hao, E.Y.; Xu, L.J.; Chen, H.; Zhou, R.Y. Effect of natural garlic essential oil on chickens with artificially infected *Eimeria tenella*. *Vet. Parasitol.* **2021**, *300*, 109614. [CrossRef]
9. Sidiropoulou, E.; Skoufos, I.; Marugan-Hernandez, V.; Giannenas, I.; Bonos, E.; Aguiar-Martins, K.; Lazari, D.; Blake, D.P.; Tzora, A. In vitro Anticoccidial Study of Oregano and Garlic Essential Oils and Effects on Growth Performance, Fecal Oocyst Output, and Intestinal Microbiota in vivo. *Front. Vet. Sci.* **2020**, *7*, 420. [CrossRef]
10. Gholami-Ahangaran, M.; Ahmadi-Dastgerdi, A.; Azizi, S.; Basiratpour, A.; Zokaei, M.; Derakhshan, M. Thymol and carvacrol supplementation in poultry health and performance. *Vet. Med. Sci.* **2022**, *8*, 267–288. [CrossRef]
11. Spisni, E.; Petrocelli, G.; Imbesi, V.; Spigarelli, R.; Azzinnari, D.; Sarti, M.D.; Campieri, M.; Valerii, M.C. Antioxidant, anti-inflammatory, and microbial-modulating activities of essential oils: Implications in colonic pathophysiology. *Int. J. Mol. Sci.* **2020**, *21*, 4152. [CrossRef] [PubMed]
12. Reza Youssefi, M.; Moghaddas, E.; Abouhosseini Tabari, M.; Moghadamnia, A.A.; Hosseini, S.M.; Hosseini Farash, B.R.; Amin Ebrahimi, M.; Nabavi Mousavi, N.; Fata, A.; Maggi, F.; et al. In Vitro and In Vivo Effectiveness of Carvacrol, Thymol and Linalool against Leishmania infantum Mohammad. *Molecules* **2019**, *24*, 2072. [CrossRef] [PubMed]
13. Dominguez-Uscanga, A.; Aycart, D.F.; Li, K.; Witola, W.H.; Andrade Laborde, J.E. Anti-protozoal activity of Thymol and a Thymol ester against Cryptosporidium parvum in cell culture. *Int. J. Parasitol. Drugs Drug Resist.* **2021**, *15*, 126–133. [CrossRef] [PubMed]
14. Bastaki, S.M.A.; Ojha, S.; Kalasz, H.; Adeghate, E. Chemical constituents and medicinal properties of *Allium* species. *Mol. Cell. Biochem.* **2021**, *476*, 4301–4321. [CrossRef] [PubMed]
15. Ali, M.; Chand, N.; Khan, R.U.; Naz, S.; Gul, S. Anticoccidial effect of garlic (*Allium sativum*) and ginger (*Zingiber officinale*) against experimentally induced coccidiosis in broiler chickens. *J. Appl. Anim. Res.* **2019**, *47*, 79–84. [CrossRef]
16. Felici, M.; Tugnoli, B.; Piva, A.; Grilli, E. In vitro assessment of anticoccidials: Methods and molecules. *Animals* **2021**, *11*, 1962. [CrossRef] [PubMed]
17. Patton, W.H. *Eimeria tenella*: Cultivation of the Asexual Stages in Cultured Animal Cells. *Science* **1965**, *150*, 767–769. [CrossRef] [PubMed]
18. Marugan-Hernandez, V.; Jeremiah, G.; Aguiar-Martins, K.; Burrell, A.; Vaughan, S.; Xia, D.; Randle, N.; Tomley, F. The Growth of *Eimeria tenella*: Characterization and Application of Quantitative Methods to Assess Sporozoite Invasion and Endogenous Development in Cell Culture. *Front. Cell. Infect. Microbiol.* **2020**, *10*, 579833. [CrossRef]

19. McDougald, L.R.; Galloway, R.B. Anticoccidial drugs: Effects on infectivity and survival intracellularly of *Eimeria tenella* sporozoites. *Exp. Parasitol.* **1976**, *40*, 314–319. [CrossRef]
20. Clark, J.D.; Billington, K.; Bumstead, J.M.; Oakes, R.D.; Soon, P.E.; Sopp, P.; Tomley, F.M.; Blake, D.P. A toolbox facilitating stable transfection of Eimeria species. *Mol. Biochem. Parasitol.* **2008**, *162*, 77–86. [CrossRef]
21. Shirley, M.W. Eimeria species and strains of chickens. In *Guidelines on Techniques in Coccidiosis Research*; Office for Official Publications of the European Communities: Luxembourg, 1995; pp. 1–25, ISBN 9282749703.
22. Pastor-Fernández, I.; Pegg, E.; Macdonald, S.E.; Tomley, F.M.; Blake, D.P.; Marugán-Hernández, V. Laboratory Growth and Genetic Manipulation of *Eimeria tenella*. *Curr. Protoc. Microbiol.* **2019**, *53*, e81. [CrossRef] [PubMed]
23. Felici, M.; Tugnoli, B.; Ghiselli, F.; Baldo, D.; Ratti, C.; Piva, A.; Grilli, E. Investigating the effects of essential oils and pure botanical compounds against *Eimeria tenella* in vitro. *Poult. Sci.* **2023**, *102*, 102898. [CrossRef] [PubMed]
24. Marugan-Hernandez, V.; Cockle, C.; Macdonald, S.; Pegg, E.; Crouch, C.; Blake, D.P.; Tomley, F.M. Viral proteins expressed in the protozoan parasite *Eimeria tenella* are detected by the chicken immune system. *Parasites Vectors* **2016**, *9*, 463. [CrossRef] [PubMed]
25. Arias-Maroto, S.; Aguiar-Martins, K.; Regidor-Cerrillo, J.; Ortega-Mora, L.; Marugan-Hernandez, V. Reduction of chickens use to perform in vitro pre-screening of novel anticoccidials by miniaturisation and increased throughput of the current *Eimeria tenella* compound-screening model. *F1000Research* **2022**, *11*, 1135. [CrossRef]
26. Thabet, A.; Alnassan, A.A.; Daugschies, A.; Bangoura, B. Combination of cell culture and qPCR to assess the efficacy of different anticoccidials on *Eimeria tenella* sporozoites. *Parasitol. Res.* **2015**, *114*, 2155–2163. [CrossRef]
27. Müller, J.; Hemphill, A. In vitro culture systems for the study of apicomplexan parasites in farm animals. *Int. J. Parasitol.* **2013**, *43*, 115–124. [CrossRef]
28. Sidiropoulou, E.; Marugán-Hernández, V.; Skoufos, I.; Giannenas, I.; Bonos, E.; Aguiar-Martins, K.; Lazari, D.; Papagrigoriou, T.; Fotou, K.; Grigoriadou, K.; et al. In Vitro Antioxidant, Antimicrobial, Anticoccidial, and Anti-Inflammatory Study of Essential Oils of Oregano, Thyme, and Sage from Epirus, Greece. *Life* **2022**, *12*, 1783. [CrossRef]
29. Singh, V.K.; Singh, D.K. Pharmacological effects of garlic (*Allium sativum* L.). *Annu. Rev. Biomed. Sci.* **2008**, *10*, 6–26. [CrossRef]
30. Long, P.L. Endogenous stages of a 'chick embryo-adapted' strain of *Eimeria tenella*. *Parasitology* **1973**, *66*, 55–62. [CrossRef]

animals

Article

Effects of Dietary Inclusion of a Proprietary Combination of *Quillaja saponaria* and *Yucca schidigera* on Intestinal Permeability and Immune Response in Broiler Chickens during a Coccidia Challenge

Kari Saddoris-Clemons *, Saheed Osho, Miriam Garcia and Brooke Humphrey

Phibro Animal Health Corporation, Teaneck, NJ 07666, USA
* Correspondence: kari.saddorisclemons@pahc.com; Tel.: +1-3176013519

Simple Summary: The implementation of antibiotic-free (ABF) and no-antibiotics-ever (NAE) poultry production has prompted interest in identifying viable options to reduce the need for medically important antibiotics (MIA) in animal feed. The Magni-Phi® Ultra (MPU) nutritional specialty product is mainly composed of natural plant extracts (*Quillaja saponaria* and *Yucca schidigera*), which are known for their numerous benefits. By evaluating the effects of feeding MPU on the immune response and intestinal permeability of broilers during coccidia challenge at the peak and recovery phases of infection, this study found that intestinal integrity was enhanced and growth performance and immune response were improved when feeding MPU.

Abstract: This study assessed the impact of Magni-Phi Ultra (MPU) inclusion on intestinal integrity and immunity in broiler chickens challenged with coccidia during peak and recovery phases. A total of 128 male Ross 708 broiler chicks were randomly allotted to one of four treatment groups (four chicks/cage). Treatments included an uninfected control (UUC); a coccidial challenge (CC) infected control (IUC); a CC fed salinomycin at 66 ppm (SAL); and a CC fed Magni-Phi Ultra at 0.11 g/kg of diet (MPU). At 16 days post-hatch, all birds in the CC groups were orally gavaged with a 3× dose of a live coccidia vaccine. At 5 dpi, the birds fed MPU and SAL showed decreased plasma FITC-d, oocyte shedding, and lesion scores and higher BWG compared to the IUC birds ($p < 0.05$). Jejunum *IL-17*, *IL-10*, and *IFN-γ* mRNA expression was higher in the IUC compared to the UUC ($p < 0.05$) group at 5 dpi. At 12 dpi, the birds fed MPU or SAL had lower plasma FITC-d and jejunum *IFN-γ* and *IL*-10 mRNA expression compared to the IUC birds ($p < 0.05$). This study indicates that MPU supports intestinal integrity and mucosal immune responses during the peak and recovery phases of infection, which may lead to improved health and performance.

Keywords: broiler chickens; coccidia; cytokines; intestinal integrity; immune response

Citation: Saddoris-Clemons, K.; Osho, S.; Garcia, M.; Humphrey, B. Effects of Dietary Inclusion of a Proprietary Combination of *Quillaja saponaria* and *Yucca schidigera* on Intestinal Permeability and Immune Response in Broiler Chickens during a Coccidia Challenge. *Animals* **2024**, *14*, 1737. https://doi.org/10.3390/ani14121737

Academic Editor: Kyung-Woo Lee

Received: 3 May 2024
Revised: 3 June 2024
Accepted: 6 June 2024
Published: 8 June 2024

1. Introduction

The implementation of antibiotic-free and no-antibiotic-ever poultry production has led to an increased prevalence of avian coccidiosis caused by *Eimeria* spp. infection. Coccidiosis is extremely costly to the poultry industry, with annual global losses estimated to be greater than USD 14.5 billion [1]. *Eimeria* spp. infections damage intestinal cells, compromise intestinal barrier function, cause inflammation, and induce oxidative stress. This damage often results in a reduction in nutrient absorption and ultimately worse growth performance and feed efficiency [2].

To help reduce the need for antibiotics, nutritional strategies such as plant extracts, essential oils, prebiotics, enzymes, minerals, probiotics, and organic acids have all been suggested as potential tools to help reduce coccidiosis-associated challenges [3–6]. These strategies are crucial in helping maintain poultry health and productivity. The phytogenic compounds, saponins, and polyphenols from *Quillaja saponaria* and *Yucca schidigera* have been shown to impact growth performance in broilers due to coccidiostat action and improvements in nutrient digestibility and intestinal permeability in both unchallenged and *Eimeria* challenge situations [2,7–12]. In addition, saponins have immunomodulatory effects [2,13]. Quillaja saponins stimulate mucosal immunity [13] and yucca saponins reduce pro-inflammatory effects [2] following immune challenges These compounds, due to their surfactant nature, disrupt protozoan cell membranes, leading to cell death, and offer antiprotozoal benefits [14]. Moreover, recent findings suggest they also increase ileal villus height and improve nutrient absorption [8], positively modulate the mRNA abundance of several tight junction proteins in the intestine [15], and alter the intestinal microflora in broiler chickens [16], pointing to their multifaceted role in avian health.

Previous studies both with and without yucca involving an *Eimeria* challenge have focused on the peak phase of *Eimeria* infection due to the high presence of oocysts and intestinal damage occurring during this period, which eventually leads to increased intestinal permeability [10,11,14,16–18]. However, little is known about the recovery phase (5 t 12 dpi) following an *Eimeria* challenge, when immune response and tissue repair are ke This phase also coincides with reduced oocyst shedding and clinical signs. Our study see to expand our understanding of the effects of feeding Magni-Phi Ultra (MPU) on broile chickens infected with *Eimeria* by assessing its effects on growth performance, intestinal health, and immunity during both the peak and recovery phases following a coccidia infection. We hypothesized that the impact of coccidia infection would be reduced in broiler chickens with dietary supplementation of MPU during both the peak and recovery phases by improving intestinal health.

2. Materials and Methods

2.1. Animal Care

All animals were cared for adhering to both the guidelines outlined in the Phibro Animal Health Animal Care and Use Policy and the Guide for the Care and Use of Agricultural Animals in Research and Teaching [19].

2.2. Product

The product tested is a commercially available product, Magni-Phi Ultra (MPU; Phibro Animal Health Corp., Teaneck, NJ, USA). The product is a blend of triterpenoid saponins composed of a unique combination of two natural, plant-based products consisting of *Quillaja saponaria* and *Yucca schidigera* formulated in a proprietary ratio, where quillaja saponins are the major component. The product is formulated to contain 8% total saponins.

2.3. Bird Husbandry, Experimental Design, and Treatments

Ross 708 broiler chicks were purchased from a commercial hatchery (Welp Hatchery, Bancroft, IA, USA). Upon arrival, a total of 128 male chicks were housed across two colony brooders (Universal Brooder Box, GQF Manufacturing, Savannah, GA, USA). Throughout the experiment, the birds were exposed to a lighting schedule of 23L:1D. The first four days, chicks were offered water and a broiler starter feed ad libitum. Both brooder and chick room temperatures decreased as the animals aged. All birds were fed a corn–soybean meal mash basal diet formulated to meet or exceed the National Research Council (1994) nutrient requirements for broiler chickens (Table 1). The basal diet was devoid of coccidiostats and antibiotics. On d 6 post-hatch, the birds were individually tagged and weighed. The birds were then assigned randomly to 1 of 4 treatment groups utilizing a randomized complete block design. A total of 32 cages were utilized, with 8 cages per treatment and 4 birds per cage. The four treatments included T1, an uninfected, unchallenged control fed the basal diet (UUC); T2, a coccidosis challenge (CC) control fed the basal diet (IUC); T3, a CC fed the basal diet supplemented with salinomycin at 66 ppm (SAL-BioCox Granular 60, Huvapharma, Peachtree City, GA, USA); and T4, a CC fed the basal diet supplemented with Magni-Phi Ultra at 0.11 g/kg of diet (MPU). The birds were fed the experimental diets for 10 days prior to coccidia challenge.

Table 1. Ingredients and nutrient composition of basal diet, g/kg as-fed basis.

Ingredients	**g/kg**
Corn	523.0
Soybean meal (47% CP)	380.0
Soybean oil	50.0
Calcium carbonate (38% Ca)	15.0
Dicalcium phosphate	15.0
Salt	4.0
Vitamin-mineral premix [1]	5.4
DL-Methionine	3.8
L-Lysine.HCl	2.8
Threonine	1.0
Total	1000.0
Calculated Nutrients and Energy Content	
CP, g/kg	220.0
ME, kcal/kg	3134.0
Ca, g/kg	9.2
Ca:Tp	1.3

[1] Supplied the following per kilogram of diet: vitamin A, 5484 IU; vitamin D_3, 2643 IU; vitamin E, 11 IU; menadione sodium bisulfite, 4.38 mg; riboflavin, 5.49 mg; D-n-pantothenic acid, 11 mg; niacin, 44.1 mg; choline chloride, 771 mg; vitamin B_{12}, 13.2 µg; biotin, 55.2 µg; thiamine mononitrate, 2.2 mg; folic acid, 990 µg; pyridoxine hydrochloride, 3.3 mg; I, 1.11 mg; Mn, 66.06 mg; Cu, 4.44 mg; Fe, 44.1 mg; Zn, 44.1 mg; Se, 300 µg.

2.4. Coccidia Infection, Lesion Scoring, and Oocyst Count

The birds in the IUC, SAL, and MPU groups were orally gavaged with a 3× dose of the live coccidia vaccine (Coccivac® B52, Merck, Madison, NJ, USA), which was prepared from anti-coccidia sensitive strains from *Eimeria maxima*, *E. acervuline*, *E. tenella*, and *E. mivati* on d 16 post-hatch (0 dpi). Each vaccine bottle contains 10,000 doses of oocysts in an unspecified proportion of the included *Eimeria* species. The recommended dosage of the vaccine is 25 doses per kg of body weight (BW), meaning a 1 kg bird would require 25 doses, per the manufacturer's instructions. At 16 days of age, a bird weighing 300 g would typically receive 7.5 doses of the vaccine. However, in order to induce a coccidia challenge, we administered 3× the recommended dosage, which was equivalent to 22.5 doses per bird. A 3× recommended dose was used to achieve a 10–15% reduction in BWG and a 2-fold increase in serum FITC-d during the peak challenge period. This vaccine dosage is similar in virulence to a live oocyst challenge model containing a mixture of 50,000 *E. acervuline* and 10,000 *E. tenella* oocysts utilized to induce a mild coccidiosis challenge [20]. The coccidia vaccine was gavaged in 1 mL at the appropriate concentration to the birds in the challenge groups (IUC, SAL, and MPU). The birds in the UUC group were orally gavaged with 1 mL of saline. Two birds per cage were randomly selected at 5 dpi and euthanized for scoring of duodenal lesions induced by coccidia infection. Duodenal lesions were scored by personnel blinded to treatment and were based on scores ranging from 0 (no gross lesion) to 4 (severe lesions) [21]. Excreta samples (~100 g) were collected at 5 dpi from each cage and stored in airtight plastic bags. Samples were homogenized and stored at 4 °C until assessed for oocyst counts. Oocyte counts were determined by dilution via a microscope utilizing a McMaster counting chamber [22] and are expressed as oocytes per gram of excreta. A description of the study sampling schematic is shown in Figure 1.

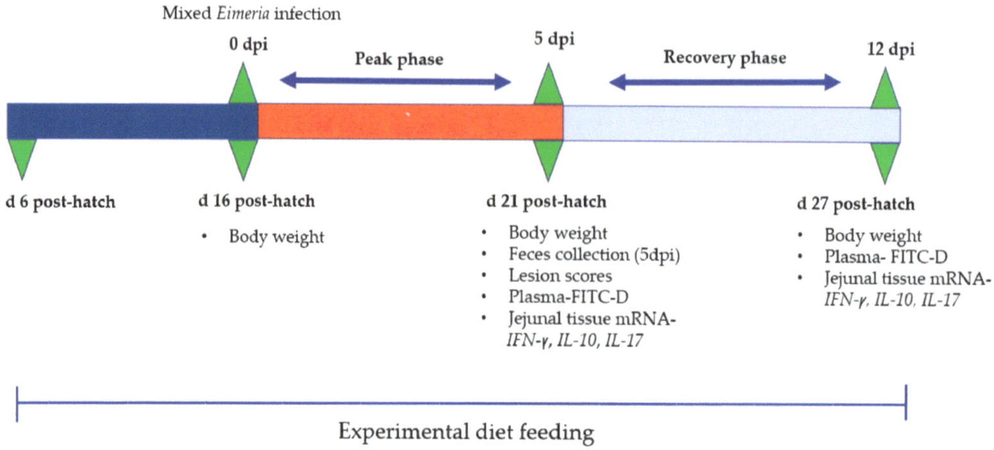

Figure 1. Schematic of study timeline and sampling. Study day 0 to challenge (d 16 post hatch) is represented by the blue bar, the peak challenge phase (d 16-d 21) is represented by the red bar, and the recovery phase (d 21-d 27) is represented by the light blue bar.

2.5. *Growth Performance Measurements*

The feed intake (FI) and BW of the birds were recorded per cage at 0, 5, and 12 dpi to evaluate body weight gain (BWG) and the feed conversion ratio (FCR) over the challenge period.

$$FCR = \text{Body Weight Gain}/\text{Feed Intake}$$

2.6. *Gastrointestinal Permeability*

Fluorescein isothiocyanate dextran (FITC-d; MW 4 kDa; Sigma-Aldrich Co., St. Louis, MO, USA) was administrated to evaluate intestinal permeability. Plasma FITC-d concentration was utilized as a marker of mucosal barrier dysfunction. At 5 and 12 dpi, two birds per cage were randomly selected and orally gavaged with 8.4 mg/kg BW of FITC-d (500 µL of FITC-d solution). The birds were euthanized by CO_2 asphyxiation at 2 h following administration of the FITC-d. Blood (6 mL) was collected into EDTA vacutainer tubes (BD Company, Franklin Lakes, NJ, USA) and stored on ice. The blood was centrifuged at 2000× *g* at 4 °C for 10 min to separate plasma. A standard solution was created by diluting FITC-d with a pool of plasma from 5 unchallenged birds at both time points. FITC-d levels in both the standard solution and plasma samples were measured by florescence at an excitation wavelength of 485 nm and an emission wavelength of 530 nm utilizing a microplate reader (BioTek Instruments, Inc., Winooski, VT, USA).

2.7. *Jejunum Collection and Real-Time Polymerase Chain Reaction Analysis*

A mid-jejunum sample was collected at 5 and 12 dpi from 2 birds per cage for a quantitative real-time Polymerase Chain Reaction (PCR) (RT-qPCR) assay to quantify the relative mRNA abundance of interleukin-17 (IL-17), interleukin-10 (IL-10), and interferon γ (*IFN-γ*). Mid-jejunal samples were flushed with PBS, and a 1 cm section was removed and immediately submerged in an RNAlater® stabilization solution (Thermo Fisher Scientific Inc., Waltham, MA, USA). Samples were held at room temperature for 1 h and then stored at −80 °C pending further analysis. RNA was extracted from the jejunal tissue utilizing a Tissue Purification Kit (NORGEN BioTek Corp, Thorold, ON, Canada), following the manufacturer's recommendations. RNA concentration and purity were measured using a NanoDrop (ThermoFisher Scientific). The purity and concentrations of RNA samples were measured using a NanoDrop (ThermoFisher Scientific). The purity for all samples was between 1.80 and 2.03 for the 260/280 ratio. RNA was analyzed for the mRNA abundance of reference genes (GADPH and YWAHZ) and target genes (*IFN-γ*, *IL10*, and *IL17A*). Primers and TaqMan probes were obtained from Applied Biosystems (TaqMan® Gene Expression Assays, Waltham, MA, USA). Primers were run in duplex pairs (I-GAPDH and *IL10-YWHAZ*) or singlets (*IFN-γ*). The lack of primer cross-hybridization was confirmed by verifying similar threshold cycle (Ct) values for the reactions in singlet and pairs.

For duplex reactions, primers were labelled with dyes with different fluorescence emission spectra (VIC or FAM). The RT-qPCR reaction volumes consisted of 5 μL of RNA (40 ng/μL), 9 (singlet) or 8 (duplex) μL of RNAse/DNase free water (ThermoFisher Scientific), 5 μL of Taqman Fast Virus 1-step Master Mix (cat no. 4444432 Applied Biosystems), and 1 μL of each primer. Each sample was set up in duplicate with specific primers and probes for chicken *IFN-γ* (assay ID number Gg03348618_m1), chicken *IL10* (assay ID number Gg03358689_m1), chicken *IL17A* (assay ID number Gg03365522_m1), and the reference chick genes *GAPDH* (assay ID number Gg03346982_m1) and *YWHAZ* (assay ID number Gg03356701_m1) (Thermo Fisher Scientific). Samples for RT-qPCR reactions were run in duplicate using a CFX96 optics unit mounted on a C1000 touch base (BioRad, Hercules, CA, USA) in the fast-cycling mode. The PCR cycle parameters were an initial denaturing step at 95 °C for 20 s followed by 40 cycles of 95 °C for 3 s and 60 °C for 30 s. Target gene Ct-values were normalized to the geometric mean of the Ct-values of the two reference genes. The $2^{-\Delta\Delta CT}$ method was used to calculate relative mRNA abundance, and the results are reported as fold-change relative to the UUC treatment group [23].

2.8. Statistical Analysis

A data analysis was conducted with the cage as the experimental unit using the GLIMMIX procedure of SAS 9.4 software (SAS Inst., Inc., Cary, NC, USA) with the random effect of block (initial BW) and the fixed effect of treatment. Data were tested for normal distribution in PROC UNIVARIATE using the Shapiro–Wilk test. In order to achieve normality, oocyst count data were natural log transformed. Differences between treatments were considered significantly different at $p \leq 0.05$.

3. Results

3.1. Growth Performance

Broiler chicks were fed the dietary treatments for 10 days prior to the coccidia challenge. During this period (d 0–10), there was no difference in growth performance between treatment groups. Table 2 summarizes the effect of dietary treatments on growth performance during the coccidia challenge. In the peak phase (0–5 dpi), the BW, FCR, and BWG of the birds was significantly reduced ($p < 0.05$) in the IUC group compared to the UUC group. However, the birds fed SAL and MPU had similar BW, FCR, and BWG compared to the birds in UUC group. The birds in the SAL group consumed more ($p < 0.05$) feed than the other treatment groups and were heavier ($p < 0.05$) than the birds in the IUC and MPU groups. However, the SAL birds had similar BWG and FCRs when compared to the birds fed MPU. During the recovery phase (6–12 dpi), there were no differences in BWG and the FCR between treatment groups. However, feed intake was higher ($p < 0.05$) for the birds in the IUC group, followed by the other CC birds (SAL, MPU), with the UUC birds having the lowest feed intake. As for final BW, the birds fed MPU had similar BW when compared to the other groups, but the birds in the SAL group were heavier than the birds in the UUC and IUC groups.

Table 2. Growth performance, lesion scores, and oocyst shedding in broiler chickens fed diets supplemented with salinomycin (SAL) or Magni-Phi Ultra (MPU) during a coccidia challenge (dpi) [1,2].

Item	Uninfected UUC	IUC	Coccidia Infection SAL	MPU	SEM	*p*-Value
0–5 dpi (Peak Phase)						
Final body weight, g	570 ^{ab}	515 ^{c}	598 ^{a}	563 ^{b}	10.7	<0.01
Body weight gain, g/bird	264 ^{a}	208 ^{b}	287 ^{a}	259 ^{a}	11.8	<0.01
Feed intake, g/bird	301 ^{b}	291 ^{b}	332 ^{a}	310 ^{b}	7.4	0.04
Feed conversion ratio	1.15 ^{b}	1.44 ^{a}	1.17 ^{b}	1.21 ^{b}	0.07	0.03
6–12 dpi (Recovery Phase)						
Final body weight, g	1025 ^{b}	1009 ^{b}	1075 ^{a}	1037 ^{ab}	16.0	0.05
Body weight gain, g/bird	454	489	476	468	19.3	0.63
Feed intake, g/bird	612 ^{c}	672 ^{a}	640 ^{b}	642 ^{b}	9.2	<0.01
Feed conversion ratio	1.37	1.40	1.35	1.38	0.05	0.96

$^{a–c}$ Means in the same row lacking a common superscript differ significantly ($p < 0.05$). [1] Data are least squares means of 8 cages per treatment. [2] Uninfected control (UUC) fed basal diet, coccidia challenge untreated control (IUC) fed basal diet, SAL (cocci challenge fed basal diet supplemented with salinomycin (66 ppm), and MPU (cocci challenge fed basal diet supplemented with Magni-Phi Ultra (0.11g/kg of diet)). On d 16 post-hatch (0 dpi), birds in the IUC, MPU, and SAL groups were orally gavaged with 3× the recommended dose of coccidia vaccine (Coccivac B52, Merck, Rahway, NJ), while birds in the UUC group were orally gavaged with 1 mL of saline.

3.2. Oocyte Count and Lesion Scores

Oocyte count and lesion scores are presented in Table 3. Dietary SAL and MPU supplementation significantly reduced ($p < 0.05$) oocyte shedding and duodenal lesion scores compared with the IUC group. The UUC birds remained free of coccidiosis infection, as indicated by the absence of duodenal lesions and oocyst shedding in that group.

Table 3. Lesion scores and oocyst shedding 5 dpi in broiler chickens fed diets supplemented with salinomycin (SAL) or Magni-Phi Ultra (MPU) during a coccidia challenge [1,2].

Item	Uninfected UUC	IUC	Coccidia Infection SAL	MPU	SEM	*p*-Value
Oocyst per gram of excreta ($\times 10^2$)						
5 dpi	0 ^{c}	81 ^{a}	30 ^{b}	28 ^{b}	6.7	<0.01
Duodenum lesion scores						
5 dpi	0 ^{c}	2.13 ^{a}	1.13 ^{b}	1.19 ^{b}	0.17	<0.01

$^{a–c}$ Means in the same row lacking a common superscript differ significantly ($p < 0.05$). [1] Data are least squares means of 8 cages per treatment. [2] Uninfected control (UUC) fed basal diet, coccidia challenge untreated control (IUC) fed basal diet, SAL (cocci challenge fed basal diet supplemented with salinomycin (66 ppm), and MPU (cocci challenge fed basal diet supplemented with Magni-Phi Ultra (0.11g/kg of diet)). On d 16 post-hatch (0 dpi), birds in the IUC, MPU, and SAL groups were orally gavaged with 3× the recommended dose of coccidia vaccine (Coccivac B52, Merck, Rahway, NJ, USA), while birds in the UUC group were orally gavaged with 1 mL of saline.

3.3. Jejunal mRNA Expression Analysis

The differences between the treatment groups' host immune responses were assessed by cytokine mRNA expression in the mid-jejunum of broiler chickens (Table 4). At the peak phase (5 dpi), the birds in the IUC group had significant upregulation ($p < 0.05$) of *IFN-γ*, *IL-17*, and *IL-10* compared to the other three groups, which had similar mRNA expression for all three measured cytokines. In the recovery phase, *IFN-γ* and IL-10 remained upregulated ($p < 0.05$) in the IUC group compared to the other three groups. However, the birds fed SAL and MPU had similar *IL-10*, *IL-17*, and *IFN-γ* mRNA expression when compared to the UUC group.

Table 4. Relative mRNA expression of cytokines in mid-jejunum tissue of broiler chicks fed diets supplemented with salinomycin (SAL) or Magni-Phi Ultra (MPU) during a coccidia challenge (dpi) [1,2].

Item	Uninfected	Coccidia Infection				
	UUC	IUC	SAL	MPU	SEM	*p*-Value
5 dpi (Peak phase)						
IL-17	1.00 [b]	2.96 [a]	1.64 [b]	0.60 [b]	0.02	0.04
IL-10	1.00 [b]	4.72 [a]	1.42 [b]	2.00 [b]	0.11	<0.01
IFN-γ	1.00 [b]	3.89 [a]	1.45 [b]	1.56 [b]	0.27	<0.01
12 dpi (Recovery Phase)						
IL-17	1.00	0.67	1.00	1.17	0.02	0.07
IL-10	1.00 [b]	2.61 [a]	1.64 [b]	1.50 [b]	0.24	0.04
IFN-γ	1.00 [b]	2.19 [a]	1.53 [b]	1.22 [b]	0.26	0.04

[a–b]Means in the same row lacking a common superscript differ significantly ($p < 0.05$). [1] Data are least squares means of 8 cages per treatment. *IL-17* = interleukin-17; *IL-10* = interleukin-10; interferon-gamma = *IFN-γ*. [2] Uninfected control (UUC) fed basal diet, coccidia challenge untreated control (IUC) fed basal diet, SAL (cocci challenge fed basal diet supplemented with salinomycin (66 ppm), and MPU (cocci challenge fed basal diet supplemented with Magni-Phi Ultra (0.11g/kg of diet)). On d 16 post-hatch (0 dpi), birds in the IUC, MPU, and SAL groups were orally gavaged with 3× the recommended dose of coccidia vaccine (Coccivac B52, Merck, Rahway, NJ, USA), while birds in the UUC group were orally gavaged with 1 mL of saline.

3.4. Intestinal Permeability

The alterations in intestinal permeability at 5 and 12 dpi are depicted in Table 5. The higher the concentration of FITC-d recovered in the plasma, the greater the intestinal permeability. Plasma FITC-d concentration was significantly higher ($p < 0.05$) in the birds in the IUC group compared to UUC birds at both 5 and 12 dpi. The birds fed SAL and MPU had significantly lower ($p < 0.05$) plasma FITC-d concentration at 5 dpi compared to the IUC birds. However, at 12 dpi, the birds fed MPU, but not SAL, had similar ($p > 0.05$) FITC-d concentrations when compared to the UUC group. As expected, 5 dpi had the most severe intestinal permeability across all treatment groups.

Table 5. Plasma fluorescein isothiocyanate dextran (FITC-d) concentration (ng/mL) of broiler chicks fed diets supplemented with salinomycin (SAL) or Magni-Phi Ultra (MPU) during a coccidia challenge (dpi) [1,2].

Item	Uninfected	Coccidia Infection				
	UUC	IUC	SAL	MPU	SEM	*p*-Value
5 dpi (Peak phase)						
FITC-d	46 [c]	94 [a]	70 [b]	63 [b]	4.79	<0.01
12 dpi (Recovery Phase)						
FITC-d	42 [b]	59 [a]	51 [ab]	42 [b]	5.59	<0.01

[a–c] Means in the same row lacking a common superscript differ significantly ($p < 0.05$). [1] Data are least squares means of 8 cages per treatment. [2] Uninfected control (UUC) fed basal diet, coccidia challenge untreated control (IUC) fed basal diet, SAL (cocci challenge fed basal diet supplemented with salinomycin (66 ppm), and MPU (cocci challenge fed basal diet supplemented with Magni-Phi Ultra (0.11g/kg of diet)). On d 16 post-hatch (0 dpi), birds in the IUC, MPU, and SAL groups were orally gavaged with 3× the recommended dose of coccidia vaccine (Coccivac B52, Merck, Rahway, NJ, USA), while birds in the UUC group were orally gavaged with 1 mL of saline.

4. Discussion

4.1. Peak Phase

In the peak phase of coccidia infection, detrimental effects were observed on the broiler chickens' health and performance, emphasizing the importance of exploring alternatives to enhance disease response without the use of antibiotics. As reported by previous studies, coccidia challenge has detrimental effects on intestinal integrity and nutrient absorption [20,24,25], which leads to altered expressions of immunoregulatory cytokines [26]. The results from this study confirm the negative impact of coccidia challenge in broiler chickens, even with the relatively lower challenge dosage utilized in this study. In addition, it investigated the effects of SAL or MPU on growth performance, intestinal permeability, and immune responses in coccidia-challenged broiler chickens. During the peak phase of coccidia infection (5 dpi), both the SAL and MPU treatments protected the broiler chickens from the negative impacts of the challenge, as evidenced by improved growth performance. The dietary inclusion of 500 mg/kg of a saponin-based product via an extract from *Yucca schidigera* effectively promotes growth in broilers infected with *Eimeria* [10]. However, the dietary inclusion of 250 or 500 mg/kg of a *Yucca schidigera*-based saponin source also had no impact on performance in broilers infected with *Eimeria* [2]. The saponin source and level of the product can vary greatly, and along with the severity of the coccidiosis challenge, impacts the efficacy. In the current study, the MPU treatment, which contained a blend of *Quillaja saponaria* and *Yucca schidigera*, was included at a much lower dosage of 110 mg/kg compared to other studies [2,10], indicating a higher potency of the *Quillaja Saponaria* compared to saponin products only from *Yucca schidigera* sources. To assess the severity of coccidia infection and its impact on the intestinal mucosa, we used intestinal lesion scores as an indicator. In this study, the birds in the IUC group exhibited a high oocyst count and duodenal lesion scores, which are indicative of an active coccidia infection. However, dietary supplementation with SAL or MPU decreased excreta oocyst counts and lesion scores. Both of these responses point to their effectiveness in reducing the replication or shedding of oocysts and protection against coccidia damage. The inclusion of *Quillaja saponaria* at 250 and 300 ppm reduced oocyst production (43% and 37%, respectively) and lesion scores following *Eimeria acervulina* and *Eimeria tenella* infection [14]. Yucca-based saponins also display coccidiostat actions, as measured by reduced oocyte production and protection against intestinal damage induced by the Eimeria in broilers [11]. Previous studies have demonstrated that saponins exhibit anti-protozoal properties through their interaction with cholesterol in protozoal cell membranes, leading to membrane damage and eventual cell lysis [27], and this effect has been observed in studies with both quillaja- and yucca-based saponin sources [10,14].

In the context of immune responses, cytokines play a crucial role in the immune system of animals [28]. T-cell-mediated immune response is an important factor in avian coccidiosis control. In this study, we chose to evaluate *IFN-γ*, a key cytokine that mediates the Th1 immune response; *IL-17*, a cytokine involved in the Th17-related immune response; and *IL-10*, a cytokine involved in the Treg-related immune response [29]. The Th17 immune response is reported to play a role in the early stages (2–4 dpi) of a coccidosis infection, while the Th1 and Treg immune response are involved in the later stages (6–8 dpi) of a coccidosis infection [29]. The peak phase (5 dpi) was characterized by a robust immune response in the IUC group relative to UUC, with marked upregulation of the cytokines *IL-17* (3×), *IL-10* (4.7×), and *IFN-γ* (3.9×), which is indicative of an active inflammatory and regulatory response to the coccidia infection. However, the SAL and MPU treatments modulated this acute response and maintained the expression of these cytokines at similar levels as the UUC group. This suggests that feeding these products may help to dampen the acute inflammatory response, potentially limiting the tissue damage associated with excessive inflammation. *Eimeria* infection, which damages the gut lining and disrupts barrier integrity, results in increased gut permeability and negative impacts on performance. This leakage, resulting from epithelial damage and tight junction disruption, can cause nutrient malabsorption and impaired immune responses, leading to secondary bacterial infections [26]. Our findings of reduced cytokine expression in the birds fed a saponin-based product during an *Eimeria* infection varies slightly from [2], who reported a reduction in *IL-1β* but no impact on *IFN-γ* or *IL-12β* in the duodenum at 7 dpi with birds fed 250 to 500 mg/kg of a *Yucca*-derived saponin. This indicates that a *Yucca*-based saponin has a lower impact on reducing pro-inflammatory cytokines involved in the Th1 immune response compared to saponin products containing *Quillaji*. *Quillaja* saponins have been suggested to act directly on T-helper cells of the mucosal immune system and induce T-cell and B-cell proliferation [30,31]. Saponin products containing both *Quillaja saponaria* and *Yucca schidigera* appear to have a greater immunomodulatory effect during an

Eimeria infection compared to products with only *Yucca*-derived saponins. In the present study, elevated levels of plasma FITC-d following a mixed *Eimeria* spp. infection and duodenal lesion scores are indicative of the presence of intestinal damage. A previous *Eimeria* study also reported the highest intestinal permeability at 5 dpi [28]. This was coupled with negative impacts on performance during 0–5 dpi compared to the UUC group. Correspondingly, the reduced plasma levels of FITC-d in the SAL and MPU groups point to a protective effect on the intestinal barrier and a reduction in leakiness that is often a result of inflammation-induced damage. In summary, during the peak phase of coccidia infection, detrimental effects on the broiler chickens' health and performance were evident, but the supplementation of the SAL and MPU treatments demonstrated promising protective effects. Feeding these products positively impacted growth performance, reduced lesion scores, modulated the acute immune response, and improved gut permeability, all of which contribute to mitigating the adverse effects of the coccidia challenge.

4.2. Recovery Phase

The current study continued to evaluate the effects of SAL and MPU treatments on broiler chickens as they progressed towards recovery from the coccidia challenge. In this phase, the birds in all feeding groups, including the IUC group, had similar growth and feed efficiencies, indicating recovery from the initial infection. However, the early protection provided by SAL and MPU may have allowed for a faster resolution of the inflammatory response and restoration of gut barrier integrity. By 12 dpi, the IUC group continued to exhibit higher expression of *IL-10* (2.6×) and *IFN-γ* (2.2×), but the degree of upregulation was about half of what was observed in the peak phase, indicating a shift from an acute response towards a resolving phase of the immune response. In all the CC birds, *IL-17*, a cytokine involved in the early immune response, had returned to levels similar to that of the UUC group. Our findings corroborate previous reports of increased mRNA expression of intestinal *IFN-γ* due to *Eimeria* infection in chickens during peak challenge (5–7 d post-infection), although the timing and magnitude of response vary considerably [32,33]. Resolution of the Th1, Th17, and Treg immune response and return of *IFN-γ*, *IL-17*, and *IL-10* cytokines to baseline levels was reported by 10 dpi in birds challenged with *E. tenella* sporulated oocytes at 18 d of age [29]. Differences in return to baseline of *IFN-γ* and *IL-10* in the current study may be due to the use of a mixed *Eimera* challenge model and the earlier age (16 d) at the time of challenge. By the recovery phase, *IL-10* and *IFN-γ* levels in the SAL and MPU groups were below that of the IUC group and similar to the UUC group. This may better enable tissue repair and the recovery of gut function, as shown by performance improvements in the recovery phase. *Yucca* and *Quillaja* saponins and polyphenols can inhibit inflammation in broilers raised under standard conditions, without any pathogen challenges, as reported by [16]. Another study also shows that *Yucca* bark and whole *Yucca* plant powder contain resveratrol, which is known for its anti-inflammatory activity [34,35]. Thus, whole-plant *Yucca* powder has powerful anti-inflammatory activity, mediated via inhibition of *NFkB* activation [27]. Dietary saponins from *Yucca schidigera* extracts included at levels as low as 100 mg/kg improve antioxidant capacity in the small intestine and increased gene expression of superoxide dismutase, glutathione peroxidase, and catylase [36]. This may help explain why a reduction of inflammation in broilers was observed during this coccidia challenge, but more research is needed, as limited research exists evaluating the immune response and antioxidant capacity of birds fed saponin-based products past the peak challenge phase.

To the best of our knowledge, publications on the effects of saponins on the intestinal permeability of broiler chickens during both the peak and recovery phases have not been well addressed in the literature. Our data suggest that MPU treatments are not only effective in reducing the impact of coccidia infection on gut barrier function during the peak phase but are also capable of supporting the recovery of gut integrity post-infection, as measured by plasma FITC-d levels. By 12 dpi, the IUC group continued to have elevated serum FITC-d levels (1.4×) compared to the UCC group. However, serum FITC-d levels of the MPU group were lower than the IUC group and had returned to levels similar to the UCC group, indicating recovery of gut barrier function.

Recovery of gut barrier function post-infection is critical, as it minimizes the risk of secondary complications associated with increased intestinal permeability, such as systemic inflammation or bacterial translocation. The observed alteration in intestinal permeability patterns may be linked to the ability of saponins to influence the permeability of intestinal cells through their interaction with cholesterol in mucosal cell membranes [37].

5. Conclusions

In conclusion, our study shows that the negative impact of a coccidia infection spans beyond the peak phase into the recovery phase, as evidenced by markers of inflammation (cytokines) and gut integrity (FITC-d levels) not yet reaching baseline (UUC) levels after 12 dpi. Also, our study provides new insights into the beneficial effects of MPU on intestinal permeability and immune responses during a coccidia challenge. We demonstrate that SAL and MPU treatments are effective not just in the acute phase but also in supporting recovery, highlighting their potential in improving intestinal integrity and growth performance. This contributes to the growing body of evidence supporting the use of saponins from both *Quillaja saponaria* and *Yucca schidigera* for their immune modulatory and gut health benefits, offering valuable alternatives to traditional antibiotic use.

Author Contributions: Conceptualization, S.O., K.S.-C., B.H., and M.G.; methodology, S.O., K.S.-C., B.H., and M.G.; formal analysis, S.O.; software, S.O.; data curation, S.O.; writing—original draft preparation, S.O.; writing—review and editing, S.O., K.S.-C., B.H., and M.G; supervision, S.O. and M.G.; project administration, K.S.-C. and B.H. All authors have read and agreed to the published version of the manuscript.

Funding: Funding for this study was provided by Phibro Animal Health Corporation.

Institutional Review Board Statement: All animals were cared for following both the Phibro Animal Health Animal Care and Use Policy and the Guide for the Care and Use of Agricultural Animals in Research and Teaching [17]. The animal study was approved by the Phibro Animal Health Corporation, Animal Care and Use Policy. The study was conducted in accordance with the local legislation (Animal and Plant Health Inspection Services, USDA) and institutional requirements (AVMA Guidelines for the Euthanasia of Animals: 2020 Edition).

Informed Consent Statement: Not applicable.

Data Availability Statement: The data presented in this study are available upon request from the corresponding author.

Conflicts of Interest: The authors were all employed at Phibro Animal Health Corporation when this work was completed. The authors declare that this study received funding from Phibro Animal Health Corporation. The funder was involved in the collection, analysis and interpretation of data and the decision to submit it for publication.

References

1. Blake, D.P.; Knox, J.; Dehaeck, B.; Huntington, B.; Rathinam, T.; Ravipati, V.; Ayoade, S.; Gilbert, W.; Adebambo, A.O.; Jatau, I.D.; et al. Re-calculating the cost of coccidiosis in chickens. *Vet. Res.* **2020**, *51*, 115. [CrossRef]
2. Oelschlager, M.; Rasheed, M.; Smith, B.; Rincker, M.; Dilger, R. Effects of *Yucca schidigera*-derived saponin supplementation during a mixed Eimeria challenge in broilers. *Poult. Sci.* **2019**, *98*, 3212–3222. [CrossRef]
3. Alagawany, M.; Elnesr, S.S.; Farag, M.R.; Tiwari, R.; Yatoo, M.I.; Karthik, K.; Michalak, I.; Dhama, K. Nutritional significance of amino acids, vitamins and minerals as nutraceuticals in poultry production and health–a comprehensive review. *Vet. Q.* **2021**, *41*, 1–29. [CrossRef]
4. Bortoluzzi, C.; Vieira, B.S.; Applegate, T.J. Influence of dietary zinc, copper, and manganese on the intestinal health of broilers under Eimeria challenge. *Front. Vet. Sci.* **2020**, *7*, 13. [CrossRef]
5. Georgieva, N.V.; Gabrashanska, M.; Koinarski, V.; Yaneva, Z. Zinc supplementation against *Eimeria acervulina*-induced oxidative damage in broiler chickens. *Vet. Med. Int.* **2011**, *2011*. [CrossRef]

6. Santos, T.S.d.; Teng, P.-Y.; Yadav, S.; Castro, F.L.d.S.; Gould, R.L.; Craig, S.W.; Chen, C.; Fuller, A.L.; Pazdro, R.; Sartori, J.R. Effects of Inorganic Zn and Cu Supplementation on Gut Health in Broiler Chickens Challenged with *Eimeria* spp. *Front. Vet. Sci.* **2020**, *7*, 230. [CrossRef]
7. Gadde, U.; Kim, W.H.; Oh, S.T.; Lillehoj, H.S. Alternatives to antibiotics for maximizing growth performance and feed efficiency in poultry: A review. *Anim. Health Res. Rev.* **2017**, *18*, 26–45. [CrossRef]
8. Bafundo, K.W.; Männer, K.; Duerr, I. The combination of quillaja and yucca saponins in broilers: Effects on performance, nutrient digestibility and ileal morphometrics. *Br. Poult. Sci.* **2021**, *62*, 589–595. [CrossRef]
9. Sahoo, S.; Kaur, D.; Sethi, A.; Saini, A.; Chandra, M. Effect of dietary supplementation of yucca schidigera extract on the performance and litter quality of broilers in winter season. *Anim. Nutr. Feed Technol.* **2016**, *16*, 475–484. [CrossRef]
10. Kozlowski, K.; Vervenne-Zetteler, P.; Konieczka, P.; Szymański, Ł.; Van Vilsteren, A. *Yucca schidigera* improves performance and lowers oocyst counts in Eimeria challenged birds. *Animals* **2022**, *12*, 1668. [CrossRef]
11. Galli, G.M.; Da Silva, A.S.; Bottari, N.B.; Biazus, A.H.; Petrolli, T.; Reis, J.H.; Morsch, V.M.; Schetinger, M.R.C.; Piva, M.M.; Baggio, R.A.; et al. Addition of yucca extract and glutamine in the diet of chicks had a protective effect against coccidiosis. *Comp. Clin. Pathol.* **2018**, *27*, 205–214. [CrossRef]
12. Stefanello, C.; Moreira, B.; Gräf, W.M.; Robalo, S.; Costa, S.T.; Vieira, I.M.; Mirandaz, D.J. Effects of a proprietary blend of Quillaja and Yucca on growth performance, nutrient digestibility, and intestinal measurements of broilers. *J. Appl. Poult. Res.* **2022**, *31*, 100251. [CrossRef]
13. Kensil, C.R. Saponins as Vaccine Adjuvants. *Crit. Rev. Ther. Drug Carr. Syst.* **1996**, *13*, 1–55.
14. Bafundo, K.W.; Johnson, A.B.; Mathis, G.F. The Effects of a Combination of *Quillaja saponaria* and *Yucca schidigera* on *Eimeria* spp. in Broiler Chickens. *Avian Dis.* **2020**, *64*, 300–304. [CrossRef] [PubMed]
15. Blue, C.E.C.; Emami, N.K.; White, M.B.; Cantley, S.; Dalloul, R.A. Inclusion of Quillaja Saponin Clarity Q Manages Growth Performance, Immune Response, and Nutrient Transport of Broilers during Subclinical Necrotic Enteritis. *Microorganisms* **2023**, *11*, 1894. [CrossRef] [PubMed] [PubMed Central]
16. Dai, Z.; Wang, H.; Liu, J.; Zhang, H.; Li, Q.; Yu, X.; Zhang, R.; Yang, C. Comparison of the Effects of Yucca saponin, *Yucca schidigera*, and *Quillaja saponaria* on Growth Performance, Immunity, Antioxidant Capability, and Intestinal Flora in Broilers. *Animals* **2023**, *13*, 1447. [CrossRef] [PubMed]
17. Alfaro, D.M.; Silva, A.V.F.; Borges, S.A.; Maiorka, F.A.; Vargas, S.; Santin, E. Use of *Yucca schidigera* Extract in Broiler Diets and Its Effects on Performance Results Obtained with Different Coccidiosis Control Methods. *J. Appl. Poult. Res.* **2007**, *16*, 248–254. [CrossRef]
18. Konieczka, P.; Tykałowski, B.; Ognik, K.; Kinsner, M.; Szkopek, D.; Wójcik, M.; Mikulski, D.; Jankowski, J. Increased arginine, lysine, and methionine levels can improve the performance, gut integrity and immune status of turkeys but the effect is interactive and depends on challenge conditions. *Vet. Res.* **2022**, *53*, 59. [CrossRef] [PubMed]
19. McGlone, J. *Guide for the Care and Use of Agricultural Animals in Teaching and Research*; Federation of Animal Science Societies: Champaign, IL, USA, 2010.
20. Jespersen, J.C.; de Paula Dorigam, J.C.; Whelan, R.; Dilger, A.C.; Oelschlager, M.L.; Sommer, K.M.; Gorenz, B.E.; White, R.R.; Dilger, R.N. Defining optimal dietary starch, oil, and amino acid inclusion levels for broilers experiencing a coccidiosis challenge. *Poult. Sci.* **2024**, *103*, 103335. [CrossRef]
21. Johnson, J.; Reid, W.M. Anticoccidial drugs: Lesion scoring techniques in battery and floor-pen experiments with chickens. *Exp. Parasitol.* **1970**, *28*, 30–36. [CrossRef]
22. Henriksen, S.A.; Aagaard, K. A simple flotation and McMaster method (author's transl). *Nord. Veterinaermed.* **1976**, *28*, 392–397.
23. Livak, K.J.; Schmittgen, T.D. Analysis of relative gene expression data using real-time quantitative PCR and the 2− ΔΔCT method. *Methods* **2001**, *25*, 402–408. [CrossRef] [PubMed]
24. Castro, F.L.S.; Tompkins, Y.H.; Pazdro, R.; Kim, W.K. The effects of total sulfur amino acids on the intestinal health status of broilers challenged with *Eimeria* spp. *Poult. Sci.* **2020**, *99*, 5027–5036. [CrossRef] [PubMed]
25. Teng, P.-Y.; Choi, J.; Yadav, S.; Tompkins, Y.H.; Kim, W.K. Effects of low-crude protein diets supplemented with arginine, glutamine, threonine, and methionine on regulating nutrient absorption, intestinal health, and growth performance of Eimeria-infected chickens. *Poult. Sci.* **2021**, *100*, 101427. [CrossRef]
26. Khatlab, A.S.; Del Vesco, A.P.; de Oliveira Neto, A.R.; Fernandes, R.P.M.; Gasparino, E. Dietary supplementation with free methionine or methionine dipeptide mitigates intestinal oxidative stress induced by *Eimeria* spp. challenge in broiler chickens. *J. Anim. Sci. Biotechnol.* **2019**, *10*, 58. [CrossRef] [PubMed]
27. Cheeke, P.R.; Piacente, S.; Oleszek, W. Anti-inflammatory and anti-arthritic effects of *Yucca schidigera*: A review. *J. Inflamm.* **2006**, *3*, 6. [CrossRef] [PubMed]
28. Akerele, G.; Al Hakeem, W.G.; Lourenco, J.; Selvaraj, R.K. The effect of necrotic enteritis challenge on production performance, cecal microbiome, and cecal tonsil transcriptome in broilers. *Pathogens* **2022**, *11*, 839. [CrossRef] [PubMed]
29. Yu, H.; Zou, W.; Mi, C.; Wang, Q.; Dai, G.; Zhang, T.; Zhang, G.; Xie, K.; Wang, J.; Shi, H. Research Note: Expression of T cell-related cytokines in chicken cecal and spleen tissues following *Eimeria tenella* infection in vivo. *Poult. Sci.* **2021**, *100*, 101161. [CrossRef] [PubMed] [PubMed Central]
30. Chavali, S.R.; Campbell, J.B. Adjuvant effects of orally administered saponins on humoral and cellular immune responses in mice. *Immunobiology* **1987**, *174*, 347–359. [CrossRef]

31. Mowat, A.M.; Donachie, A.M.; Reid, G.; Jarrett, O. Immune-stimulating complexes containing Quil A and protein antigen prime Class I MHC-restricted T lymphocytes in-vivo and are immunogenic by the oral route. *Immunology* **1991**, *72*, 317–322.
32. Rochell, S.J.; Helmbrecht, A.; Parsons, C.M.; Dilger, R.N. Influence of dietary amino acid reductions and *Eimeria acervulina* infection on growth performance and intestinal cytokine responses of broilers fed low crude protein diets. *Poult. Sci.* **2016**, *95*, 2602–2614. [CrossRef]
33. Yun, C.H.; Lillehoj, H.S.; Choi, K.D. *Eimeria tenella* infection induces local gamma interferon production and intestinal lymphocyte subpopulation changes. *Infect. Immun.* **2000**, *68*, 1282–1288. [CrossRef]
34. Bertelli, A.A.; Migliori, M.; Panichi, V.; Origlia, N.; Filippi, C.; Das, D.K.; Giovannini, L. Resveratrol, a component of wine and grapes, in the prevention of kidney disease. *Ann. N. Y. Acad. Sci.* **2002**, *957*, 230–238. [CrossRef]
35. Bhat, K.P.; Pezzuto, J.M. Cancer chemopreventive activity of resveratrol. *Ann. N. Y. Acad. Sci.* **2002**, *957*, 210–229. [CrossRef]
36. Sun, D.; Jin, X.; Camerlink, I.; Tong, M.; Su, J.; Zhao, F.; Yan, S.; Shi, B. Effects of *Yucca schidigera* extract on growth performance and antioxidative function of small intestine in broilers. *J. Anim. Physiol. Anim. Nutr.* **2019**, *103*, 738–746. [CrossRef]
37. Johnson, I.T.; Gee, J.M.; Price, K.; Curl, C.; Fenwick, G.R. Influence of saponins on gut permeability and active nutrient transport in vitro. *J. Nutr.* **1986**, *116*, 2270–2277. [CrossRef]

animals

MDPI

Article

Effect of an *Alliaceae* Encapsulated Extract on Growth Performance, Gut Health, and Intestinal Microbiota in Broiler Chickens Challenged with *Eimeria* spp.

Gonzalo Villar-Patiño [1,2], María del Carmen Camacho-Rea [3,*], Myrna Elena Olvera-García [2], Julio César Baltazar-Vázquez [4], Gabriela Gómez-Verduzco [5], Guillermo Téllez [6], Aurora Labastida [7] and Aurora Hilda Ramírez-Pérez [5,*]

1 Programa de Doctorado en Ciencias de la Salud y de la Producción Animal, Facultad de Medicina Veterinaria y Zootecnia, Universidad Nacional Autónoma de México, Avenida Universidad 3000, Coyoacán, Ciudad de Mexico 04510, Mexico; gvillar@gponutec.com
2 Grupo Nutec, Avenida de las Fuentes No. 14, Parque Industrial Bernardo Quintana, El Marqués 76246, Querétaro, Mexico; molvera@gponutec.com
3 Instituto Nacional de Ciencias Médicas y Nutrición "Salvador Zubirán", Departamento de Nutrición Animal, Tlalpan, Ciudad de Mexico 14080, Mexico
4 Specialized Animal Nutrition Research Network, Grupo Nutec, La Valla, San Juan del Río 76814, Querétaro, Mexico; jbaltazar@gponutec.com
5 Facultad de Medicina Veterinaria y Zootecnia, Universidad Nacional Autónoma de México, Avenida Universidad 3000, Coyoacán, Ciudad de Mexico 04510, Mexico; gagove@unam.mx
6 Department of Poultry Science, University of Arkansas, Fayetteville, AR 72701, USA; gtellez@uark.edu
7 OMICs Analysis, Camino a Xilotepetl No. 45, Tepoztlán 62520, Morelos, Mexico; a.labastida@omicsanalysis.mx
* Correspondence: carmen.camachor@incmnsz.mx (M.d.C.C.-R.); rpereza@unam.mx (A.H.R.-P.); Tel.: +52-55-1234-6028 (M.d.C.C.-R.); +52-55-5622-5906 (A.H.R.-P.)

Citation: Villar-Patiño, G.; Camacho-Rea, M.d.C.; Olvera-García, M.E.; Baltazar-Vázquez, J.C.; Gómez-Verduzco, G.; Téllez, G.; Labastida, A.; Ramírez-Pérez, A.H. Effect of an *Alliaceae* Encapsulated Extract on Growth Performance, Gut Health, and Intestinal Microbiota in Broiler Chickens Challenged with *Eimeria* spp. *Animals* **2023**, *13*, 3884. https://doi.org/10.3390/ani13243884

Academic Editor: Kenneth Bafundo

Received: 25 October 2023
Revised: 13 December 2023
Accepted: 15 December 2023
Published: 18 December 2023

Simple Summary: Coccidiosis is caused by an intracellular parasite that damages the intestinal integrity, negatively affecting the digestion and absorption of nutrients and consequently worsening weight gain, feed efficiency, and pigmentation of birds, even causing mortality. Therefore, it has a negative impact on the economy of the poultry industry. Currently, the disease is mainly treated by using anticoccidials drugs added to the diet. The drug resistance, as well as the residue of drugs in the meat, has prompted the development of natural alternatives to combat coccidiosis. The purpose of this research was to determine whether an *Alliaceae* encapsulated extract added to the broiler chickens diet decreased the number of oocysts excreted in feces and the harm caused to the intestinal mucosa, consequently improving the productive performance of broiler chickens challenged with *Eimeria* spp. Under our experimental conditions, both the inclusion of *Alliaceae* extract, as well as the use of conventional anticoccidials (nicarbazin/narasin/salinomycin), diminished the detrimental effect of *Eimeria* spp. Moreover, the *Alliaceae* extract favored the abundance of acid butyric bacteria (*Ruminococcus* spp. and *Intestinimonas* spp.) in the cecum, related to intestinal health. Based on the current findings, the *Alliaceae* extract could be a natural additive used to lessen the effects of coccidiosis infections.

Abstract: This study analyzed the effects of an *Alliaceae* encapsulated extract (AE-e) on daily gain (ADG), feed intake (ADFI), feed conversion ratio (FCR), oocysts per gram of feces (OPG), intestinal lesion (LS), and microbiota composition in broilers challenged with *Eimeria* spp. A total of 4800 one day Cobb-500 were allotted into 10 treatment groups with 12 replicates of 40 birds in a 2 × 4 + 2 factorial arrangement. The first factor was non-challenged (NC) or challenged (C), the second was four levels of AE-e added in the basal diet, 0 (AE0), 250 (AE250), 500 (AE500), and 750 mg·kg^{-1} (AE750), plus two ionophore controls, non-challenged (NC-Ion) and challenged (C-Ion). No interactions were observed between factors (NC0, NC250, NC500, NC750, C0, C250, C500, and C750), while C-Ion improved FCR at 21 d. The challenge affected negatively ADG and FCR and promoted enteropathogens in cecum. AE750 improved FCR in the finisher and cumulative phases, while C-Ion had fewer total

OPG than C0 and C250. Likewise, at 21d, C250, C500, and C-Ion had fewer LS than C0, while at 28 d, C750 showed lower than C-Ion. In the cecum microbiota, C500 had more *Ruminococcus, Firmicutes b*, and *Intestinimonas* than C-Ion. In summary, AE-e showed beneficial results in broilers infected with *Eimeria* spp.

Keywords: broiler chickens; coccidiosis; *Alliaceae* extract; ionophore; microbiota intestinal

1. Introduction

Currently, coccidiosis disease continues to be one of the most serious problems in the commercial broiler poultry industry, resulting in great economic loss all over the world [1]. *Eimeria acervulina, Eimeria maxima,* and *Eimeria tenella* are the main species that cause disease in broiler chickens, impairing both intestinal function and growth performance [2]. To date, coccidiosis prevention has been through the addition of synthetic anticoccidials such as nicarbazin, decoquinate, and zoalene, as well as ionophores [3] such as monensin, lasalocid, salinomycin, narasin, etc., to either the poultry diet or drinking water, or the use of vaccines [4]. However, *Eimeria* spp. has developed drug resistance, causing a loss of effectiveness in the anticoccidials [5]. In addition, there is a global concern about drug resistance and the presence of residual drugs in meat [6] that has prompted the study and development of natural alternatives to prevent or control coccidiosis [7], such as phytochemicals, which are a suitable alternative due to their favorable effects against *Eimeria* spp. [8].

Several works have demonstrated that extracts and the essential oil of *Allium cepa* and *Allium sativum* improved the average daily feed intake (ADFI), average daily gain (ADG), and feed conversion ratio (FCR), as well as intestinal health and carcass quality in broiler chickens [9–11]. Likewise, blends of the genus *Allium* with oregano essential oil and eugenol [12], as well as organic acids [8,13], have shown benefits in broiler performance and health. The genus *Allium* has different sulfur compounds that have been studied as additives in animal nutrition [14]. These compounds are conformed by sulfur atoms attached to a cyanate group in cyclic or non-cyclic forms [15]. The most studied are the sulfoxydes s-methyl-L-cysteine (methionine), s-allil-L-cysteine (alliin), s-propenyl-L-cysteine (isoalliin), and s-propyl-L-cysteine (propiin). Propyl propane thiosulfinate (PTS) and propyl propane thiosulfonate (PTSO) are derived from the natural degradation of propiin [16,17]. Unlike other sulfur compounds from the genus *Allium*, PTSO is chemically stable but is insoluble in water. For this reason, it should be provided with a water-soluble carrier to increase its availability and absorption [18,19]. In addition, PTS and PTSO have shown to reduce the viability of *E. acervulina* sporozoites, improving the innate immune response [20], and inhibit the sporulation of *E. tenella* [21], and they are able to promote the growth of beneficial bacteria and decrease intestinal pathobionts [22] in broiler chickens. Recently, Aguinaga-Casanas, et al. [19] conducted a study in vitro that showed that PTSO inhibit the capability of *E. acervulina* sporozoites to penetrate Madin-Darby bovine kidney cells (MDBK cells), concluding that PTSO is a promising alternative to coccidiosis treatment. The *Allium* spp. could be used as an alternative to anticoccidials in broiler production due to its proven benefits. Nevertheless, it is necessary to carry out more studies to understand the mechanism by which they exert their favorable effects, as well as determine the appropriate inclusion level in the poultry diet and the period of use to enhance broiler performance [23].

Based on the above information, we hypothesize that an *Alliaceae (A. cepa* and *A. sativum)* encapsulated extract (AE-e) used as a feed additive decreases the oocysts per gram in feces (OPG), reduces the intestinal lesion score (LS), and improves productive performance, as well as modulates positively the intestinal microbiota in *Eimeria* challenged birds. Therefore, the aim of this study was to examine the effects of increasing levels of AE-e on ADG, ADFI, FCR, OPG, LS, and intestinal microbiota composition in broiler chickens challenged with a mixture of *Eimeria* spp.

2. Materials and Methods

2.1. Ethical Standard

The present study was approved on 5 October 2020, by the Animal Welfare and Experimentation Ethics Committee of the National Autonomous University of México SICUAE-DC-2020/3-6 in compliance with the Mexican Official Norm NOM-062-ZOO-1999.

2.2. Housing, Animals and Experimental Design

A total of 4800 one-day-old Cobb-500 broiler chickens were housed in 1204 m^2 pens separated by wire mesh partitions and new wood shaving litter throughout the period of study of 49 d. The facility temperature was set as 30 °C during the first week using thermostatically controlled propane gas heaters, reducing 2.5 °C each week. After the fourth week, the temperature was controlled through curtains and kept between 18–21 °C. The first 4 d after reception, the chickens had access to 23 h of light; after that, a natural photoperiod was maintained throughout the study. The birds were assigned in a completely randomized experimental design with a $2 \times 4 + 2$ factorial arrangement. One factor was the challenge level, formed by a non-challenged group (NC) and a challenged group (C) with *Eimeria* spp. The second was 4 levels of AE-e (0, 250, 500, and 750 mg AE-e per each kg of feed). In addition, two ionophore controls, C-Ion and NC-Ion, were used to contrast the challenged and non-challenged AE-e treatments, respectively. The experimental unit was the pen. To prevent cross contamination, the C and NC birds were housed in separate but identical buildings.

2.3. Alliaceae Encapsulated Extract Supplementation

We used a concentrated liquid commercial *Alliaceae* extract (Garlicon™; DOMCA S.A.U., Granada, Spain), which has shown positive effects on bird productivity [20,24,25]. It was encapsulated into a dextrin–lecithin matrix and validated by the presence of PTSO, which has a concentration of 12 $g \cdot kg^{-1}$, as determined by gas chromatography–mass spectrometry (Gas chromatograph model 7890A Agilent Technologies Inc., coupled to a simple quadrupole mass detector model 5975C Agilent Technologies Inc., Santa Clara, CA, USA). The retention time of the chromatography peak was indicated for the PTSO according to the databases of the NIST/EPA/NIH Mass Spectra Library, version 1.7 (Gaithersburg, MD, USA). The analysis was carried out in the laboratory of the Center for Research in Applied Sciences and Advanced Technology of the National Polytechnic Institute (IPN, Querétaro, Mexico).

2.4. Diets and Experimental Groups

A corn–soybean meal basal diet was formulated to meet the nutritional specifications for the Cobb-500 lineage™ (Table 1). The basal diet was split into five portions to be mixed with the experimental doses of AE-e or ionophore and then were pelleted at 80 °C for 30 s. The broiler chickens were on two phases of feeding, a starter phase (1–21 day of age) and a finisher phase (22–49 day of age). The feed was restricted from 15:00 h to 08:00 h to avoid ascites syndrome, and the water was provided ad libitum.

The birds were assigned to either a NC or C group and feed with the basal diet containing 4 different levels of AE-e as follows: basal diet without AE-e (AE0); basal diet with added 250 $mg \cdot kg^{-1}$ of AE-e (AE250); basal diet with added 500 $mg \cdot kg^{-1}$ of AE-e (AE500); and basal diet with added 750 $mg \cdot kg^{-1}$ of AE-e (AE750). Two positive control treatments, non-challenged ionophore control (NC-Ion) and a challenged ionophore control (C-Ion), both formed by a basal diet with 50 ppm of nicarbazin and 50 ppm of narasin added for the starter phase and 60 ppm of salinomycin for the finisher phase, summarizing a total of 10 treatments (NC0, NC250, NC500, NC750, C0, C250, C500, C750, NC-Ion, C-Ion) with 12 replicates of 40 birds for each one.

Table 1. Ingredients (kg·t^{-1} of feed) and calculated chemical composition (% as fed) and metabolizable energy (EM, kcal·kg^{-1}) in the basal diets.

Ingredient	Starter (0–21 d)	Finisher (22–49 d)
Corn	513.9	552.5
Soybean meal	406.0	360.0
Soybean oil	40.5	51.0
Limestone	14.7	13.6
Calcium orthophosphate	9.1	7.0
Sodium carbonate	4.9	4.7
Methionine DL	3.6	3.2
Xanthophylls.		2.4
Salt	2.0	1.6
L-Lysine HCL	2.2	1.5
Threonine	1.1	0.8
Vitamin-and mineral trace Premix [1]	0.9	0.8
Betaine anhydrous	0.6	0.4
L-valine	0.2	
Biocholine	0.2	0.17
Tryptophan		0.16
Phytase 5000	0.1	0.12
Calculated nutrient levels (%)		
Humidity	11.61	11.66
EM (kcal/kg)	3156	3245
Crude protein CP	23.90	21.60
Ether extract	6.21	7.34
Ashes	5.83	5.27
Crude fiber	2.34	2.47
Total phosphorous	0.59	0.53
Total calcium	1.00	0.90
Sodium	0.23	0.21
Lysine	1.50	1.29
Xanthophylls	0.0008	0.0080

[1] Content per kilogram: vitamin A (retinol acetate), 12,000 international units (IU); vitamin D3, 5000 IU; vitamin E (DL-α-tocopherol acetate), 50 IU; vitamin K, 3 mg; thiamine, 3 mg; riboflavin, 9 mg; pantothenic acid, 15 mg; pyridoxine, 4 mg; biotin, 0.2 mg; folic acid, 2 mg; vitamin B12, 0.02 mg; manganese, 100 mg; zinc 100 mg; iron, 40 mg; copper, 15 mg; iodine, 1 mg; selenium, and 0.35 mg.

2.5. Productive Performance

Each pen was monitored for body weight (BW), weight gain (WG), and feed intake (FI) at 0, 21, and 49 d of age. On the same day as the event took place, we recorded the age and weight of dead birds to determine (a) ADG: [(mean final BW of live birds in the pen) − (mean initial BW of all birds in that pen)]/days of testing. (b) ADFI: (total feed consumed in a pen)/(birds alive × days on test in the pen + days dead birds on test in that pen). (c) FCR: (total feed consumption in a pen)/(WG of birds alive + WG of dead birds in the same pen) [26].

2.6. *Eimeria* Challenge

At 12 d of age, the broiler chickens from the challenged group (C) were inoculated directly into the crop with 0.5 mL of a mixture of sporulated oocysts of *E. acervulina* 1×10^5, *E. maxima* 2×10^4, and *E. tenella* 2×10^4 using sterile plastic syringes, while the birds from the non-challenged group (NC) received a sham 0.5 mL of distilled water.

The *Eimeria* mixture was obtained from a non-governmental laboratory of parasitology, Morelos, México, and it was assessed by counting the oocysts sporulated from the different species of *Eimeria* at the National Autonomous University of México (UNAM) and by PCR Sanger sequencing at the Faculty of Chemistry, Querétaro University, México (UAQ).

2.7. *Eimeria Oocysts Count*

On d 9, 16, and 23 post-inoculation (p.i.), approximately 10 g of fresh fecal material was collected from each pen and mixed thoroughly in a plastic bag and kept at 4 °C until the total count of oocysts. Five grams of each sample was homogenized in a saturated NaCl solution (400 g·L^{-1}) and filtered through a 300-mesh sieve. The filtrate was centrifuged at 800× *g* for 2 min and an aliquot of the supernatant was poured into a Mc Master Chamber and counted at 10× magnification on a compound microscope following the technique described by Long, et al. [27]. The morphological characteristics of the sporulated oocysts were used to identify *E. acervulina*, *E. maxima*, and *E. tenella*; the number of oocysts was expressed as OPG. The total OPG is the sum of OPG for all three species.

2.8. *Intestinal Lesion Score (LS)*

The LS in the duodenum, jejunum, and cecum were evaluated at 9- and 16-day p.i., twenty-four birds from each treatment were randomly selected and humanely killed by cervical dislocation [28,29]. The gastrointestinal tract was removed and opened; the scores for macroscopic lesions for *E. acervulina, E. maxima, and E. tenella* were determined according to the scale of Johnson and Reid [30]. A score of "0" represented no visual lesions, "1" was minimal lesions, "2" was moderate lesions, "3" was severe lesions, and "4" was extremely severe lesions.

2.9. *Anticoccidial Index (ACI)*

The relative ratio weight gain (rBWG), survival rate (SR), total mean lesion score (TMLS), and OPG value are necessary to calculate the anticoccidial index (ACI) and are recognizes as good indicators of the efficacy of the anticoccidial compounds. The ACI was calculated for each group according to the following equation proposed by Merk, et al. [31]:

$$ACI = (rBWG + SR) \times 100 - (TMLS \times 10 + OPG\ value)$$

The variables were calculated as follows:

- rBWG: BWG rate of the challenged unmedicated control or drug treated group/BWG rate of unchallenged unmedicated control group × 100.
- BWG rate: (Final BW − initial BW)/initial BW × 100.
- SR: Number of final birds alive/ number of total initial birds × 100.
- TMLS: Sum of the LS caused by all the *Eimeria* spp.
- OPG value: OPG in unchallenged unmedicated control or challenged drug-treated group/OPG in infected/unmedicated control group × 100 [32].

2.10. *Intestinal Microbiota Samples*

At 21 d of age (9 d p.i.), 6 chickens from 5 treatments, NC0, C0, NC500, C500, and C-Ion, were randomly selected and sacrificed by the manual cervical dislocation method [29,33]. The gastrointestinal tract was dissected, the ileum and cecum contents were scraped carefully and collected in cryogenic vials, snap frozen in liquid nitrogen, and stored at −80 °C until the microbiota composition analysis [34,35].

2.11. *DNA Extraction, 16s rRNA Gene Amplification, and Library Preparation for Sequencing*

The bacterial DNA from the ileal and cecal contents was extracted using the ZymoBIOMICS™ DNA Miniprep kit (D4300 Zymo Research, Irvine, CA, USA), according to the manufacturer recommendations, it was quantified by fluorometry using Qubit chemistry (Invitrogen, Waltham, MA, USA), while its integrity was assessed by spectrometry (NanoDrop, Thermo Fisher Scientific, Whaltam, MA, USA). The libraries were made following the two-step polymerase chain reaction (PCR) protocol suggested by Illumina (Illumina Part# 15044223 Rev.B, San Diego, CA, USA) to sequence a single segment comprising the 16S rRNA V3-V4 region [36]. The libraries were quantified by fluorometry, pooled at 4nM with 10% PhiX sequencing control, and sequenced using the Illumina MiSeq platform to obtain 300 paired-end reads following the manufacturer´s instructions (Illumina, San Diego, CA, USA).

2.12. Bioinformatic Analysis

The paired-end raw reads were analyzed with Cutadapt v1.15 to eliminate any traces of 16S-rRNA amplification primers or Illumina adapter sequences and then scanned with Trimmomatic v36 [37] to filter out the lower quality reads. The forward and reverse reads of each pair were then overlapped into single fragments using FLASH v1.2.11 software, employing an expected fragment length of 409 $\pm$ 20 bp and an expected read length of 279 bp and were further filtered with DADA2 (included in the QIIME2 v2020.8 suite) [38]; to eliminate reads where 2 or more sequencing errors were expected, groups of reads were produced by experimental errors (noise removal) and chimeric fragments.

To assign a taxonomic classification to the pre-processed sequences, we used the naive Bayes classifier [39], as implemented in the QIIME2 suite [38]. The classifier was trained with the annotations of the SILVA 138 ribosomal reference database [40] using the sequences grouped to a similarity of 99%.

The Shannon and Simpson indexes, as well as the total OTU counts, were obtained for each sample to study the intestinal microbial α-diversity. β-diversity was assessed by measuring the Bray–Curtis dissimilarity of each pair of samples, followed by nonmetric multi-dimensional scaling (NMDS) to observe the clustering of the different sample groups. The α and β-diversity profiles were visualized through box-plots and NMDS scaling plots [41]. To obtain the relative abundance of the OTUs, the number of reads per OTU was normalized by library size. Only the abundance changes in genus with a significance $p \leq 0.05$ were represented in a heatmap.

2.13. Statistical Analyses

The statistical analyses were performed using the JMP statistical software v 17.0.0 (SAS Institute Inc., Cary, NC, USA) and R version 4.0.2. The data normality and variance homogeneity among groups were tested using the Shapiro–Wilk and Levene´s tests, respectively. The variables with non-normal distributions were analyzed by nonparametric statistics. The significance level was set at $p \leq 0.05$, and a trend was set between $p > 0.05$ and ≤ 0.10.

For the analysis of productive performance, the initial BW was included as a covariate. The ADFI, ADG, and FCR were analyzed by 2-way ANOVA. The ACI was analyzed by one-way ANOVA. Post hoc Tukey tests were performed. The experimental design considered two controls with ionophores to contrast the treatments. Contrast A was NC-Ion vs. NC0, NC250, NC500, and NC750; contrast B was C-Ion vs. C0, C250, C500, and C750; and contrast C was NC-Ion vs. C-Ion. Additionally, the AE-e factor was analyzed to determine if the effects of different doses of AE-e had a linear trend. Furthermore, AE-e treatments included an analysis of the polynomial orthogonal contrast trend in the variable ACI.

Since there were no detected oocyst or coccidia lesions in nonchallenged birds, the OPG and LS were analyzed only in the challenged birds using the Kruskal–Wallis test and, as post hoc, the Steel–Dwass test, the medians, and quantiles q25 and q75 were reported. Correlations were carried out between ADG, FCR, OPG, and LS using Spearman Rho analysis.

The α-diversity changes among groups were assessed with the Kruskal–Wallis test, while the β-diversity was assessed by the NMDS and PerMANOVA tests to identify significant differences in the clustering position of the groups. To obtain the diversity measures and the corresponding statistical tests, we used the R phyloseq v1.38 package [41]. We used the DESeq2 package to estimate the differential abundance of specific clades between group sample pairs, using the Wald test and adjusting the p-values through the Benjamini–Hochberg multiple sampling correction.

3. Results

3.1. Productive Performance

The following tables provide comprehensive data on the interactions between the *Eimeria* challenge and AE-e supplementation (Table 2) and subsequently present independently the effects of the *Eimeria* challenge (Table 3) and the effects of AE-e supplementation (Table 4) on ADG, ADFI, and FCR. No significant differences ($p > 0.05$) were found in the interaction of factors on ADG, ADFI, and FCR in starter and finisher phases or in the cumulative period of study (Table 2). As well as the orthogonal contrast (contrast A), NC-Ion vs. NC0, NC250, NC500, and NC750 did not show differences in ADFI, ADG, and FCR ($p > 0.05$) in both growing phases. On the other hand, the comparison of contrast B, C-Ion vs. C0, C250, C500 and C750, did not show differences in ADFI and ADG throughout the study period ($p > 0.05$). However, at 21 d, C-Ion showed lower FCR, 1.30 vs. 1.39, respectively, ($p < 0.01$) and a trend ($p = 0.07$) to improve ADG. There was no effect on FCR in the finisher or cumulative period ($p > 0.05$). Finally, in contrast C, NC-Ion vs. C-Ion, there was a trend in the starter phase in which the challenged group (C-Ion) had lower ADG ($p = 0.06$) and higher FCR ($p = 0.07$) than NC-Ion.

Table 2. Initial body weight (BW), average daily gain (ADG), average daily feed intake (ADFI), and feed conversion ratio (FCR) of broilers chickens fed a corn–soybean diet supplemented with different doses of *Alliaceae* encapsulated extract (AE-e) [1] or anticoccidial drugs (Ion) [2] under challenge with *Eimeria* spp. [3].

	Non-Challenged (NC)				Challenged (C)								Contrasts [4]		
													A	B	C
	NC0	NC250	NC500	NC750	C0	C250	C500	C750	SEM	*p* val.	NC-Ion	C-Ion	*p* val.	*p* val.	*p* val.
BW 0 d	42.3	41.9	41.7	41.9	42.1	42.3	42.2	42.0	0.21	0.30	42.0	42.2	0.71	0.69	0.53
Starter phase (1–21 d)															
ADG (g)	48.2	47.0	47.4	47.5	41.3	41.7	43.5	39.9	1.09	0.33	46.5	43.7	0.41	0.07	0.06
ADFI (g)	58.4	58.1	58.1	58.2	58.7	59.9	58.7	57.2	0.81	0.36	58.1	57.3	0.90	0.15	0.47
FCR	1.21	1.24	1.22	1.21	1.41	1.43	1.34	1.41	0.03	0.47	1.23	1.30	0.75	0.01	0.07
Finisher phase (22–49 d)															
ADG (g)	79.5	81.5	84.0	82.4	78.2	80.6	82.2	82.7	1.99	0.96	84.1	82.6	0.29	0.42	0.58
ADFI (g)	156.6	159.1	158.6	149.6	157.0	156.2	151.9	151.7	3.95	0.70	150.9	153.1	0.20	0.87	0.67
FCR	2.16	2.06	2.12	2.01	2.23	2.08	1.97	1.93	0.08	0.47	1.99	2.00	0.21	0.58	0.92
Cumulative study (1–49 d)															
ADG (g)	66.1	66.7	68.3	67.4	62.4	63.9	65.6	64.4	1.10	0.97	68.0	66.0	0.46	0.11	0.17
ADFI (g)	114.5	115.8	115.5	110.5	114.8	115.0	112.0	111.2	2.32	0.80	111.1	112.1	0.20	0.70	0.76
FCR	1.77	1.75	1.75	1.70	1.91	1.84	1.74	1.76	0.03	0.22	1.69	1.74	0.17	0.19	0.29

[1] NC0, Non-challenged + basal diet; NC250, Non-challenged + basal diet + 250 $mg \cdot kg^{-1}$ AE-e; NC500, Non-challenged + basal diet + 500 $mg \cdot kg^{-1}$ AE-e; NC750, Non-challenged + basal diet + 750 $mg \cdot kg^{-1}$ AE-e; C0, Challenged + basal diet; C250, Challenged + basal diet + 250 $mg \cdot kg^{-1}$ AE-e; C500, Challenged + basal diet + 500 $mg \cdot kg^{-1}$ AE-e; C750, Challenged + basal diet + 750 $mg \cdot kg^{-1}$ AE-e; [2] NC-Ion, Non-challenged + basal diet + 50 ppm nicarbazin–50 ppm narasin 1–21 d/ salinomycin 60 ppm 22–49 d. C-Ion, Challenged + basal diet + 50 ppm nicarbazin–50 ppm narasin 1–21 d/ salinomycin 60 ppm 22–49 d. [3] Challenge with *E. acervulina* 1×10^5, *E. maxima* 2×10^4, and *E. tenella* 2×10^4, [4] Contrast A: NC-Ion vs. NC0, NC250, NC500, and NC750; Contrast B: C-Ion vs. C0, C250, C500, and C750; Contrast C: NC-Ion vs. C-Ion.

At the starter phase, the challenge affected the productive performance. Broiler chickens from the challenged group (C) had a reduction of 12.4% in ADG compared to those from the non-challenged (NC) group ($p < 0.0001$); moreover, the FCR was also deteriorated by 12.8%, 1.40 vs. 1.22 ($p < 0.0001$), respectively. Despite not finding changes in the finisher phase ($p > 0.05$), the negative effect observed in the starter phase continued in the cumulative period, ADG ($p < 0.001$) and FCR ($p = 0.01$) in broiler chickens from C group compared to NC group. However, the ADFI was not affected by the challenge ($p > 0.05$; Table 3).

Table 3. Effects of the challenge with *Eimeria* spp. [1] on average daily gain (ADG), average daily feed intake (ADFI), and feed conversion ratio (FCR) of broiler chickens fed a corn–soybean diet.

	Challenge Factor			
	NC	**C**	**SEM**	***p* Value**
Body weight (0 d)	42.00	42.10	0.11	0.23
Starter phase				
ADG (g)	47.50 [a]	41.60 [b]	0.54	<0.001
ADFI (g)	58.20	58.60	0.40	0.44
FCR	1.22 [a]	1.40 [b]	0.01	<0.001
Finisher phase				
ADG (g)	81.80	80.90	0.99	0.53
ADFI (g)	156.00	154.20	1.97	0.52
FCR	2.09	2.05	0.04	0.51
Cumulative study				
ADG (g)	67.10 [a]	64.10 [b]	0.55	<0.001
ADFI (g)	114.10	113.20	1.17	0.61
FCR	1.74 [a]	1.81 [b]	0.02	0.01

[ab] Means within rows not sharing a common superscript differ at the *P* value reported. [1] NC Non-challenged; C, challenged with *E. acervulina* 1×10^5, *E. maxima* 2×10^4, and *E. tenella* 2×10^4.

Table 4. Initial body weight (BW), average daily gain (ADG), average daily feed intake (ADFI), and feed conversion ratio (FCR) of broiler chickens fed a corn–soybean diet supplemented with different doses of *Alliaceae* encapsulated extract (AE-e) [1].

	AE-e Factor						Linear Trend [2]	
	AE0	**AE250**	**AE500**	**AE750**	**SEM**	***p* Value**	***p* Value**	**R^2**
BW 0 d	42.20	42.10	42.00	42.00	0.15	0.65		
Starter phase								
ADG (g)	44.80	44.40	45.54	43.70	0.77	0.41	0.61	0.002
ADFI (g)	58.60	59.00	58.40	57.70	0.57	0.45	0.38	0.008
FCR	1.31	1.34	1.28	1.31	0.02	0.23	0.82	0.005
Finisher phase								
ADG (g)	78.90	81.00	83.10	82.60	1.40	0.15	0.05	0.04
ADFI (g)	156.70	157.70	155.30	150.70	2.78	0.31	0.16	0.02
FCR	2.20 [a]	2.07 [ab]	2.05 [ab]	1.97 [b]	0.05	0.03	0.006	0.08
Cumulative study								
ADG (g)	64.20	65.30	67.00	65.90	0.78	0.10	0.14	0.02
ADFI (g)	114.70	115.40	113.80	110.80	1.64	0.23	0.13	0.02
FCR	1.84 [a]	1.79 [ab]	1.75 [ab]	1.73 [b]	0.03	0.01	0.004	0.08

[ab] Means within rows not sharing a common superscript differ at *p* value reported. [1] AE0, basal diet; AE250, basal diet + 250 mg·kg^{-1} AE-e; AE500, basal diet + 500 mg·kg^{-1} AE-e; AE750, basal diet + 750 mg·kg^{-1} AE-e; [2] Linear trend AE0, AE250, AE500, and AE750.

Table 4 shows that during the starter phase, the inclusion of AE-e did not influence performance ($p > 0.05$). Nevertheless, in the finisher phase, the FCR was better in AE750, 1.97 compared to AE0, 2.20 ($p = 0.03$). Moreover, in the cumulative period, AE750 continued to show better FCR, 1.73 regarding 1.84 from AE0 ($p = 0.01$); however, the AE250 and AE500 inclusion did not show a difference in FCR ($p > 0.05$). In addition, no significant differences in ADG and ADFI were observed between treatments during the study ($p > 0.05$); however, AE500 displayed a trend to improve ADG in the cumulative study ($p = 0.10$). On the other hand, a positive linear trend was observed in ADG ($p = 0.05$) and FCR ($p = 0.006$) in the finisher phase, as well as FCR in the cumulative period ($p = 0.004$).

3.2. Oocysts Shedding

Table 5 describes the effect of feed supplementation with AE-e or ionophores on the OPG of *Eimeria*-challenged broiler chickens (C0, C250, C500, C750, and C-Ion), because of the absence of OPG in non-challenged treatments, NC0, NC250, NC500, NC750, and

NC-Ion were omitted from analysis. The differences between treatments were observed at 9 d p.i., corresponding to 21 d of age. The higher OPG values of *E. acervulina* and *E. tenella* were observed in C0, 7.75 and 22.26-fold changes, respectively, regarding C-Ion ($p < 0.05$).

Table 5. Medians and ranges (Q25–Q75) of oocysts per gram of feces (OPG) in broiler chickens supplemented with different doses of *Alliaceae* encapsulated extract (AE-e) [1] or anticoccidial drugs (Ion) [2], under challenge with *Eimeria* spp. [3].

		Challenge Treatments						
OPG	**Age (d)**	**C0**	**C250**	**C500**	**C750**	**C-Ion**	**H [4] (Ji^2)**	***p* Value**
Total OPG	21	95,575 ^{a} (35,712–186,537)	109,725 ^{a} (28,150–125,350)	48,900 ^{ab} (35,887–66,337)	35,350 ^{ab} (24,462–107,225)	17,975 ^{b} (4862–56,237)	14.22	<0.01
	28	3525 (112–6787)	1575 387–17,612)	700 (0–12,262)	1150 (0–3000)	9225 (812–17,275)	3.76	0.44
	35	0 (0–0)	0 (0–187)	0 (0–37)	0 (0–0)	0 (0–225)	5.50	0.24
Eacervulina	21	78,700 ^{a} (30,537–142,175)	62,850 ^{ab} (12,925–86,512)	32,900 ^{ab} (24,300–46,575)	31,700 ^{ab} (18,550–83,950)	10,150 ^{b} (1562–49,687)	12.16	0.02
	28	2625 (50–5350)	1025 (50–15,350)	375 (0–9000)	400 (0–1200)	7950 (200–16,400)	5.20	0.26
	35	0 (0–0)	0 (0–0)	0 (0–0)	0 (0–0)	0 (0–100)	10.36	0.35
E. maxima	21	3750 (387–16,550)	850 (25–11,312)	3925 (62–14,450)	1550 (162–6037)	3500 (1437–7462)	2.77	0.60
	28	0 (0–137)	25 (0–187)	0 (0–175)	0 (0–250)	100 (0–725)	3.11	0.53
	35	0 (0–0)	0 (0–187)	0 (0–0)	0 (0–0)	0 (0–37)	2.79	0.60
E. tenella	21	21,150 ^{a} (3550–35,862)	15,450 ^{ab} (0–54,025)	3875 ^{ab} (287–17,425)	2575 ^{ab} (637–10,700)	950 ^{b} (62–5675)	10.33	0.04
	28	700 (25–1462)	325 (212–1225)	225 (0–8712)	100 (0–2400)	425 (25–1950)	0.51	0.97
	35	0 (0–0)	0 (0–0)	0 (0–0)	0 (0–0)	0 (0–0)	2.80	0.59

[ab] Medians within rows not sharing a common superscript differ at *P value* reported. [1] C0, Challenged + basal diet; C250, Challenged + basal diet + 250 mg·kg^{-1} AE-e; C500, Challenged + basal diet + 500 mg·kg^{-1} AE-e; C750, Challenged + basal diet + 750 mg·kg^{-1} AE-e; [2] C-Ion, Challenged + basal diet + 50 ppm nicarbazin–50 ppm narasin 1–21 d/ salinomycin 60 ppm 22–49 d. [3] Challenged with *E. acervulina* 1×10^5, *E. maxima* 2×10^4 and *E. tenella* 2×10^4. [4] H, test statistic of Kruskal–Wallis.

While C250, C500, and C750 had a similar OPG shedding in both species compared to C-Ion and C0 ($p > 0.05$), in addition, *E. maxima* did not show differences between treatments ($p > 0.05$). Moreover, the total OPG values were significantly higher (+5-fold) in C0 and C250 than C-Ion ($p < 0.01$), whereas C500 and C750 had a similar OPG shedding compared to C-Ion, C0, and C250 ($p > 0.05$). The decrease in OPG excretion from 21 to 28 d of age was more evident in the groups C250 (−69.66-fold), C500 (−69.85-fold), and C750 (−30.73-fold) than C-Ion (−1.94-fold) and C0 (−27.11-fold). At 28 and 35 day of age, *E. acervulina, E. maxima,* and *E. tenella* OPG differences between treatments were not detected ($p > 0.05$).

3.3. Intestinal Lesion Score

Due to the absence of injuries in the NC groups, NC0, NC250, NC500, NC750, and NC-Ion, we only show the analysis of the challenged groups, C0, C250, C500, C750, and C-Ion. The LS of duodenum, jejunum, and cecum, as well as TMLS, are described in Table 6. At 21 days of age (9 d p.i.), the LS observed in the duodenum reveled that the birds given the C0 diet showed a more severe LS of 1.5, which was significantly ($p < 0.0001$) higher than the 0.5, 1, and 0 scores from birds given the C250, C500, and C-Ion diets, respectively. Remarkably, the LS observed in broilers from the C250 diet were similar to those observed in the broilers on the C-Ion diet. In addition, no significant differences ($p > 0.05$) were detected among the C250, C500, and C750 diets. However, in the jejunum, significant differences were identified ($p = 0.03$) between the C0 and C750 vs. the C250 and C-Ion diets; despite the LS medians being equal in the groups, the range q25–q75 was higher in

C0 and C750, (0-1), than in C250 and C-Ion, (0-0). In addition, in this intestinal section, the C500 diet had similar effects to the others on LS ($p > 0.05$). In the cecum, the LS in the broilers given the C-Ion diet were lower than those observed in birds fed the C0, C250, and C750 diets ($p < 0.001$) but not compared to C500 ($p > 0.05$). Furthermore, on day 21, the total mean lesion score (TMLS) was significantly higher ($p < 0.001$) in birds from diet C0 compared to C250, C500, and C-Ion. Moreover, the chicken on the C-Ion diet had lower TMLS compared to other treatments at 21 d of age ($p < 0.0001$). On the other hand, on day 28, the LS caused by *E. acervuline*, *E. maxima*, and *E. tenella* were not different between all treatments ($p > 0.05$). However, C-Ion showed higher TMLS compared to the C750 diet ($p = 0.02$), while the C0, C250, and C500 diets did not show significant differences ($p > 0.05$).

Table 6. Medians and ranges (Q25–Q75) in intestinal lesion scores (LS) in broiler chickens supplemented with different doses of *Alliaceae* encapsulated extract (AE-e) [1] or anticoccidial drugs (Ion) [2], under challenge with *Eimeria* spp.[3].

		Challenge Treatments						
LS	Age (d)	C0	C250	C500	C750	C-Ion	H [5] (Ji²)	p Value
Duodenum	21	1.5 c (1–2)	0.5 ab (0–1)	1 b (0–1)	1 bc (1–1)	0 a (0–0)	49.7	<0.0001
	28	0 (0–1)	0 (0–0)	0 (0–0)	0 (0–0)	0 (0–0.7)	9.2	0.06
Jejunum	21	0 b (0–1)	0 a (0–0)	0 ab (0–0.7)	0 b (0–1)	0 a (0–0)	10.7	0.03
	28	0 (0–0)	0 (0–0)	0 (0–0)	0 (0–0)	0 (0–0)	9.7	0.09
Cecum	21	1 c (1–2)	1 bc (0–3)	1 abc (0–1)	1 bc (0–2)	0 a (0–0)	31.4	<0.0001
	28	0.5 (0–2)	0.5 (0–2)	1 (0–1)	0 (0–0.7)	1.5 (0–2)	9.2	0.06
TMLS [4]	21	3 c (2–4.7)	2 b (1–3)	2 b (1–3)	2 bc (1.2–3.7)	0 a (0–0.7)	54.0	<0.0001
	28	1 ab (1–2.7)	1 ab (0–2)	1 ab (0–2)	0 a (0–1.7)	2 b (1–2)	12.0	0.02

abc Medians within rows not sharing a common superscript differ at p value reported. [1] C0, Challenged + basal diet; C250, Challenged + basal diet + 250 mg·kg^{-1} AE-e; C500, Challenged + basal diet + 500 mg·kg^{-1} AE-e; C750, Challenged + basal diet + 750 mg·kg^{-1} AE-e; [2] C-Ion, Challenged + basal diet + 50 ppm nicarbazin–50 ppm narasin 1–21 d/salinomycin 60 ppm 22–49 d. [3] Challenged with *E. acervulina* 1 × 10^5, *E. maxima* 2 × 10^4 and *E. tenella* 2 × 10^4; [4] TMLS Total mean lesion score. [5] H, test statistic of Kruskal–Wallis.

3.4. Anticoccidial Index (ACI)

Table 7 describes the ACI of the different treatments, NC0 had superior ACI compared to C0, C250, C500, and C750 ($p < 0.0001$). C-Ion performed better than C0 and C250 ($p < 0.0001$). The groups C500 and C750 had similar results compared to C-Ion, C0, and C250 ($p > 0.05$). There were linear ($p < 0.04$) and quadratic ($p < 0.03$) positive trends in ACI when comparing the responses to C0, C250, C500 and C750.

Table 7. Calculated anticoccidial index (ACI) in 21 d broiler chickens supplemented with different doses of *Alliaceae* encapsulated extract (AE-e) [1] or anticoccidial drugs (Ion) [2], under challenge with *Eimeria* spp. [3], as well as orthogonal polynomial contrast comparison of AE-e [4].

	Treatments								Linear Trend		Quadratic Trend	
Means [5]	NC0	C0	C250	C500	C750	C-Ion	SEM	p Value	p Value	R²	p Value	R²
rBWG %	100	85.8	86.1	90.0	82.7	91.4						
SR %	95.0	91.4	92.1	91.0	93.8	94.8						
TMLS × 10	0	34.2	21.7	19.2	25.8	2.5						
OPG value %	0	100	79.1	54.3	59.3	28.0						
ACI	195 a	43.1 c	77.5 c	107.6 bc	91.3 bc	155.7 ab	21.99	<0.0001	0.03	0.09	0.04	0.13

abc Means within rows not sharing a common superscript differ at p value reported. [1] NC0, Non-challenged + basal diet; C0, Challenged + basal diet; C250, Challenged + basal diet + 250 mg·kg^{-1} AE-e; C500, Challenged + basal diet + 500 mg·kg^{-1} AE-e; C750, Challenged + basal diet + 750 mg·kg^{-1} AE-e; [2] C-Ion, Challenged + basal diet + 50 ppm nicarbazin–50 ppm narasin 1–21 d. [3] Challenged with *E. acervulina* 1 × 10^5, *E. maxima* 2 × 10^4, and *E. tenella* 2 × 10^4; [4] Polynomic trends, treatments C0, C250, C500, and C750. [5] rBWG, relative ratio growth gain; SR, survival rate; TMLS, total mean lesion score; OPG value, oocyst per gram of feces value.

Spearman correlations (rho) were calculated between OPG, TMLS, ADG, and FCR; the coefficient showed that the presence of OPG at 21 d of age had a positive correlation of 0.80 and 0.71 with TMLS at 21 and 28 d of age, respectively ($p < 0.0001$). In addition, this OPG had also a positive correlation of 0.73 ($p < 0.0001$) with the FCR at 21 d. On the other hand, there was a negative correlation (−0.54) with ADG at 21 d of age and −0.29 at 49 d of age ($p < 0.01$). At 28 d of age, OPG and TMLS showed a positive correlation of 0.42 ($p < 0.001$). The TMLS at 21 and 28 d were positively correlated (0.67). TMLS at 21d showed a positive correlation of 0.72 with the FCR in the starter phase ($p < 0.001$), as well as a positive correlation of 0.22 with the cumulative period ($p < 0.05$), while showing a negative correlation (−0.51) with the ADG at the same age ($p < 0.001$) and of −0.18 with ADG at 49 d ($p < 0.05$). Therefore, TMLS at 28 d was negatively correlated (−0.18) with ADG at 49 d of age ($p < 0.05$).

3.5. *Analysis of Bacterial Composition, 16s rRNA*

To evaluate the effects of 5 treatments, NC0, NC500, C0, C500, and C-Ion, on the intestinal microbiota composition in broiler chickens, we studied the ileum and cecum microbial community using 16S rRNA sequencing. We obtained a total of 2,683,053 high-quality sequences of the V3–V4 region of the 16S rRNA gene; a total of 10,710 operational taxonomic units (OTUs) at the 99% sequence similarity level were identified in all samples.

3.6. *Alpha- and Beta-Diversity*

We detected no differences in the α-diversity indexes (Shannon and Simpson) or in the OTU count of different treatment pairs in the ileum and cecum ($p > 0.05$). In addition, the β-diversity assessed by NMDS from Bray–Curtis distance matrices and a permutational analysis of variance in the ileum section showed no differences in the microbiome between the treatments ($p > 0.05$). On the other hand, in the cecum, there were significant differences in the clustering position in the groups at the taxonomic levels of family, Figure 1a ($p < 0.01$), and genus, Figure 1b ($p = 0.01$). The results showed that NC0 and NC500 were different from C0 and C500, while C-Ion showed no differences against any treatment, except in the taxonomic category family, which was different from NC0, Figure 1.

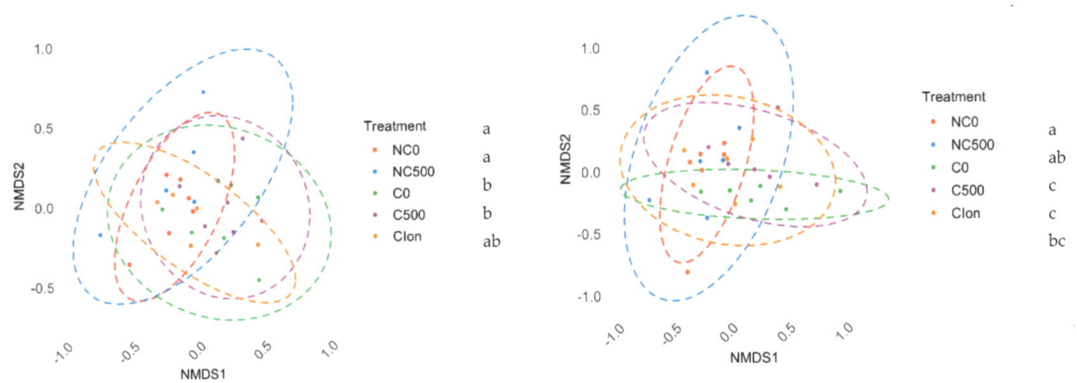

(**a**) Family cecum; ($p < 0.01$)

(**b**) Genus cecum; ($p = 0.01$)

Figure 1. β-diversity comparison between treatments [1]. NMDS of Bray-Curtis distances in (**a**) Family cecum and (**b**) Genus cecum. [abc] The treatments not sharing superscript are different ($p < 0.05$). NC0, Non-challenged + basal diet; NC500, Non-challenged + basal diet + 500 mg·kg^{-1} AE-e; C0, Challenged + basal diet; C500, Challenged + basal diet + 500 mg·kg^{-1} AE-e; C-Ion, Challenged + basal diet + 50 ppm nicarbazin–50 ppm narasin. Challenged with *E. acervulina* 1×10^5, *E. maxima* 2×10^4, and *E. tenella* 2×10^4.

3.7. Relative Intestinal Microbiota Abundance

Figure 2 depicts the analysis of the relative abundance of different bacterial clades in the cecum at the phylum and genus level for each group of samples. In the cecum microbiota, *Firmicutes* was the dominant phylum (Figure 2a), with a relative abundance ranging from 57.8% (C0 treatment) to 79.4% (NC0 treatment). At the genus level (Figure 2b), *Bacteroides* was the taxon with the highest relative abundance (9.7–22.1%), highlighting their presence in the groups challenged, C0, C500, and C-Ion.

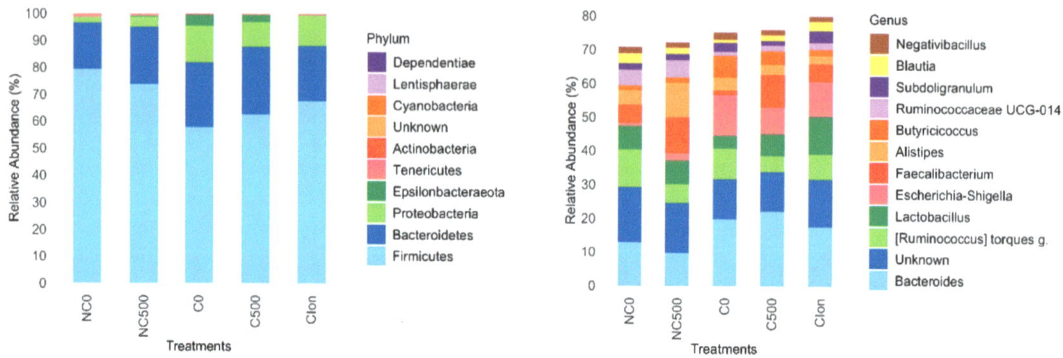

(**a**) Phylum cecum (**b**) Genus cecum

Figure 2. Relative abundance in the intestinal microbiota in broiler chickens at 21 d in (**a**) Phylum cecum, (**b**) Genus cecum. Treatments: NC0, Non-challenged + basal diet; NC500, Non-challenged + basal diet + 500 mg·kg^{-1} AE-e; C0, Challenged + basal diet; C500, Challenged + basal diet + 500 mg·kg^{-1} AE-e; C-Ion, Challenged + basal diet+50 ppm nicarbazin–50 ppm narasin. Challenged with *E. acervulina* 1×10^5, *E. maxima* 2×10^4, and *E. tenella* 2×10^4. The relative abundances were obtained after normalizing the reads per clade counts by sequencing the library size and obtaining the average for each clade across treatments. The 12 most abundant clades are shown.

To further explore the effect of the treatments on the cecum bacterial communities, we searched for bacterial genus abundance changes among pairs of treatments, C0 vs. NC0; C500 vs. NC500; C-Ion vs. C0; NC500 vs. NC0; C500 vs. C0; and C500 vs. C-Ion (Figure 3). The results are presented in a heat map from a hierarchical clustering analysis based only on significant changes ($p < 0.05$). The results from this study showed that the pair comparisons C0 vs. NC0 and C500 vs. NC500 had more bacterial change-in-abundance differences than the other comparisons studied: C-Ion vs. C0; NC500 vs. NC0; C500 vs. C0; and C500 vs. C-Ion. There was a significantly higher abundance ($p < 0.05$) of genus *Ruminococcus 2* in birds from C500 compared to NC500, C0, and C-Ion (+23.36 log2-fc, +8.0 log2-fc, and +23.08 log2-fc, respectively). Moreover, the abundance change in *Intestinimonas* in the C500 group was +1.78 log2-fc higher than the C-Ion group. However, *Escherichia-Shigella*, another predominant bacterium in broiler chicken intestines, was also found to be abundant (+4.2 log2fc) in C0 with respect to the NC0 ($p < 0.05$). Other bacteria, such as *Tyzzerella*, *Eggerthella*, *Clostridium innocuum* g., *Ruminococcaceae* UCG-009, u. *Clostridia* b., and *Ruminococcus* 1., show a higher abundance under the *Eimeria* challenge conditions. On the other hand, NC0 has higher numbers ($p < 0.05$) of *Bacillus*, *Hydrogenoanaerobacterium*, and *Ruminococus 1* than its challenged counterparts.

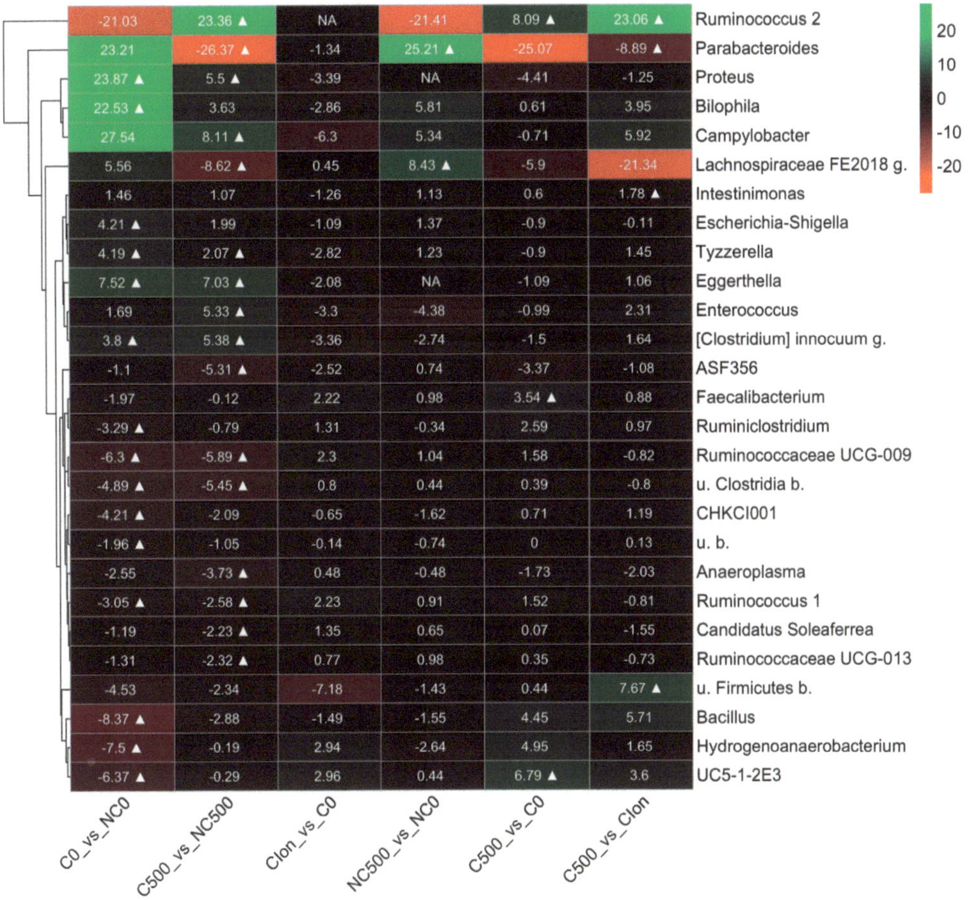

Figure 3. Heat map of the bacterial genus from the cecum of broiler chickens at 21d (9d p.i.). The color block in the heatmap indicates the normalized abundance log2 fold change (log2-fc) of each pair of groups at the bacterial genus: C0 vs. NC0; C500 vs. NC500; C-Ion vs. C0; NC500 vs. NC0; C500 vs. C0; and C500 vs. C-Ion. The positive and negative changes are indicated by the intensity of the green and red color, respectively. Significant changes (with an adjusted $p \leq 0.05$) are marked with a symbol Δ. The non-available (NA) label indicates that the corresponding comparison was not performed as there were no counts for most of the samples of the group pair. A hierarchical clustering tree based on the log2-fc signals of the clades is shown to the left of the heatmap. The abbreviations u. b. and g. in the genus names stand for "unknown", "bacterium", and "genus", respectively.

4. Discussion

In our study, the *Eimeria* challenge caused a negative impact on growth performance parameters, being more evident at 9 d p.i. (21 days of age). The decrease in the ADG in infected birds, as well as the increases in the FCR, has been reported in other works [8,26,42,43]. In our study, the inclusion of AE750 improved the FCR of broilers in the finisher phase, and in the cumulative study regarding AE0, these results are in agreement with Peinado, et al. [44], who also include two levels of PTSO in the broiler chicken diet and did not find any effect on ADFI, whereas ADG improved using 45 mg per PTSO kg^{-1} of the diet. Moreover, the FCR was enhanced with both doses tested, 45 and 90 mg per PTSO kg^{-1} of diet. Similar results were obtained by Kim, et al. [20], who reported that broilers challenged with *E. acervuline* and fed with 10 ppm of PTSO (67%) and PTS (33%) showed a better growth than the birds that were not supplemented ($p < 0.05$). Furthermore, Kairalla, et al. [11] reported that some feed additives, particularly garlic (*A. sativum*), have shown to improve FCR, as was also previously demonstrated by Aarti and Khusro [45]. This result is important since *Eimeria* infection destroys epithelial cells and affects intestinal villi, causing poor nutrient digestion and severe damage to the host intestinal mucosa, resulting in clinical or subclinical symptoms [46]. In this regard, in a previous study, we demonstrated that organosulfur compounds from garlic, particularly PTSO in 250 $g{\cdot}t^{-1}$ of feed, improved the amino acid and energy digestibility in broiler chickens fed with a soybean meal–yellow corn diet [47].

ADG or FCR are not the only good indicators for measuring the effectiveness of anticoccidial drug, but LS and OPG are also considered complementary indicators [48] to the performance. In our study, the higher OPG excretion detected was 9 d p.i., then OPG gradually decreased over time; at 16 d p.i., it is barely noticeable and almost disappears at 23 d p.i. This trend after infection was reported by You [49]. In addition to the treatment effect, the reduction in OPG may be due in part to self-limitation of parasitosis and the immune response developed by the host [50]. We found that OPG shedding was decreased in the challenged birds supplemented with anticoccidial treatment (C-Ion), while 500 and 750 $mg{\cdot}kg^{-1}$ of AE-e tended to reduce OPG shedding. A similar effect was observed on the OPG of broiler chickens challenged with *Eimeria* and supplemented with garlic extracts (*A. sativum*) [51,52], their active derivative compounds [20], or a premix of garlic and oregano essential oils [53]. The ACI results were better for C-Ion than for C0 and C250. A positive linear and quadratic trend in ACI suggests that generally, birds challenged with *Eimeria* spp. would perform better when 500 or 750 $mg{\cdot}kg^{-1}$ of AE-e is included in their diet.

The correlation analysis in our study indicates that the reduction in the LS was supported by the decreasing OPG. Similar results were found by Elkhtam, et al. [51], where OPG, the clinical symptoms of the disease, and the LS decreased with the addition of garlic extract (*A. sativum*) to the diet of the broiler chickens challenged with *Eimeria* spp. The LS could explain the negative effect on ADG and FCR at 21 and 28 d of age, which was previously noticed by Reid and Johnson [54], who contrasted the LS of birds infected with *E. acervulina* with their weight reached, finding that at higher LS, a lower ADG was obtained, mostly by one week after the challenge; however, LS are not always correlated in coccidia infections, as reported by Ringenier, et al. [55], who did not find a relationship between LS and FCR in broilers at 28 d of age, concluding that broiler chickens are able to cope with a certain level of gut damage before it influences the overall performance. Other researchers such as Conway, et al. [56] concluded that the correlation between OPG, LS, ADG, and FCR depends on the type of *Eimeria* and the use or not of anticoccidials.

On the other hand, it has been reported that diet composition [57,58] and phytochemicals modulate intestinal microbiota [44,53]. It is well known that a healthy and functional intestinal microbiome is related to a positive productive performance of the chickens [59,60]. It was reported that when growing broiler chickens that feed on diets supplemented with *Allium* derivatives, PTSO or PTS are able to influence intestinal microbiota composition [22,61], decreasing enteropathogens and increasing the nutrient absorption in the intestine [22,62–64]. In the current study, neither AE-e nor anticoccidials modified the α- or β-diversity in the ileum or the α-diversity in the cecum. Similar findings were noted by Abdelli, et al. [13], who found that dietary supplementation with natural compounds such as organic acids and essential oils does not always result in changes in diversity in microbial populations within the gastrointestinal tract. Nevertheless, in the cecum, we found changes in β-diversity between clades at the taxonomic levels of family and genus, where the challenged groups showed a different spatial distribution to those not challenged, concluding that the infection with *Eimeria* spp. influences β-diversity, as reported by other researchers [33,65,66]. The analysis of the microbiota clearly showed that the groups under challenge of *Eimeria* spp., regardless of the presence of anticoccidial drugs or AE-e in the diet, had a higher number of enteropathogens belonging to the *Enterobacteriaceae* family such as *Proteus*, *Escherichia*, and *Shigella*, as well as other opportunistic pathogens, including the genera *Tyzerella*, *Eggertella*, and *Biophila*.

In our study, we found that in the cecum, at the genus level, C500 had a higher abundance of *Ruminococcus*, *Firmicutes* b., and *Intestinimonas* than C-Ion, considering that *Ruminococcus* bacteria synthesize digestive enzymes such as cellulases, xylanases, and cellobioses [67–69], which contributes to the hydrolysis and fermentation of non-structural carbohydrates, producing butyric acid. It has also been reported that some species of *Ruminococcus* produce bacteriocins that contribute to controlling undesirable bacterial populations, enhancing the growth of *Lactobacillus* and promoting intestinal health. All this could explain the improvement in nutrient digestibility and the productive behavior of birds when *Ruminococcus* is present in abundance. On the other hand, species of the *Intestinimonas* genus are producers of short-chain fatty acids increasing butyric fatty acid from simple sugars and amino acids such as lysine [70,71]. It connects two important metabolic characteristics, butyric acid production and amino acid fermentation in the intestinal tract. Thus, in this study, the beneficial effects of AE-e on the modulation of the intestinal microbiota were consistent with the results of Vezza, et al. [57], who demonstrated in a murine model of metabolic syndrome that PTSO supplementation at doses of 0.1, 0.5, and 1 mg·kg^{-1}·day^{-1} counteracts intestinal dysbiosis.

5. Conclusions

In summary, the dose of 750 mg·kg^{-1} of an *Alliaceae* encapsulated extract added to the diet of broiler chickens improved the feed conversion ratio in the finisher phase compared to broiler chickens fed a diet without additives. Furthermore, during the finisher phase and cumulative study, its addition to the diet resulted in a positive linear trend in average daily gain and feed conversion ratio.

The anticoccidial index showed a quadratic trend in which the dose of 500 mg·kg^{-1} of the *Alliaceae* encapsulated extract displayed the best response. Moreover, it promoted the abundance of some butyrate-producing bacteria such as *Intestinimonas* and *Ruminococcus* in the cecum.

However, the best anticoccidial index, feed conversion ratio, oocyst shedding, and intestinal lesion score was observed in the group fed with the anticoccidial ionophore program.

Further research is required to explain the mode of action, as well as determine the optimal dose of the *Alliaceae* encapsulated extract in the diet of broiler chickens to lessen or control the detrimental effects of coccidiosis under industrial conditions. It is necessary to verify whether using an *Alliaceae* encapsulated extract in dual programs combined with ionophores or vaccines is feasible.

Author Contributions: Conceptualization, G.V.-P. and M.E.O.-G.; methodology, G.V.-P., M.E.O.-G. and J.C.B.-V.; validation, M.E.O.-G. and J.C.B.-V.; formal analysis, G.V.-P., M.d.C.C.-R., M.E.O.-G. and A.L.; investigation, G.V.-P. and M.E.O.-G.; resources, G.V.-P.; data curation, G.V.-P. and A.L.; writing—original draft preparation, G.V.-P.; writing—review and editing A.H.R.-P. and M.d.C.C.-R.; visualization, A.H.R.-P. and M.d.C.C.-R.; supervision, G.G.-V., G.T. and A.H.R.-P.; project administration, G.V.-P., M.E.O.-G. and J.C.B.-V.; funding acquisition, G.V.-P. All authors have read and agreed to the published version of the manuscript.

Funding: This research was funded by Grupo Nutec (Avenida del Marqués 32, Parque Industrial Bernardo Quintana, El Marqués, Querétaro, México, 76246). "https://www.gponutec.com/ (accessed on 16 October 2023)", funding number iiia 2-19 L-A.

Institutional Review Board Statement: This experiment was approved on 5 October 2020 by the Institutional Subcommittee for the Care and Use of Experimental Animals (protocol SICUAE. DC-2020/3-6) of the School of Veterinary Medicine and Zootechnics of the National Autonomous University of México (UNAM), in compliance with the Mexican Official Norm NOM-062-ZOO-1999.

Informed Consent Statement: The study was conducted at the facilities of the SANUREN (Specialized Animal Nutrition Research Network) experimental farm owned by the Grupo Nutec company. Informed consent was obtained from all subjects involved in the study.

Data Availability Statement: The data presented in this study are available on request from the corresponding authors. The data are not publicly available due to privacy.

Acknowledgments: The authors wish to thank all staff members of the Specialized Animal Nutrition Research Network (SANUREN, Querétaro, México) and Carlos Vázquez-Hernández staff member of OMICs who contributed to this study.

Conflicts of Interest: At the time of submitting this manuscript, G. Villar-Patiño, M.E. Olvera-García, and J.C. Baltazar-Vázquez declare a potential direct conflict of interest as they work at Grupo NUTEC. https://www.gponutec.com/ (accessed on 24 October 2023). M.C. Camacho-Rea, G. Gómez-Verduzco, G. Téllez, A. Labastida, and A.H. Ramírez-Pérez have no conflicts of interest.

References

1. Britez, J.D.; Rodriguez, A.E.; Di Ciaccio, L.; Marugán-Hernandez, V.; Tomazic, M.L. What do we know about surface proteins of chicken parasites *Eimeria*? *Life* **2023**, *13*, 1295. [CrossRef] [PubMed]
2. Liu, J.; Liu, L.; Li, L.; Tian, D.; Li, W.; Xu, L.; Yan, R.; Li, X.; Song, X. Protective immunity induced by *Eimeria* common antigen 14–3-3 against *Eimeria tenella*, *Eimeria acervulina* and *Eimeria maxima*. *BMC Vet. Res.* **2018**, *14*, 337. [CrossRef] [PubMed]
3. Noack, S.; Chapman, H.D.; Selzer, P.M. Anticoccidial drugs of the livestock industry. *Parasitol. Res.* **2019**, *118*, 2009–2026. [CrossRef] [PubMed]
4. Zaheer, T.; Abbas, R.Z.; Imran, M.; Abbas, A.; Butt, A.; Aslam, S.; Ahmad, J. Vaccines against chicken coccidiosis with particular reference to previous decade: Progress, challenges, and opportunities. *Parasitol. Res.* **2022**, *121*, 2749–2763. [CrossRef] [PubMed]
5. Mesa-Pineda, C.; Navarro-Ruíz, J.L.; López-Osorio, S.; Chaparro-Gutiérrez, J.J.; Gómez-Osorio, L.M. Chicken coccidiosis: From the parasite lifecycle to control of the disease. *Front. Vet. Sci.* **2021**, *8*, 787653. [CrossRef] [PubMed]
6. Martins, R.R.; Silva, L.J.G.; Pereira, A.M.P.T.; Esteves, A.; Duarte, S.C.; Pena, A. Coccidiostats and poultry: A comprehensive review and current legislation. *Foods* **2022**, *11*, 2738. [CrossRef] [PubMed]
7. Wunderlich, F.; Al-Quraishy, S.; Steinbrenner, H.; Sies, H.; Dkhil, M.A. Towards identifying novel anti-*Eimeria* agents: Trace elements, vitamins, and plant-based natural products. *Parasitol. Res.* **2014**, *113*, 3547–3556. [CrossRef]
8. Stefanello, C.; Rosa, D.P.; Dalmoro, Y.K.; Segatto, A.L.; Vieira, M.S.; Moraes, M.L.; Santin, E. Protected blend of organic acids and essential oils improves growth performance, nutrient digestibility, and intestinal health of broiler chickens undergoing an intestinal challenge. *Front. Vet. Sci.* **2019**, *6*, 491. [CrossRef]
9. Ur Rahman, S.; Khan, S.; Chand, N.; Sadique, U.; Khan, R.U. In vivo effects of *Allium cepa* L. on the selected gut microflora and intestinal histomorphology in broiler. *Acta Histochem.* **2017**, *119*, 446–450. [CrossRef]
10. Malematja, E.; Manyelo, T.G.; Ng'ambi, J.W.; Nemauluma, M.F.D.; Kolobe, S.D. Effects of onion extracts (*Allium cepa*) inclusion in diets on growth performance, carcass characteristics, and bone morphometric of broiler chickens. *Anim. Biosci.* **2023**, *36*, 1075–1082. [CrossRef]
11. Kairalla, M.A.; Alshelmani, M.I.; Aburas, A.A. Effect of diet supplemented with graded levels of garlic (*Allium sativum* L.) powder on growth performance, carcass characteristics, blood hematology, and biochemistry of broilers. *Open Vet. J.* **2022**, *12*, 595–601. [CrossRef] [PubMed]
12. Kumar, A.; Sharma, N.K.; Kheravii, S.K.; Keerqin, C.; Ionescu, C.; Blanchard, A.; Wu, S.B. Potential of a mixture of eugenol and garlic tincture to improve performance and intestinal health in broilers under necrotic enteritis challenge. *Anim. Nutr.* **2022**, *8*, 26–37. [CrossRef] [PubMed]
13. Abdelli, N.; Perez, J.F.; Vilarrasa, E.; Cabeza Luna, I.; Melo-Duran, D.; D'Angelo, M.; Sola-Oriol, D. Targeted-release organic acids and essential oils improve performance and digestive function in broilers under a necrotic enteritis challenge. *Animals* **2020**, *10*, 259. [CrossRef] [PubMed]
14. Kothari, D.; Lee, W.D.; Niu, K.M.; Kim, S.K. The genus *Allium* as poultry feed additive: A review. *Animals* **2019**, *9*, 1032. [CrossRef] [PubMed]
15. Putnik, P.; Gabric, D.; Roohinejad, S.; Barba, F.J.; Granato, D.; Mallikarjunan, K.; Lorenzo, J.M.; Bursac Kovacevic, D. An overview of organosulfur compounds from *Allium* spp.: From processing and preservation to evaluation of their bioavailability, antimicrobial, and anti-inflammatory properties. *Food Chem.* **2019**, *276*, 680–691. [CrossRef] [PubMed]
16. Subramanian, M.S.; Nandagopal, M.G.; Amin Nordin, S.; Thilakavathy, K.; Joseph, N. Prevailing knowledge on the bioavailability and biological activities of sulphur compounds from *Alliums*: A potential drug candidate. *Molecules* **2020**, *25*, 4111. [CrossRef] [PubMed]
17. Sorlozano-Puerto, A.; Albertuz-Crespo, M.; Lopez-Machado, I.; Ariza-Romero, J.J.; Banos-Arjona, A.; Exposito-Ruiz, M.; Gutierrez-Fernandez, J. In Vitro antibacterial activity of propyl-propane-thiosulfinate and Propyl-propane-thiosulfonate derived from *Allium* spp. against gram-negative and gram-positive multidrug-resistant bacteria isolated from human samples. *Biomed Res. Int.* **2018**, *2018*, 7861207. [CrossRef]
18. Abad, P.; Arroyo-Manzanares, N.; Rivas-Montoya, E.; Ochando-Pulido, J.M.; Guillamon, E.; García-Campaña, A.M.; Martinez-Ferez, A.; Plaizier, J. Effects of different vehiculization strategies for the allium derivative propyl propane thiosulfonate during dynamic simulation of the pig gastrointestinal tract. *Can. J. Anim. Sci.* **2019**, *99*, 244–253. [CrossRef]

19. Aguinaga-Casanas, M.A.; Mut-Salud, N.; Falcon-Pineiro, A.; Alcaraz-Martinez, A.; Guillamon, E.; Banos, A. In vitro antiparasitic activity of propyl-propane-thiosulfinate (PTS) and propyl-propane-thiosulfonate (PTSO) from *Allium cepa* against *Eimeria acervulina* sporozoites. *Microorganisms* **2022**, *10*, 2040. [CrossRef]
20. Kim, D.K.; Lillehoj, H.S.; Lee, S.H.; Lillehoj, E.P.; Bravo, D. Improved resistance to *Eimeria acervulina* infection in chickens due to dietary supplementation with garlic metabolites. *Br. J. Nutr.* **2013**, *109*, 76–88. [CrossRef]
21. Pourali, M.; Kermanshahi, H.; Golian, A.; Razmi, G.; Soukhtanloo, M. Antioxidant and anticoccidial effects of garlic powder and sulfur amino acids on *Eimeria*-infected and uninfected broiler chickens. *Iran. J. Vet. Res.* **2014**, *15*, 227–232.
22. Peinado, M.J.; Ruiz, R.; Echávarri, A.; Aranda-Olmedo, I.; Rubio, L.A. Garlic derivative PTS-O modulates intestinal microbiota composition and improves digestibility in growing broiler chickens. *Anim. Feed. Sci. Technol.* **2013**, *181*, 87–92. [CrossRef]
23. Khan, R.; Nikousefat, Z.; Tufarelli, V.; Naz, S.; Javdani, M.; Laudadio, V. Garlic (*Allium sativum*) supplementation in poultry diets, effects on production and physiology. *World's Poult. Sci. J.* **2012**, *68*, 417–424. [CrossRef]
24. Rabelo-Ruiz, M.; Ariza-Romero, J.J.; Zurita-Gonzalez, M.J.; Martin-Platero, A.M.; Banos, A.; Maqueda, M.; Valdivia, E.; Martinez-Bueno, M.; Peralta-Sanchez, J.M. *Allium*-based phytobiotic enhances egg production in laying hens through microbial composition changes in ileum and cecum. *Animals* **2021**, *11*, 448. [CrossRef] [PubMed]
25. Abad, P.; Arroyo-Manzanares, N.; Ariza, J.J.; Baños, A.; García-Campaña, A.M. Effect of *Allium* extract supplementation on egg quality, productivity, and intestinal microbiota of laying hens. *Animals* **2021**, *11*, 41. [CrossRef] [PubMed]
26. Holdsworth, P.A.; Conway, D.P.; McKenzie, M.E.; Dayton, A.D.; Chapman, H.D.; Mathis, G.F.; Skinner, J.T.; Mundt, H.C.; Williams, R.B.; World Association for the Advancement of Veterinary, P. World Association for the Advancement of Veterinary Parasitology (WAAVP) guidelines for evaluating the efficacy of anticoccidial drugs in chickens and turkeys. *Vet. Parasitol.* **2004**, *121*, 189–212. [CrossRef] [PubMed]
27. Long, P.L.; Millard, B.J.; Joyner, L.P.; Norton, C.C. A guide to laboratory techniques used in the study and diagnosis of avian coccidiosis. *Folia Vet. Lat.* **1976**, *6*, 201–217.
28. Kraieski, A.L.; Hayashi, R.M.; Sanches, A.; Almeida, G.C.; Santin, E. Effect of aflatoxin experimental ingestion and *Eimeria* vaccine challenges on intestinal histopathology and immune cellular dynamic of broilers: Applying an Intestinal Health Index. *Poult. Sci.* **2017**, *96*, 1078–1087. [CrossRef]
29. Moraes, P.O.; Cardinal, K.M.; Gouvêa, F.L.; Schroeder, B.; Ceron, M.S.; Lunedo, R.; Frazzon, A.P.G.; Frazzon, J.; Ribeiro, A.M.L. Comparison between a commercial blend of functional oils and monensin on the performance and microbiota of coccidiosis-challenged broilers. *Poult. Sci.* **2019**, *98*, 5456–5464. [CrossRef]
30. Johnson, J.; Reid, W.M. Anticoccidial drugs: Lesion scoring techniques in battery and floor-pen experiments with chickens. *Exp. Parasitol.* **1970**, *28*, 30–36. [CrossRef]
31. Merk Sharp; Dohome Laboratory. *Anticoccidial Index*; Merk Company: Kenilworth, NJ, USA, 1976.
32. Wang, L.; Guo, W.; Haq, S.U.; Guo, Z.; Cui, D.; Yang, F.; Cheng, F.; Wei, X.; Lv, J. Anticoccidial activity of Qinghao powder against *Eimeria tenella* in broiler chickens. *Front. Vet. Sci.* **2021**, *8*, 709046. [CrossRef] [PubMed]
33. Choi, J.; Kim, W. Interactions of Microbiota and Mucosal Immunity in the Ceca of Broiler Chickens Infected with *Eimeria tenella*. *Vaccines* **2022**, *10*, 1941. [CrossRef] [PubMed]
34. Bae, Y.; Koo, B.-S.; Lee, S.; Mo, J.; Oh, K.; Mo, I. Bacterial diversity and its relationship to growth performance of broilers. *Korean J. Vet. Res.* **2017**, *57*, 159–167. [CrossRef]
35. Yang, C.; Kennes, Y.; Lepp, D.; Yin, X.; Wang, Q.; Yu, H.; Yang, C.; Gong, J.; Diarra, M. Effects of encapsulated cinnamaldehyde and citral on the performance and cecal microbiota of broilers vaccinated or not vaccinated against coccidiosis. *Poult. Sci.* **2020**, *99*, 936–948. [CrossRef] [PubMed]
36. Klindworth, A.; Pruesse, E.; Schweer, T.; Peplies, J.; Quast, C.; Horn, M.; Glöckner, F.O. Evaluation of general 16S ribosomal RNA gene PCR primers for classical and next-generation sequencing-based diversity studies. *Nucleic Acids Res.* **2012**, *41*, e1. [CrossRef]
37. Bolger, A.M.; Lohse, M.; Usadel, B. Trimmomatic: A flexible trimmer for Illumina sequence data. *Bioinformatics* **2014**, *30*, 2114–2120. [CrossRef] [PubMed]
38. Bolyen, E.; Rideout, J.R.; Dillon, M.R.; Bokulich, N.A.; Abnet, C.C.; Al-Ghalith, G.A.; Alexander, H.; Alm, E.J.; Arumugam, M.; Asnicar, F.; et al. Author Correction: Reproducible, interactive, scalable and extensible microbiome data science using QIIME 2. *Nat. Biotechnol.* **2019**, *37*, 1091. [CrossRef]
39. Wang, Q.; Garrity, G.M.; Tiedje, J.M.; Cole, J.R. Naïve Bayesian classifier for rapid assignment of rRNA sequences into the new bacterial taxonomy. *AEM* **2007**, *73*, 5261–5267. [CrossRef]
40. Glockner, F.O.; Yilmaz, P.; Quast, C.; Gerken, J.; Beccati, A.; Ciuprina, A.; Bruns, G.; Yarza, P.; Peplies, J.; Westram, R.; et al. 25 years of serving the community with ribosomal RNA gene reference databases and tools. *J. Biotechnol.* **2017**, *261*, 169–176. [CrossRef]
41. McMurdie, P.J.; Holmes, S. Phyloseq: An R package for reproducible interactive analysis and graphics of microbiome census data. *PLoS ONE* **2013**, *8*, e61217. [CrossRef]
42. Ritzi, M.M.; Abdelrahman, W.; Mohnl, M.; Dalloul, R.A. Effects of probiotics and application methods on performance and response of broiler chickens to an *Eimeria* challenge. *Poult. Sci.* **2014**, *93*, 2772–2778. [CrossRef] [PubMed]
43. Osho, S.O.; Xiao, W.W.; Adeola, O. Response of broiler chickens to dietary soybean bioactive peptide and coccidia challenge. *Poult. Sci.* **2019**, *98*, 5669–5678. [CrossRef] [PubMed]

44. Peinado, M.J.; Ruiz, R.; Echavarri, A.; Rubio, L.A. Garlic derivative propyl propane thiosulfonate is effective against broiler enteropathogens in vivo. *Poult. Sci.* **2012**, *91*, 2148–2157. [CrossRef] [PubMed]
45. Aarti, C.; Khusro, A. Role of garlic (*Allium sativum*) as feed supplements in poultry industries: An overview. *WNOFNS* **2020**, *29*, 151–161.
46. Blake, D.P.; Knox, J.; Dehaeck, B.; Huntington, B.; Rathinam, T.; Ravipati, V.; Ayoade, S.; Gilbert, W.; Adebambo, A.O.; Jatau, I.D.; et al. Re-calculating the cost of coccidiosis in chickens. *Vet. Res.* **2020**, *51*, 115. [CrossRef] [PubMed]
47. Villar-Patiño, G.; Camacho-Rea, M.d.C.; Olvera-García, M.E.; Soria-Soria, A.; Baltazar-Vázquez, J.C.; Gómez-Verduzco, G.; Solano, L.; Téllez, G.; Ramírez-Pérez, A.H. The effect of encapsulated Propyl propane thiosulfonate (PTSO) on apparent ileal digestibility and productive performance in broiler chickens. *Animals* **2023**, *13*, 1123. [CrossRef] [PubMed]
48. Chasser, K.M.; Duff, A.F.; Wilson, K.M.; Briggs, W.N.; Latorre, J.D.; Barta, J.R.; Bielke, L.R. Research Note: Evaluating fecal shedding of oocysts in relation to body weight gain and lesion scores during *Eimeria* infection. *Poult. Sci.* **2020**, *99*, 886–892. [CrossRef]
49. You, M.-J. Detection of four important *Eimeria* species by multiplex PCR in a single assay. *Parasitol. Int.* **2014**, *63*, 527–532. [CrossRef]
50. Kim, W.H.; Chaudhari, A.A.; Lillehoj, H.S. Involvement of T cell immunity in avian coccidiosis. *Front. Immunol.* **2019**, *10*, 2732. [CrossRef]
51. Elkhtam, A.; Shata, A.; El-Hewaity, M.H. Efficacy of turmeric (*Curcuma longa*) and garlic (*Allium sativum*) on *Eimeria* species in broilers. *IJBAS* **2014**, *3*, 349. [CrossRef]
52. Ali, M.; Chand, N.; Khan, R. Anticoccidial effect of garlic (*Allium sativum*) and ginger (*Zingiber officinale*) against experimentally induced coccidiosis in broiler chickens. *J. Appl. Anim. Res.* **2019**, *1*, 79–84. [CrossRef]
53. Sidiropoulou, E.; Skoufos, I.; Marugan-Hernandez, V.; Giannenas, I.; Bonos, E.; Aguiar-Martins, K.; Lazari, D.; Blake, D.P.; Tzora, A. In vitro anticoccidial study of oregano and garlic essential oils and effects on growth performance, fecal oocyst output, and intestinal microbiota in vivo. *Front. Vet. Sci.* **2020**, *7*, 420. [CrossRef] [PubMed]
54. Reid, W.M.; Johnson, J. Pathogenicity of *Eimeria acervulina* in light and heavy coccidial infections. *Avian Dis.* **1970**, *14*, 166–171. [CrossRef] [PubMed]
55. Ringenier, M.; Caekebeke, N.; De Meyer, F.; Van Limbergen, T.; Eeckhaut, V.; Ducatelle, R.; Van Immerseel, F.; Dewulf, J. A field study on correlations between macroscopic gut health scoring, histological measurements and performance parameters in broilers. *Avian Pathol.* **2021**, *50*, 500–506. [CrossRef] [PubMed]
56. Conway, D.P.; McKenzie, M.E.; Dayton, A.D. Relationship of coccidial lesion scores and weight gain in infections of *Eimeria acervulina, E. maxima* and *E. tenella* in broilers. *Avian Pathol.* **1990**, *19*, 489–496. [CrossRef] [PubMed]
57. Vezza, T.; Garrido-Mesa, J.; Diez-Echave, P.; Hidalgo-García, L.; Ruiz-Malagón, A.J.; García, F.; Sánchez, M.; Toral, M.; Romero, M.; Duarte, J.; et al. Allium-derived compound Propyl propane thiosulfonate (PTSO) attenuates metabolic alterations in mice fed a high-fat diet through its anti-inflammatory and prebiotic properties. *Nutrients* **2021**, *13*, 2595. [CrossRef] [PubMed]
58. Beam, A.; Clinger, E.; Hao, L. Effect of diet and dietary components on the composition of the gut microbiota. *Nutrients* **2021**, *13*, 2795. [CrossRef]
59. Aruwa, C.E.; Pillay, C.; Nyaga, M.M.; Sabiu, S. Poultry gut health-microbiome functions, environmental impacts, microbiome engineering and advancements in characterization technologies. *J. Anim. Sci. Biotechnol.* **2021**, *12*, 119. [CrossRef]
60. Mohammadi, R.; Hosseini-Safa, A.; Ehsani Ardakani, M.J.; Rostami-Nejad, M. The relationship between intestinal parasites and some immune-mediated intestinal conditions. *Gastroenterol. Hepatol. Bed Bench* **2015**, *8*, 123–131.
61. Kumar, S.; Sharadamma, K.C.; Radhakrish, P.M. Effects of a garlic active based growth promoter on growth performance and specific pathogenic intestinal microbial counts of broiler chicks. *Int. J. Poult. Sci.* **2010**, *9*, 244–246. [CrossRef]
62. Pelicano, E.R.; Souza, P.A.; Souza, H.B.A.; Figueiredo, D.F.; Boiago, M.M.; Carvalho, S.R.; Bordon, V.F. Intestinal mucosa development in broiler chickens fed natural growth promoters. *Braz. J. Poult. Sci.* **2005**, *7*, 221–229. [CrossRef]
63. Kim, G.B.; Seo, Y.M.; Kim, C.H.; Paik, I.K. Effect of dietary prebiotic supplementation on the performance, intestinal microflora, and immune response of broilers. *Poult. Sci.* **2011**, *90*, 75–82. [CrossRef] [PubMed]
64. Frick, J.S.; Autenrieth, I.B. The gut microflora and its variety of roles in health and disease. *Curr. Top. Microbiol. Immunol.* **2013**, *358*, 273–289. [CrossRef] [PubMed]
65. Campos, P.M.; Miska, K.B.; Jenkins, M.C.; Yan, X.; Proszkowiec-Weglarz, M. Effects of *Eimeria acervulina* infection on the luminal and mucosal microbiota of the duodenum and jejunum in broiler chickens. *Front. Microbiol.* **2023**, *14*, 1147579. [CrossRef] [PubMed]
66. Jebessa, E.; Guo, L.; Chen, X.; Bello, S.F.; Cai, B.; Girma, M.; Hanotte, O.; Nie, Q. Influence of *Eimeria maxima* coccidia infection on gut microbiome diversity and composition of the jejunum and cecum of indigenous chicken. *Front. Immunol.* **2022**, *13*, 994224. [CrossRef] [PubMed]
67. Flint, H.J.; McPherson, C.A.; Martin, J. Expression of two xylanase genes from the rumen cellulolytic bacterium *Ruminococcus flavefaciens* 17 cloned in pUC13. *Microbiology* **1991**, *137*, 123–129. [CrossRef] [PubMed]
68. Pettipher, G.L.; Latham, M.J. Characteristics of enzymes produced by *Ruminococcus flavefaciens* which degrade plant cell walls. *Microbiology* **1979**, *110*, 21–27. [CrossRef]
69. Saburi, W.; Yamamoto, T.; Taguchi, H.; Hamada, S.; Matsui, H. Practical preparation of epilactose produced with cellobiose 2-epimerase from *Ruminococcus albus* NE1. *Biosci. Biotechnol. Biochem.* **2010**, *74*, 1736–1737. [CrossRef]

70. Bui, T.P.N.; Ritari, J.; Boeren, S.; de Waard, P.; Plugge, C.M.; de Vos, W.M. Production of butyrate from lysine and the Amadori product fructoselysine by a human gut commensal. *Nat. Commun.* **2015**, *6*, 10062. [CrossRef]
71. Zhang, Q.; Zhang, S.; Wu, S.; Madsen, M.H.; Shi, S. Supplementing the early diet of broilers with soy protein concentrate can improve intestinal development and enhance short-chain fatty acid-producing microbes and short-chain fatty acids, especially butyric acid. *J. Anim. Sci. Biotechnol.* **2022**, *13*, 97. [CrossRef]

animals

MDPI

Article

Alterations in Ileal Microbiota and Fecal Metabolite Profiles of Chickens with Immunity to *Eimeria mitis*

Zhongchuang Wang [1,2,†], Peiyao Shang [1,†], Xingju Song [2], Minghui Wu [3], Tong Zhang [3], Qiping Zhao [1], Shunhai Zhu [1], Yu Qiao [1], Fanghe Zhao [1], Ruiting Zhang [1,2], Jinwen Wang [1], Yu Yu [1], Hongyu Han [1] and Hui Dong [1,*]

[1] Key Laboratory of Animal Parasitology of Ministry of Agriculture, Shanghai Veterinary Research Institute, Chinese Academy of Agricultural Sciences, Minhang, Shanghai 200241, China; zxclkjhf1999601@163.com (Z.W.); 15928747313@163.com (P.S.); zqp@shvri.ac.cn (Q.Z.); zhushunhai@shvri.ac.cn (S.Z.); qiaoyu540447420@163.com (Y.Q.); xzfh824@163.com (F.Z.); 2218302044@st.gxu.edu.cn (R.Z.); 82101221311@caas.cn (J.W.); yy21058@163.com (Y.Y.); hhysh@shvri.ac.cn (H.H.)

[2] College of Animal Science and Technology, Guangxi University, Nanning 530004, China; songxingju@gxu.edu.cn

[3] Beijing Yuanda Spark Medical Technology Co., Ltd., Beijing 102615, China; 19355320229@163.com (M.W.); 19159913684@163.com (T.Z.)

* Correspondence: donghui@shvri.ac.cn

† These authors contributed equally to this work.

Simple Summary: This study is the first to investigate the effects of *E. mitis* on the ileal microbiota and fecal metabolites of chickens. Non-targeted metabolic technology and 16S rRNA sequencing technology were used to study changes in the ileal microbiota and fecal metabolites after repeated infections with *E. mitis*. The results indicated that chickens developed solid immunity against a high dose of *E. mitis* infection after repeated infections with low-dose *E. mitis* and that repeated low-dose infections of *E. mitis* disrupted the ileal microbiota balance.

Abstract: Coccidiosis, caused by different species of *Eimeria* parasites, is an economically important disease in poultry and livestock worldwide. This study aimed to investigate the changes in the ileal microbiota and fecal metabolites in chickens after repeated infections with low-dose *E. mitis*. The chickens developed solid immunity against a high dose of *E. mitis* infection after repeated infections with low-dose *E. mitis*. The composition of the ileal microbiota and the metabonomics of the *Eimeria*-immunized group and the control group were detected using 16S rRNA sequencing and liquid chromatography–mass spectrometry (LC-MS). The relative abundance of *Neisseria*, *Erysipelotrichaceae*, *Incertae sedis*, *Coprobacter*, *Capnocytophaga*, *Bifidobacterium*, and the *Ruminococcus torques* group declined in the *Eimeria*-immunized chickens, whereas *Alloprevotella*, *Staphylococcus*, *Haemophilus*, and *Streptococcus* increased. Furthermore, 286 differential metabolites (including N-undecylbenzenesulfonic acid, 1,25-dihydroxyvitamin D3, gluconic acid, isoleucylproline, proline, and 1-kestose) and 19 significantly altered metabolic pathways (including galactose metabolism, ABC transporters, starch and sucrose metabolism, the ErbB signaling pathway, and the MAPK signaling pathway) were identified between the *Eimeria*-immunized group and the control group. These discoveries will help us learn more about the composition and dynamics of the gut microbiota as well as the metabolic changes in chickens infected with *Eimeria* spp.

Keywords: Coccidia; *Eimeria mitis*; gut microbiota; fecal metabolites

Citation: Wang, Z.; Shang, P.; Song, X.; Wu, M.; Zhang, T.; Zhao, Q.; Zhu, S.; Qiao, Y.; Zhao, F.; Zhang, R.; et al. Alterations in Ileal Microbiota and Fecal Metabolite Profiles of Chickens with Immunity to *Eimeria mitis*. *Animals* **2024**, *14*, 3515. https://doi.org/10.3390/ani14233515

Academic Editors: Kenneth Bafundo and Juan David Latorre

Received: 24 October 2024
Revised: 19 November 2024
Accepted: 3 December 2024
Published: 5 December 2024

1. Introduction

Avian coccidiosis is an intestinal disease caused by infection with *Eimeria* spp. It causes severe diarrhea, weight loss, poor nutrient absorption, and increased susceptibility to other pathogens [1]. The global cost of poultry coccidiosis is estimated to exceed USD

14 billion each year, resulting in economic losses through increased drug costs [2]. Seven main species of *Eimeria (E. acervulina, E. brunetti, E. maxima, E. mitis, E. necatrix, E. praecox,* and *E. tenella*) are recognized worldwide in poultry and vary by geographic region and poultry production system [3]. These *Eimeria* species differ in terms of their invasion of distinct sites within the intestine, pathogenicity, and type of lesion they produce.

The gut, which contains trillions of microorganisms, plays a key role in digestion, the absorption of nutrients, and the regulation of physiological processes, including immune responses and pathogen elimination [4–6]. Coccidiosis infection alters the intestinal microbiota's composition and integrity, resulting in elevated susceptibility to diseases that threaten the overall health and productivity of chickens [7–10]. It was reported that *E. tenella* infection altered the composition and diversity of cecal microbiota, significantly reducing Proteobacteria and Firmicutes (*Enterococcus*) [11]. In another study, the abundance of *Bacteroides* and *SMB53*, as well as an unclassified genus in the family of *Coriobacteriaceae*, was significantly reduced by *E. maxima* infection [12]. A recently published study showed that infection with mixed *Eimeria* spp. caused a greater abundance of *Clostridium sordellii, Clostridium butyricum*, unclassified *Ruminococcaceae*, and unclassified *Bacteroides* in the ileum and a higher abundance of *Clostridium perfringens* and unclassified *Enterobacteriaceae* in the cecum [13].

Metabolomics facilitates a comprehensive examination of metabolic alterations in reaction to external stimuli or disturbances and provides new understandings about disease pathogenesis and infection responses [14,15]. In an experiment with mice infected with *Trypanosoma crudii*, the authors found reductions in glucose import, reductions in fatty acid and phospholipid synthesis in plasma, and an increase in heart tissue [16]. In previous reports, a total of 389 significant metabolites were identified in the liver of acutely infected mice with type II *T. gondii* by using metabolomics, while 368 metabolites were identified in the liver of chronically infected mice compared to the control group [17]. It was reported that after rabbits were infected with *E. magna*, 13 metabolites were changed and 7 metabolic pathways were dysregulated [18]. Another similar study demonstrated that after rabbits were infected with *E. intestinalis*, 20 metabolites and 2 metabolic pathways were altered, with lipid metabolism as the major disrupted metabolic pathway [19].

At present, the changes in the gut microbiota and metabolites caused by *Eimeria* spp. infection are mainly studied under the condition of severe gut damage caused by high-dose oocysts. However, until now, no study has been conducted on the gut microbiota and metabolites of chickens with strong immunity to *Eimeria* spp. *E. mitis* is less pathogenic and parasitizes in the small intestine, and its negative influence is often disregarded in the poultry industry [20,21]. There has been no study on the effects of *E. mitis* on the gut microbiome and metabolites of chickens. In the present study, the chickens were immunized four times with low-dose *E. mitis*, resulting in complete immunity against *E. mitis* infection. Then, the changes in the gut microbiota and metabolites of chickens were analyzed by 16S rRNA sequencing technology and non-targeted metabolic technology, respectively. This study will provide a new insight into the composition and dynamics of the gut microbiota as well as the metabolic changes in chickens infected with *Eimeria* spp.

2. Materials and Methods

2.1. Parasites and Experimental Chickens

E. mitis sporulated oocysts were preserved by the Shanghai Veterinary Research Institute, Chinese Academy of Agricultural Sciences. The oocysts were acquired, purified, and sporulated following established protocols [22]. One-day-old Sanhuang chickens, purchased from Shanghai Minyou Breeding Cooperative, were raised in isolators and maintained at 34–35 °C in a standard animal house without coccidia.

2.2. Infection of Chickens

An *Eimeria*-immunized group of 12 chickens was set up, along with a healthy control group of 12 chickens. All chickens freely consumed a non-antibiotic commercial feed.

The environment temperature was maintained at 25–30 °C. Seven-day-old chickens were infected using oral gavage with low-dose *E. mitis* (2×10^4) sporulated oocysts (*Eimeria*-immunized group). The uninfected chickens received an equal volume of PBS (control group). A total of four infections were performed throughout the course at 7-day intervals. On the sixth to eighth days after each infection, feces were collected daily from each group for oocysts per gram of feces (OPGs) using McMaster's counting technique.

2.3. Evaluation of Immunization Effectiveness

On the seventh day after the final infection, six chickens from the *Eimeria*-immunized group and six chickens from the control group were selected and challenged with a high dose of *E. mitis* (2×10^5) sporulated oocysts. Fecal oocyst numbers for each group were determined on days 6–7 post-challenge. The remaining *Eimeria*-immunized group (n = 6) and the control group (n = 6) were kept for subsequent analyses of the gut microbiota and metabolite profiles.

2.4. The 16S rDNA Amplification and Bioinformatic Analysis

The fresh contents of the ileum from the remaining chickens were collected, and the total DNA was isolated using a Hipure Soil DNA Kit (Magen, Guangzhou, China) following the manufacturer's instructions. DNA concentration and integrity were measured using a NanoDrop 2000 (Thermo Fisher Scientific, Waltham, MA, USA) and agarose gel electrophoresis. For bacterial diversity analysis, 16S rRNA genes were amplified with the universal primers 343F (5′-TACGGRAGGCAGCAG-3′) and 798R (5′-AGGGTATCTAATCCT-3′) for the V3-V4 region [23]. After purifying the PCR products with AMPure XP beads, they underwent another round of PCR. The final amplicon was then quantified using the Qubit dsDNA Assay Kit (Thermo Fisher Scientific, Waltham, MA, USA). The concentrations were modified prior to sequencing and subsequently processed using an Illumina NovaSeq 6000 system by OE Biotech Co., Ltd. (Shanghai, China).

Raw sequencing data were in FASTQ format. Paired-end reads were then preprocessed using Cutadapt software to detect and cut off the adapter. After trimming, paired-end reads were filtered with low-quality sequences, denoised, and merged, and the chimera reads were detected and cut off using DADA2 with the default parameters of QIIME2 [24]. At last, the software output the representative reads and the ASV abundance table. The representative read of each Amplicon Sequence Variant (ASV) was selected using the QIIME2 package. All representative reads were annotated and blasted against the Silva database using the q2-feature-classifier with the default parameters.

QIIME2 software was used for alpha and beta diversity analysis. The unweighted UniFrac distance matrix performed by the R package was used for an unweighted UniFrac principal coordinate analysis (PCoA) to estimate the beta diversity [25]. To test for significance in UniFrac distances, the nonparametric permutational analysis of variance (PERMANOVA) test was used. The Wilcoxon statistical test was conducted using the R package to assess significant differences between the *Eimeria*-immunized and the control group. The research utilized the linear discriminant analysis effect size (LEfSe) method to analyze significant biomarker taxa [26].

2.5. LC-MS Sample Preparation and Bioinformatic Analysis

Fresh feces from the remaining chickens were collected and snap-frozen in liquid nitrogen. Frozen samples were thawed slowly at room temperature. The extraction and analysis of samples were performed according to previously described methods [27]. The obtained supernatants were filtered using 0.22 μm microfilters and analyzed by LC-MS (OE Biotech Company, Shanghai, China). The samples were subjected to analysis using the Agilent 7890B gas chromatography system and the Agilent 5977B MSD system, according to the established parameters as previously outlined [28].

The raw data were transferred to the analysis base file (ABF) format via the Analysis Base File Converter software. An Orthogonal Partial Least Squares-Discriminant Analysis

(OPLS-DA) was used to evaluate the different metabolites between the *Eimeria*-immunized group and the control group. In order to evaluate the quality of the OPLS-DA and avoid overfitting, we utilized a 7-fold cross-validation and 200 response permutation testing (RPT). Variable Importance in the Projection (VIP) obtained from the OPLS-DA model was utilized to assess the contribution of each metabolite to the classification. A two-tailed Student's t-test was employed to assess the significance of metabolite differences between the *Eimeria*-immunized group and the control group. The selection of important metabolites was chosen based on VIP values > 1.0 and *p*-values < 0.05. Furthermore, we utilized the KEGG database for a comprehensive analysis of the metabolic pathways associated with the identified metabolites.

2.6. Statistical Analysis

Data are presented as the mean ± standard error of the mean (SEM). A Student's *t*-test was performed to determine the significance between the two groups for OPGs; $p < 0.05$ was considered significant.

3. Results

3.1. Fecal OPG Counts During Immunizations

A higher OPG count was produced on days 6 to 8 after the first infection and relatively low OPGs on the sixth day after the second infection (Table 1). No oocysts were detected from the seventh day after the second infection to the eighth day after the fourth infection.

Table 1. Output of oocysts after a total of four infections performed with low doses (2×10^4) of *E. mitis* oocysts at 7-day intervals.

Infection	OPGs		
	6 dpi	**7 dpi**	**8 dpi**
1st	166,500 ± 19,200	102,000 ± 13,800	51,000 ± 7200
2nd	1500 ± 300	0	0
3rd	0	0	0
4th	0	0	0

OPGs: oocysts per gram feces; dpi: day(s) post-infection.

3.2. Immunization Effectiveness

After four infections, the chickens were challenged with a high dose of *E. mitis*. The OPG results are shown in Table 2. Our results demonstrated that the *Eimeria*-immunized group had no oocysts discharged 6–7 days after the challenge. However, the control group had a large number of oocysts discharged. These results indicated that the chickens had developed immunity against a high dose of *E. mitis* infection after repeated infections with low-dose *E. mitis*.

Table 2. OPG values after a high dose of *E. mitis* (2×10^5) sporulated oocysts.

Group	OPG		
	Day 6	**Day 7**	**Means**
A	0	0	0 [b]
B	306,000 ± 38,400	93,000 ± 12,600	199,500 ± 25,800 [a]

OPG: oocysts per gram of feces; A: the *Eimeria*-immunized group was challenged with a high dose of *E. mitis*; B: the control group was challenged with a high dose of *E. mitis*. Different letters indicate a statistically significant difference.

3.3. *E. mitis* Changed Ileal Microbiota Diversity

We assessed the alpha diversity using the Chao1 and Abundance-based Coverage Estimator (ACE) indices to measure microbiota richness and the Shannon and Simpson indices for species diversity. The analysis showed no significant variation in the four indices between individuals within the *Eimeria*-immunized group and the control group (Figure 1a). The evaluation of beta diversity was conducted using an unweighted UniFrac distance analysis. The PCoA result showed that there were significant differences ($p < 0.05$) between the *Eimeria*-immunized group and the control group (Figure 1b).

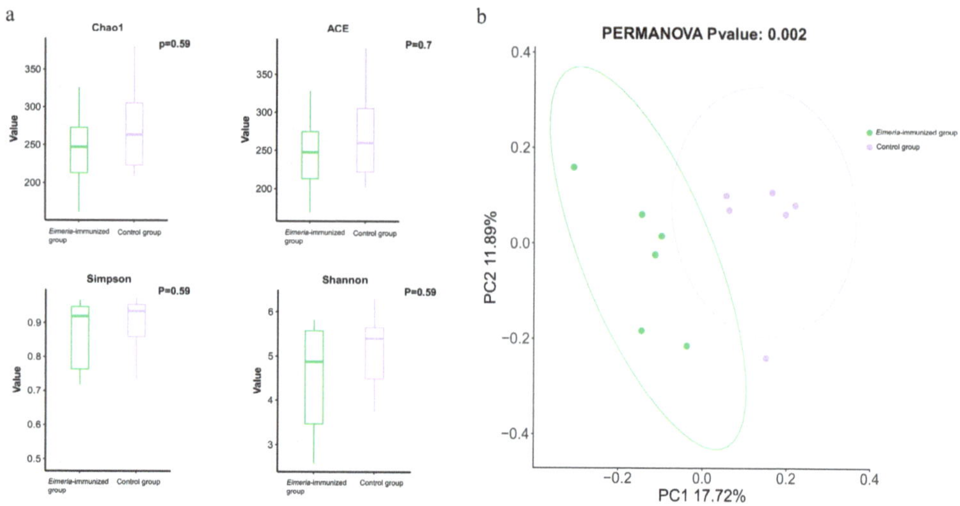

Figure 1. Diversity analysis of the gut microbiota: (**a**) analysis of the alpha diversity between the *Eimeria*-immunized group and the control group; (**b**) PCoA analysis based on the unweighted UniFrac distance.

3.4. *E. mitis* Altered Ileal Microbiota Composition

In accordance with Figure 2, our analysis displayed the predominant bacterial phyla and genera based on their relative abundance. The dominant phyla included Firmicutes, Proteobacteria, and Bacteroidetes. In the control group, the relative abundance of Firmicutes, Proteobacteria, and Bacteroidetes was 76.09%, 12.47%, and 9.17%, respectively. In the *Eimeria*-immunized group, the relative abundance of Firmicutes, Proteobacteria, and Bacteroidetes was 77.48%, 14.99%, and 5.71%, respectively. We further compared the bacterial composition of all experimental treatments at the genus level. In the control group, the relative abundance of *Lactobacillus*, *Romboutsia*, *Faecalibacterium*, and *Bacteroides* was 23.61%, 12.41%, 8.46%, and 8.01%, respectively. In the *Eimeria*-immunized group, the relative abundance of *Lactobacillus*, *Romboutsia*, *Faecalibacterium*, and *Bacteroides* was 55.11%, 2.22%, 5.30%, and 4.84%, respectively.

We detected significant taxonomic variations between the *Eimeria*-immunized group and the control group by linear discriminant analysis (LDA) scores above 3.0 (Figure 3). At the genus level, the control group displayed a significant increase in the abundance of *Neisseria*, *Erysipelotrichaceae*, *Incertae sedis*, *Coprobacter*, *Capnocytophaga*, *Butyricimonas*, *Bifidobacterium*, the *Eubacterium xylanophilum* group, *Helicobacter*, the *Eubacterium coprostanoligenes* group, the *Ruminococcus torques* group, *Oscillibacter*, *Clostridia UCG-014*, the *Clostridia vadin BB60* group, *Colidextribacter*, and *UCG-005* compared to the *Eimeria*-immunized group. The *Eimeria*-immunized group displayed a significant increase in the abundance of *Alloprevotella*, *Streptococcus*, *Staphylococcus*, and *Haemophilus* compared to the control group.

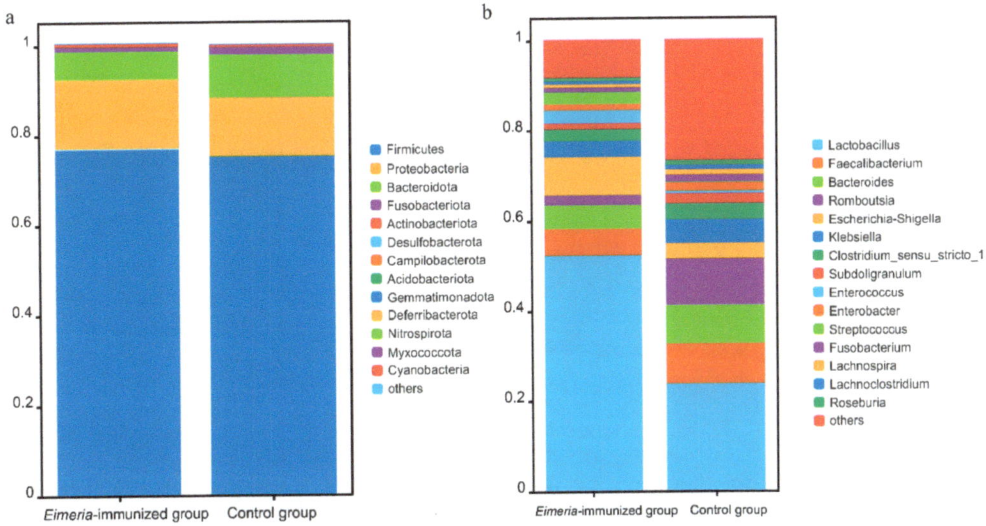

Figure 2. The dominant abundant bacterial phyla and genera are shown. Other phyla and genera were all assigned as "Others". Gut bacterial composition at the phylum (**a**) and genus (**b**) levels in different experiment groups.

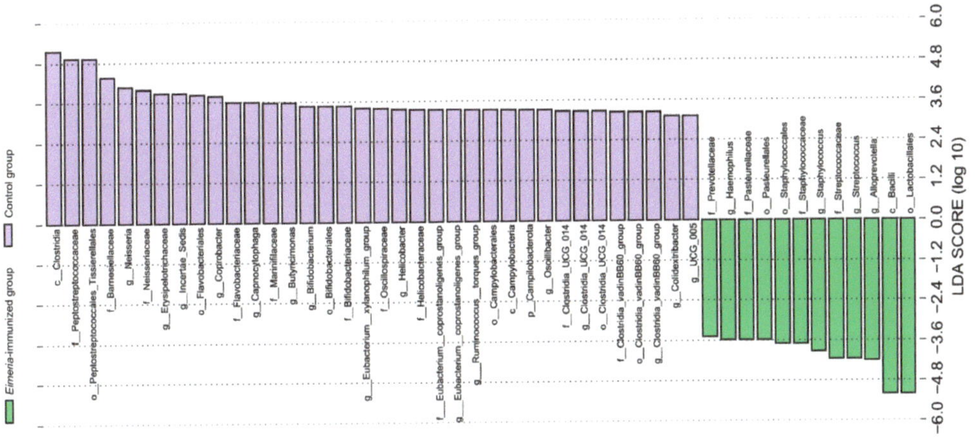

Figure 3. The linear discriminant analysis (LDA) effect size (LEfSe) analysis. Histogram of the linear discriminant analysis (LDA) scores.

3.5. *E. mitis* Perturbed Fecal Metabolites

In order to determine the impact of *E. mitis* on the chickens' metabolites, an LC-MS untargeted metabolomics analysis was performed. The OPLS-DA score scatter plots showed that the *Eimeria*-immunized group and the control group clearly separated into distinct clusters according to their metabolic differences (Figure 4a). The permutation test was applied to validate the fit accuracy and predictability of the OPLS-DA model (Figure 4b). Between the *Eimeria*-immunized group and the control group, the model's parameters were R2Y(cum) = 0.959 and Q2(cum) = 0.244. Generally, R2Y(cum) > 0.5 is considered to be good, indicating that the OPLS-DA model does not exceed the fitting.

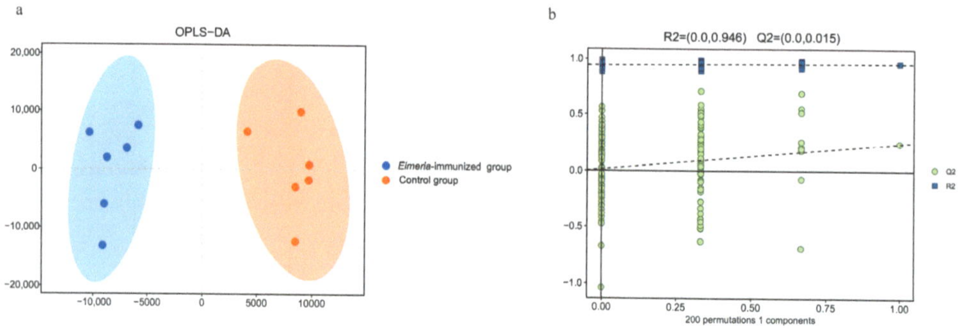

Figure 4. OPLS-DA score scatter plot of the *Eimeria*-immunized group vs. the control group (**a**) and permutation test of the *Eimeria*-immunized group and the control group (**b**).

The difference in screening results shown by the volcano maps (Figure 5) illustrated 286 differential metabolites, 73 of which were upregulated and 213 downregulated in the *Eimeria*-immunized group compared to the control group. A heat map was drawn using the top 50 differential metabolites between the *Eimeria*-immunized group and the control group (Figure 6). Metabolites such as trehalose, gluconic acid, 1-kestose, and ginsenoside Rg1 were significantly reduced in the *Eimeria*-immunized group compared to the control group. Meanwhile, metabolites such as N-undecylbenzenesulfonic acid, 1,25-dihydroxyvitamin D3, and isoleucylproline were significantly increased in the *Eimeria*-immunized group compared to the control group.

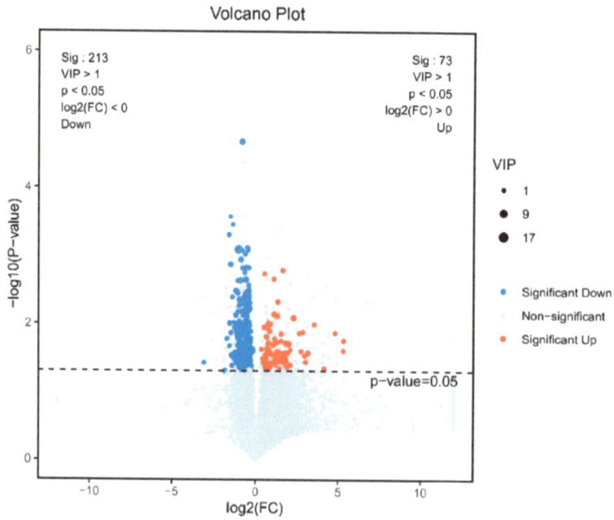

Figure 5. Volcano map of different metabolites: blue represents the downregulated differential metabolites, red represents the upregulated differential metabolites, and gray represents metabolites without differences.

Furthermore, we conducted classification and annotation of the detected differential metabolites utilizing the KEGG database. We identified 19 significantly altered metabolic pathways ($p < 0.05$), including galactose metabolism, ABC transporters, starch and sucrose metabolism, the ErbB signaling pathway, the MAPK signaling pathway, the GnRH signaling pathway, and the adipocytokine signaling pathway (Figure 7).

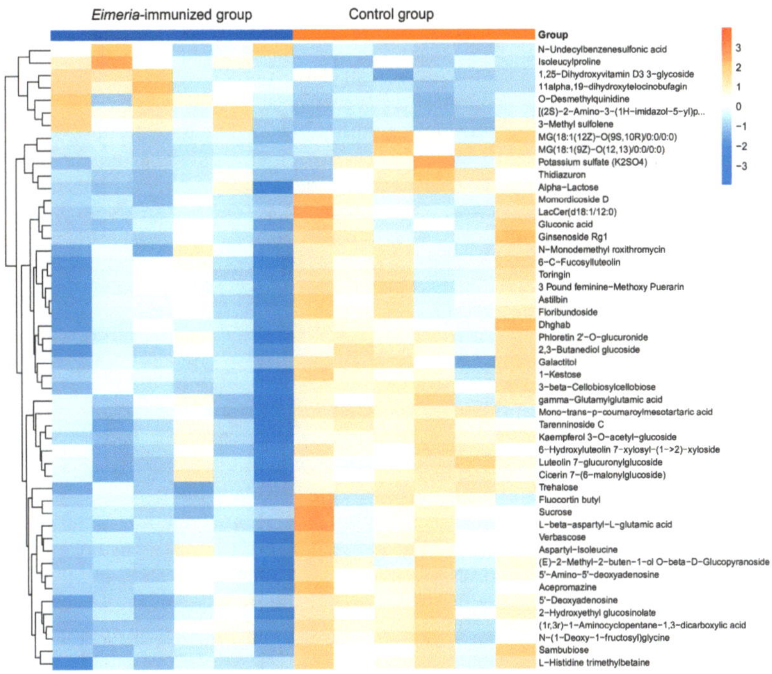

Figure 6. Cluster analysis of the top 50 differential metabolites: each row in the figure represents a differential metabolite, and each column represents a sample.

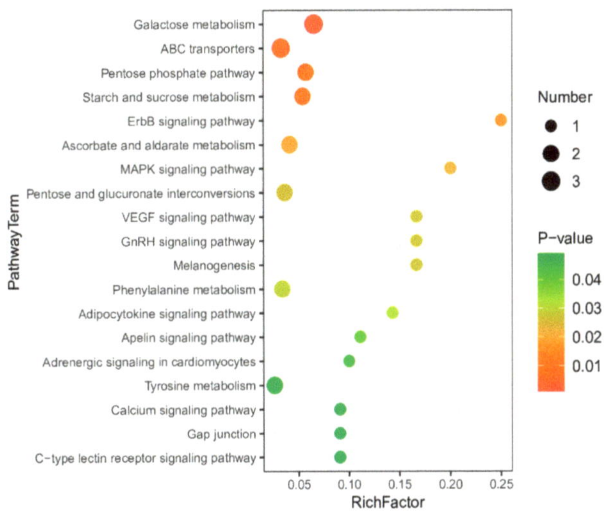

Figure 7. Differentially altered metabolic pathways, as visualized using bubble plots ($p < 0.05$); the bubble size represents the number of metabolites. Coloring from green to red indicates a decrease in the p-value; the larger the point, the more metabolites are enriched in the pathway.

4. Discussion

In this experiment, chickens were repeatedly infected with a low dose of *E. mitis*. After four infections, the chickens were challenged with a high dose of *E. mitis*. No oocysts were excreted in their stool, indicating the chickens had immunity to *E. mitis* infection. Non-targeted metabolic technology and 16S rRNA sequencing technology were used to study changes in the ileal microbiota and fecal metabolites in chickens with immunity to *E. mitis* infection.

At the phylum level, we found Firmicutes, Bacteroidetes, and Proteobacteria to be the most abundant. However, no significant difference was observed between the two groups. The Firmicutes phylum is prevalent in the intestines of poultry and involved in harvesting energy from the diet in many animals [29]. Bacteroidetes contribute to host intestinal energy metabolism through the breakdown of complex polysaccharides to produce propionic and butyric acids [30]. Proteobacteria encompass a variety of famous opportunistic pathogens, including *Escherichia coli*, *Salmonella*, and *Campylobacter*, which could have adverse effects on the health of the host [31]. Contrary to our results, the bacterial communities that were significantly increased in abundance in *E. tenella*-infected birds were Firmicutes and Proteobacteria, while Bacteroidetes was decreased [32]. We considered that this may be related to the different *Eimeria* species (*E. tenella* vs. *E. mitis*), the location of infection (cecum vs. ileum), and the infected dose (high vs. low).

At the genus level, the *Eimeria*-immunized group significantly increased the abundance of *Alloprevotella*, *Staphylococcus*, *Haemophilus*, and *Streptococcus*. Among them, *Alloprevotella* is a bacterium that produces short-chain fatty acids (SCFAs) and positively correlates with the anti-inflammatory cytokine IL-10 [33,34]. SCFAs provide energy and carbon for broilers, while also playing a role in blood flow regulation and the stimulation of intestinal cell growth [35]. However, *Staphylococcus*, *Haemophilus*, and *Streptococcus* are generally opportunistic pathogens [36,37]. Song et al. found that broilers challenged with necrotic enteritis (NE) had a significantly increased relative abundance of *Staphylococcus* in the ileum [38]. Chen et al. documented that the abundance of *Streptococcus* increased over time after *E. tenella* infection [7]. After repeated infections with a low dose of *E. mitis*, the increase in opportunistic bacterial pathogens led to gut microbiota dysbiosis. This may be related to the damage of the gut mucosal barrier and the immune response triggered by pathogen infections in the gut [39].

In contrast, the *Eimeria*-immunized group significantly decreased the levels of *Erysipelotrichaceae*, *Coprobacter*, *Bifidobacterium*, and the *Ruminococcus torques* group. *Erysipelotrichaceae* is associated with better feed conversion in broilers [40]. *Coprobacter* contributes to the maintenance of intestinal health and propionic acid production [41,42]. Earlier studies indicated that *Bifidobacterium* possesses multiple physiological functions, including modulating animals' microflora and immunity enhancement [43]. In broilers, supplementing with *Bifidobacterium*-based probiotics increases the goblet cell cup area in the gut, significantly upregulates the expression of mucin, and increases the production of mucin glycoprotein [44]. *Ruminococcus torques* are bacteria that can produce SCFAs and are considered a dominant player in the degradation of diverse polysaccharides and fibers [45]. We found that infection with a low dose of *E. mitis* had similar effects on gut microbiota as the live vaccines currently used [46]. It has been reported that live vaccines of avian coccidiosis may have a deleterious effect on the early growth of broilers and potentially increase the susceptibility of broilers to secondary infection, although it can effectively prevent coccidiosis [47].

The gut microbiota generates multiple metabolites that are essential for regulating the intricate interactions between the gut microbiota and host physiology [48]. Metabolites derived from the intestinal tract provide valuable information for discovering potential biomarkers related to intestinal homeostasis [49]. In the present study, we found that N-undecylbenzenesulfonic acid and 1,25-dihydroxyvitamin D3 were significantly increased after chickens had been infected repeatedly with low-dose *E. mitis*. As far as we know, N-undecylbenzenesulfonic acid is an antibacterial compound against *S. aureus*, *B. subtilis*, *E. coli*, and *Klebsiella pneumoniae* [50]. The 1,25-dihydroxyvitamin D3 plays a major role in regulating immune responses, which helps the maturation of monocytes into macrophages [51]. Rajapakse et al. reported that 1,25(OH)2D3 inhibited *T. gondii* parasite growth in vivo by limiting tachyzoite invasion of cells and its proliferation [52]. After rabbits were infected with *E. magna*, a fecal metabolomics analysis also found a significant increase in

1,25(OH)2D3 metabolites [18]. Therefore, the increasing 1,25-dihydroxyvitamin D3 may help protect against *E. mitis* infection.

In addition, we noted significant reductions in trehalose, gluconic acid, 1-kestose, and ginsenoside Rg1 after the chickens had been infected repeatedly with low-dose *E. mitis*. Previous research found that trehalose improved growth and boosted immunity by moderating inflammatory cytokines in the chicken intestines [53]. Gluconic acid is generated via starch fermentation and subsequently converted into butyrate, which serves as a vital energy source for colonocytes and promotes bacterial proliferation in the host intestine [54]. Previous studies have shown that a wide variety of bacteria are capable of metabolizing glucose into gluconic acid, including various strains of the AAB genus and other genera, such as *Pseudomonas* and *Zymomonas* [55]. 1-Kestose is a fructo-oligosaccharide, produced from sucrose by the action of fructosyltransferases in plants, bacteria, yeast, and fungi [56]. One study showed that 1-kestose supplementation promoted beneficial gut microbiota in rat cecal contents and increased acetate, butyrate, and lactate levels in a dose-dependent manner [57]. A prior study demonstrated that ginsenoside Rg1 enhanced the growth and intestinal health of broilers by improving inflammatory and oxidative activities [58]. These altered metabolites further suggest that repeated low-dose *E. mitis* infections can lead to ileal microbiota dysbiosis.

5. Conclusions

This study demonstrated that repeated low-dose *E. mitis* infections can lead to ileal microbiota dysbiosis. This perturbation was dominated by an increased relative abundance of the pathobionts *Staphylococcus* and *Haemophilus* and an apparent decrease in the levels of nonpathogenic bacteria, including *Bifidobacterium* and *Coprobacter*. The metabolomics analysis showed that *E. mitis* infection altered 286 metabolites and 19 metabolic pathways. These findings enhanced our understanding of gut microbiota and fecal metabolic profiles of *Eimeria* spp.-infected chickens.

Author Contributions: Conceptualization, H.D., P.S., and Y.Y.; methodology, H.D., P.S., and Y.Y.; software, Z.W. and J.W.; validation, H.D.; formal analysis, Z.W., Y.Q., R.Z., and F.Z.; investigation, M.W., T.Z., Q.Z., and S.Z.; resources, Q.Z. and S.Z.; data curation, Z.W. and J.W.; writing—original draft preparation, Z.W., J.W., and X.S.; writing—review and editing, H.H. and H.D.; visualization, Z.W.; supervision, Q.Z. and S.Z.; project administration, M.W., T.Z., Q.Z., and S.Z.; funding acquisition, H.D. All authors have read and agreed to the published version of the manuscript.

Funding: This work was supported by the National Key Research and Development Program of China (Grant No. 2023YFD1802400), the Key Research and Development of Science and Technology Plan in the Tibet Autonomous Region (XZ202401ZY0052), the National Natural Science Foundation of China (Grant No. 32373038), and the National Parasitic Resources Center (NPRC-2019-194-30).

Institutional Review Board Statement: All animal studies were approved and authorized by the Animal Care and Use Committee of the Shanghai Veterinary Research Institute, Chinese Academy of Agricultural Sciences (SV-20231215-Y11).

Informed Consent Statement: Not applicable.

Data Availability Statement: The 16S rRNA sequencing data in this study were submitted to the NCBI Sequence Read Archive (SRA) under BioProject accession number PRJNA1129660.

Acknowledgments: We appreciate the technical support provided by OE Biotech Co., Ltd. (Shanghai, China).

Conflicts of Interest: Author Minghui Wu and Tong Zhang were employed by the company Beijing Yuanda Spark Medical Technology Co., Ltd. The remaining authors declare that the research was conducted in the absence of any commercial or financial relationships that could be construed as a potential conflict of interest.

References

1. Kadykalo, S.; Roberts, T.; Thompson, M.; Wilson, J.; Lang, M.; Espeisse, O. The value of anticoccidials for sustainable global poultry production. *Int. J. Antimicrob. Agents* **2018**, *51*, 304–310. [CrossRef]
2. Blake, D.P.; Knox, J.; Dehaeck, B.; Huntington, B.; Rathinam, T.; Ravipati, V.; Ayoade, S.; Gilbert, W.; Adebambo, A.O.; Jatau, I.D.; et al. Re-calculating the cost of coccidiosis in chickens. *Vet. Res.* **2020**, *51*, 115. [CrossRef] [PubMed]
3. Chapman, H.D.; Barta, J.R.; Blake, D.; Gruber, A.; Jenkins, M.; Smith, N.C.; Suo, X.; Tomley, F.M. A selective review of advances in coccidiosis research. *Adv. Parasitol.* **2013**, *83*, 93–171. [PubMed]
4. Brestoff, J.R.; Artis, D. Commensal bacteria at the interface of host metabolism and the immune system. *Nat. Immunol.* **2013**, *14*, 676–684. [CrossRef]
5. Dominguez-Bello, M.G.; Godoy-Vitorino, F.; Knight, R.; Blaser, M.J. Role of the microbiome in human development. *Gut* **2019**, *68*, 1108–1114. [CrossRef]
6. Li, A.; Liu, B.; Li, F.; He, Y.; Wang, L.; Fakhar-E-Alam Kulyar, M.; Li, H.; Fu, Y.; Zhu, H.; Wang, Y.; et al. Integrated bacterial and fungal diversity analysis reveals the gut microbial alterations in diarrheic giraffes. *Front. Microbiol.* **2021**, *12*, 712092. [CrossRef]
7. Chen, H.L.; Zhao, X.Y.; Zhao, G.X.; Huang, H.B.; Li, H.R.; Shi, C.W.; Yang, W.T.; Jiang, Y.L.; Wang, J.Z.; Ye, L.P.; et al. Dissection of the cecal microbial community in chickens after *Eimeria tenella* infection. *Parasit. Vectors* **2020**, *13*, 56. [CrossRef]
8. Huang, G.; Tang, X.; Bi, F.; Hao, Z.; Han, Z.; Suo, J.; Zhang, S.; Wang, S.; Duan, C.; Yu, Z.; et al. *Eimeria tenella* infection perturbs the chicken gut microbiota from the onset of oocyst shedding. *Vet. Parasitol.* **2018**, *258*, 30–37. [CrossRef] [PubMed]
9. Liu, P.Y.; Liaw, J.; Soutter, F.; Ortiz, J.J.; Tomley, F.M.; Werling, D.; Gundogdu, O.; Blake, D.P.; Xia, D. Multi-omics analysis reveals regime shifts in the gastrointestinal ecosystem in chickens following anticoccidial vaccination and *Eimeria tenella* challenge. *mSystems* **2024**, *9*, e0094724. [CrossRef]
10. Tomal, F.; Sadrin, G.; Gaboriaud, P.; Guitton, E.; Sedano, L.; Lallier, N.; Rossignol, C.; Larcher, T.; Rouille, E.; Ledevin, M.; et al. The caecal microbiota promotes the acute inflammatory response and the loss of the intestinal barrier integrity during severe *Eimeria tenella* infection. *Front. Cell. Infect. Microbiol.* **2023**, *13*, 1250080. [CrossRef]
11. Zhou, B.H.; Jia, L.S.; Wei, S.S.; Ding, H.Y.; Yang, J.Y.; Wang, H.W. Effects of *Eimeria tenella* infection on the barrier damage and microbiota diversity of chicken cecum. *Poult. Sci.* **2020**, *99*, 1297–1305. [CrossRef] [PubMed]
12. Lu, M.; Li, R.W.; Zhao, H.; Yan, X.; Lillehoj, H.S.; Sun, Z.; Oh, S.; Wang, Y.; Li, C. Effects of *Eimeria maxima* and *Clostridium perfringens* infections on cecal microbial composition and the possible correlation with body weight gain in broiler chickens. *Res. Vet. Sci.* **2020**, *132*, 142–149. [CrossRef] [PubMed]
13. Graham, D.; Petrone-Garcia, V.M.; Hernandez-Velasco, X.; Coles, M.E.; Juarez-Estrada, M.A.; Latorre, J.D.; Chai, J.; Shouse, S.; Zhao, J.; Forga, A.J.; et al. Assessing the effects of a mixed *Eimeria* Spp. challenge on performance, intestinal integrity, and the gut microbiome of broiler chickens. *Front. Vet. Sci.* **2023**, *10*, 1224647. [CrossRef]
14. Jacot, D.; Waller, R.F.; Soldati-Favre, D.; MacPherson, D.A.; MacRae, J.I. Apicomplexan energy metabolism: Carbon source promiscuity and the quiescence hyperbole. *Trends Parasitol.* **2016**, *32*, 56–70. [CrossRef]
15. Yang, J.; Zhao, X.; Liu, X.; Wang, C.; Gao, P.; Wang, J.; Li, L.; Gu, J.; Yang, S.; Xu, G. High performance liquid chromatography−mass spectrometry for metabonomics: potential biomarkers for acute deterioration of liver function in chronic hepatitis B. *J. Proteome. Res.* **2006**, *5*, 554–561. [CrossRef]
16. Gironès, N.; Carbajosa, S.; Guerrero, N.A.; Poveda, C.; Chillón-Marinas, C.; Fresno, M. Global metabolomic profiling of acute myocarditis caused by *Trypanosoma cruzi* infection. *PLoS Negl. Trop. Dis.* **2014**, *8*, e3337. [CrossRef] [PubMed]
17. Chen, X.Q.; Elsheikha, H.M.; Hu, R.S.; Hu, G.X.; Guo, S.L.; Zhou, C.X.; Zhu, X.Q. Hepatic metabolomics investigation in acute and chronic murine Toxoplasmosis. *Front. Cell. Infect. Microbiol.* **2018**, *8*, 189. [CrossRef]
18. Yuan, X.; Hu, X.; Yang, S.; Zhong, S.; Yang, T.; Zhou, Y.; Li, Y. Fecal metabolomic analysis of rabbits infected with *Eimeria intestinalis* and *Eimeria magna* based on LC-MS/MS technique. *Microb. Pathog.* **2022**, *162*, 105357. [CrossRef]
19. Yuan, X.; Liu, J.; Hu, X.; Yang, S.; Zhong, S.; Yang, T.; Zhou, Y.; Zhao, G.; Jiang, Y.; Li, Y. Alterations in the jejunal microbiota and fecal metabolite profiles of rabbits infected with *Eimeria intestinalis*. *Parasit. Vectors* **2022**, *15*, 231. [CrossRef]
20. Jorgensen, W.K.; Stewart, N.P.; Jeston, P.J.; Molloy, J.B.; Blight, G.W.; Dalgliesh, R.J. Isolation and pathogenicity of Australian strains of *Eimeria praecox* and *Eimeria mitis*. *Aust. Vet. J.* **1997**, *75*, 592–595. [CrossRef]
21. Xu, L.; Xiang, Q.; Li, M.; Sun, X.; Lu, M.; Yan, R.; Song, X.; Li, X. Pathogenic Effects of Single or Mixed Infections of *Eimeria mitis*, *Eimeria necatrix*, and *Eimeria tenella* in Chickens. *Vet. Sci.* **2022**, *9*, 657. [CrossRef] [PubMed]
22. Long, P.L.; Millard, B.J.; Joyner, L.P.; Norton, C.C. A guide to laboratory techniques used in the study and diagnosis of avian coccidiosis. *Folia Vet. Lat.* **2014**, *6*, 201–217.
23. Nossa, C.W.; Oberdorf, W.E.; Yang, L.; Aas, J.A.; Paster, B.J.; Desantis, T.Z.; Brodie, E.L.; Malamud, D.; Poles, M.A.; Pei, Z. Design of 16S rRNA gene primers for 454 pyrosequencing of the human foregut microbiome. *World J. Gastroenterol.* **2010**, *16*, 4135. [CrossRef] [PubMed]
24. Callahan, B.J.; McMurdie, P.J.; Rosen, M.J.; Han, A.W.; Johnson, A.J.A.; Holmes, S.P. DADA2: High-resolution sample inference from Illumina amplicon data. *Nat. Methods* **2016**, *13*, 581–583. [CrossRef]
25. Lozupone, C.; Lladser, M.E.; Knights, D.; Stombaugh, J.; Knight, R. UniFrac: An effective distance metric for microbial community comparison. *ISME J.* **2010**, *5*, 169–172. [CrossRef]
26. Segata, N.; Izard, J.; Waldron, L.; Gevers, D.; Miropolsky, L.; Garrett, W.S.; Huttenhower, C. Metagenomic biomarker discovery and explanation. *Genome Biol.* **2011**, *12*, R60. [CrossRef]

27. Hu, W.; Xu, D.; Zhou, Z.; Zhu, J.; Wang, D.; Tang, J. Alterations in the gut microbiota and metabolic profiles coincide with intestinal damage in mice with a bloodborne *Candida albicans* infection. *Microb. Pathog.* **2021**, *154*, 104826. [CrossRef] [PubMed]
28. Liu, Z.; Liu, H.; Chen, Z.; Deng, C.; Zhou, L.; Chen, S.; Kang, J.; Chen, Y.; He, S.; Zhou, Z. Identification of a novel plasma metabolite panel as diagnostic biomarker for hepatocellular carcinoma. *Clin. Chim. Acta.* **2023**, *543*, 117302. [CrossRef]
29. Hou, Q.; Kwok, L.Y.; Zheng, Y.; Wang, L.; Guo, Z.; Zhang, J.; Huang, W.; Wang, Y.; Leng, L.; Li, H.; et al. Differential fecal microbiota are retained in broiler chicken lines divergently selected for fatness Traits. *Sci. Rep.* **2016**, *6*, 37376. [CrossRef]
30. Gibiino, G.; Lopetuso, L.R.; Scaldaferri, F.; Rizzatti, G.; Binda, C.; Gasbarrini, A. Exploring Bacteroidetes: Metabolic key points and immunological tricks of our gut commensals. *Dig. Liver Dis.* **2018**, *50*, 635–639. [CrossRef]
31. Moon, C.D.; Young, W.; Maclean, P.H.; Cookson, A.L.; Bermingham, E.N. Metagenomic insights into the roles of Proteobacteria in the gastrointestinal microbiomes of healthy dogs and cats. *MicrobiologyOpen* **2018**, *7*, e00677. [CrossRef] [PubMed]
32. Zhou, Z.; Nie, K.; Huang, Q.; Li, K.; Sun, Y.; Zhou, R.; Wang, Z.; Hu, S. Changes of cecal microflora in chickens following *Eimeria tenella* challenge and regulating effect of coated sodium butyrate. *Exp. Parasitol.* **2017**, *177*, 73–81. [CrossRef] [PubMed]
33. Qu, W.; Yuan, X.; Zhao, J.; Zhang, Y.; Hu, J.; Wang, J.; Li, J. Dietary advanced glycation end products modify gut microbial composition and partially increase colon permeability in rats. *Mol. Nutr. Food Res.* **2017**, *61*, 1700118. [CrossRef] [PubMed]
34. Li, A.L.; Ni, W.W.; Zhang, Q.M.; Li, Y.; Zhang, X.; Wu, H.Y.; Du, P.; Hou, J.C.; Zhang, Y. Effect of cinnamon essential oil on gut microbiota in the mouse model of dextran sodium sulfate-induced colitis. *Microbiol. Immunol.* **2020**, *64*, 23–32. [CrossRef]
35. Clavijo, V.; Flórez, M.J.V. The gastrointestinal microbiome and its association with the control of pathogens in broiler chicken production: A review. *Poult. Sci.* **2018**, *97*, 1006–1021. [CrossRef]
36. Abranches, J.; Zeng, L.; Kajfasz, J.K.; Palmer, S.R.; Chakraborty, B.; Wen, Z.T.; Richards, V.P.; Brady, L.J.; Lemos, J.A. Biology of oral *Streptococci*. *Microbiol. Spectr.* **2018**, *6*, 426. [CrossRef]
37. Wen, S.; Feng, D.; Chen, D.; Yang, L.; Xu, Z. Molecular epidemiology and evolution of *Haemophilus influenzae*. *Infect. Genet. Evol* **2020**, *80*, 104205. [CrossRef]
38. Song, B.; Li, P.; Yan, S.; Liu, Y.; Gao, M.; Lv, H.; Lv, Z.; Guo, Y. Effects of dietary astragalus polysaccharide supplementation on the Th17/Treg balance and the gut microbiota of broiler chickens challenged with necrotic enteritis. *Front. Immunol.* **2022**, *13*, 781934. [CrossRef] [PubMed]
39. Hargis, B.; Tellez, G.; Bielke, L.R. Gut health and susceptibility to enteric bacterial diseases in poultry. In *Achieving Sustainable Production of Poultry Meat*; Applegate, T., Ed.; Burleigh Dodds Science Publishing: London, UK, 2017; pp. 13–37.
40. Stanley, D.; Hughes, R.J.; Geier, M.S.; Moore, R.J. Bacteria within the gastrointestinal tract microbiota correlated with improved growth and feed conversion: Challenges presented for the identification of performance enhancing probiotic bacteria. *Front. Microbiol.* **2016**, *7*, 187. [CrossRef]
41. Biasato, I.; Ferrocino, I.; Grego, E.; Dabbou, S.; Gai, F.; Gasco, L.; Cocolin, L.; Capucchio, M.T.; Schiavone, A. Gut microbiota and mucin composition in female broiler chickens fed diets including yellow mealworm (*Tenebrio molitor*, L.). *Animals* **2019**, *9*, 213. [CrossRef]
42. Kim, Y.B.; Park, J.; Lee, H.G.; Song, J.Y.; Kim, D.H.; Ji, W.; Joo, S.S.; Kim, M.; Jung, J.Y.; Kim, M.; et al. Dietary probiotic *Lacticaseibacillus paracasei* NSMJ56 modulates gut immunity and microbiota in laying hens. *Poult. Sci.* **2024**, *103*, 103505. [CrossRef]
43. Abd El-Hack, M.E.; El-Saadony, M.T.; Shafi, M.E.; Qattan, S.Y.A.; Batiha, G.E.; Khafaga, A.F.; Abdel-Moneim, A.E.; Alagawany, M. Probiotics in poultry feed: A comprehensive review. *J. Anim. Physiol. Anim. Nutr.* **2020**, *104*, 1835–1850. [CrossRef] [PubMed]
44. Smirnov, A.; Perez, R.; Amit-Romach, E.; Sklan, D.; Uni, Z. Mucin dynamics and microbial populations in chicken small intestine are changed by dietary probiotic and antibiotic growth promoter supplementation. *J. Nutr.* **2005**, *135*, 187–192. [CrossRef] [PubMed]
45. Zhou, Q.; Lan, F.; Li, X.; Yan, W.; Sun, C.; Li, J.; Yang, N.; Wen, C. The spatial and temporal characterization of gut microbiota in broilers. *Front. Vet. Sci.* **2021**, *8*, 712226. [CrossRef] [PubMed]
46. Wang, X.; Farnell, Y.Z.; Kiess, A.S.; Peebles, E.D.; Wamsley, K.G.S.; Zhai, W. Effects of *Bacillus subtilis* and coccidial vaccination on cecal microbial diversity and composition of Eimeria-challenged male broilers. *Poult Sci.* **2019**, *98*, 3839–3849. [CrossRef] [PubMed]
47. Soutter, F.; Werling, D.; Tomley, F.M.; Blake, D.P. Poultry coccidiosis: Design and interpretation of vaccine studies. *Front. Vet. Sci.* **2020**, *7*, 101. [CrossRef]
48. Sharon, G.; Garg, N.; Debelius, J.; Knight, R.; Dorrestein, P.C.; Mazmanian, S.K. Specialized metabolites from the microbiome in health and diseases. *Cell Metab.* **2014**, *20*, 719–730. [CrossRef]
49. Lamichhane, S.; Sen, P.; Dickens, A.M.; Orešič, M.; Bertram, H.C. Gut metabolome meets microbiome: A methodological perspective to understand the relationship between host and microbe. *Methods* **2018**, *149*, 3–12. [CrossRef]
50. Ali, S.M.; Khan, N.A.; Sagathevan, K.; Anwar, A.; Siddiqui, R. Biologically active metabolite(s) from haemolymph of red-headed centipede *Scolopendra subspinipes* possess broad spectrum antibacterial activity. *AMB Express* **2019**, *9*, 95. [CrossRef]
51. Shanmugasundaram, R.; Morris, A.; Selvaraj, R.K. Effect of 25-hydroxycholecalciferol supplementation on turkey performance and immune cell parameters in a coccidial infection model. *Poult. Sci.* **2019**, *98*, 1127–1133. [CrossRef]
52. Rajapakse, R.; Uring-Lambert, B.; Andarawewa, K.L.; Rajapakse, R.P.; Abou-Bacar, A.; Marcellin, L.; Candolfi, E. 1,25(OH)2D3 inhibits in vitro and in vivo intracellular growth of apicomplexan parasite *Toxoplasma gondii*. *J. Steroid Biochem. Mol. Biol.* **2007**, *103*, 811–814. [CrossRef]

53. Kikusato, M.; Nanto, F.; Mukai, K.; Toyomizu, M. Effects of trehalose supplementation on the growth performance and intestinal innate immunity of juvenile chicks. *Br. Poult. Sci.* **2016**, *57*, 375–380. [CrossRef] [PubMed]
54. Tsukahara, T.; Koyama, H.; Okada, M.; Ushida, K. Stimulation of butyrate production by gluconic acid in batch culture of pig cecal digesta and identification of butyrate-producing bacteria. *J. Nutr.* **2002**, *132*, 2229–2234. [CrossRef] [PubMed]
55. Cañete-Rodríguez, A.M.; Santos-Dueñas, I.M.; Jiménez-Hornero, J.E.; Ehrenreich, A.; Liebl, W.; García-García, I. Gluconic acid: Properties, production methods and applications—An excellent opportunity for agro-industrial by-products and waste bio-valorization. *Process. Biochem.* **2016**, *51*, 1891–1903. [CrossRef]
56. Bali, V.; Panesar, P.S.; Bera, M.B.; Panesar, R. Fructo-oligosaccharides: Production, purification and potential applications. *Crit. Rev. Food Sci. Nutr.* **2013**, *55*, 1475–1490. [CrossRef] [PubMed]
57. Tochio, T.; Kitaura, Y.; Nakamura, S.; Sugawa, C.; Takahashi, M.; Endo, A.; Shimomura, Y. An alteration in the cecal microbiota composition by feeding of 1-Kestose results in a marked increase in the cecal butyrate content in rats. *PLoS ONE* **2016**, *11*, e0166850. [CrossRef]
58. Song, Z.; Xie, K.; Zhang, Y.; Xie, Q.; He, X.; Zhang, H. Effects of dietary ginsenoside Rg1 supplementation on growth performance, gut health, and serum immunity in broiler chickens. *Front. Nutr.* **2021**, *8*, 705279. [CrossRef]

animals

MDPI

Article

Temporal Changes in Jejunal and Ileal Microbiota of Broiler Chickens with Clinical Coccidiosis (*Eimeria maxima*)

Katarzyna B. Miska [1,*], Philip M. Campos [1], Sara E. Cloft [2], Mark C. Jenkins [3] and Monika Proszkowiec-Weglarz [1]

1 Animal Biosciences and Biotechnology Laboratory (ABBL), Beltsville Agricultural Research Center (BARC), United States Department of Agriculture—Agricultural Research Service (USDA—ARS), Beltsville, MD 20705, USA; philip.campos@usda.gov (P.M.C.); monika.weglarz@usda.gov (M.P.-W.)

2 Animal Sciences Department, Purdue University, West Lafayette, IN 47907, USA; scloft@purdue.edu

3 Animal Parasitic Diseases Laboratory (ABBL), Beltsville Agricultural Research Center (BARC), United States Department of Agriculture—Agricultural Research Service (USDA—ARS), Beltsville, MD 20705, USA; mark.jenkins@usda.gov

* Correspondence: kate.miska@usda.gov; Tel.: +1-(301)-504-5596

Simple Summary: Coccidiosis affects broiler chickens, causing economic losses since infected chickens grow less efficiently than healthy birds. Coccidiosis is caused by a parasite *Eimeria* that infects the gastrointestinal tract. In the current study, changes in the bacteria of the jejunum and ileum following infection with *Eimeria maxima* were determined. Samples were taken between 0 and 14 days post-infection. The infection resulted in decreased body weight gain, increased feed conversion ratio, compromised gut morphology, and decreased plasma carotenoid levels. Microbiota composition was determined by sequencing a portion of the 16s rDNA gene. The results indicate that infection affected the diversity within and between bacterial communities, particularly at the height of infection. In samples from infected birds, species of bacteria that can be opportunistic pathogens were more abundant than in healthy birds. In uninfected birds, bacteria that produce short-chain fatty acids and are associated with improved growth were more abundant. *Eimeria maxima* affects the ecology of the small intestine by disrupting its integrity and changes the types of bacteria present to those that could be opportunistic pathogens. Coccidiosis is a complex gastrointestinal disease, and understanding its effects on the host will help find effective control methods in the future.

Abstract: Coccidiosis in broiler chickens continues to be a major disease of the gastrointestinal tract, causing economic losses to the poultry industry worldwide. The goal of this study was to generate a symptomatic *Eimeria maxima* (1000 oocysts) infection to determine its effect on the luminal and mucosal microbiota populations (L and M) in the jejunum and ileum (J and IL). Samples were taken from day 0 to 14 post-infection, and sequencing of 16S rRNA was performed using Illumina technology. Infected birds had significantly ($p < 0.0001$) lower body weight gain (BWG), higher feed conversion ratio (FCR) ($p = 0.0015$), increased crypt depth, and decreased villus height ($p < 0.05$). The significant differences in alpha and beta diversity were observed primarily at height of infection (D7). Analysis of taxonomy indicated that J-L and M were dominated by *Lactobacillus*, and in IL-M, changeover from *Candidatus Arthromitus* to *Lactobacillus* as the major taxon was observed, which occurred quicky in infected animals. LEfSe analysis found that in the J-M of infected chickens, *Lactobacillus* was significantly more abundant in infected (IF) chickens. These findings show that *E. maxima* infection affects the microbiota of the small intestine in a time-dependent manner, with different effects on the luminal and mucosal populations.

Keywords: coccidiosis; *Eimeria maxima*; microbiota; broilers; 16S rRNA sequencing; gut morphology

Citation: Miska, K.B.; Campos, P.M.; Cloft, S.E.; Jenkins, M.C.; Proszkowiec-Weglarz, M. Temporal Changes in Jejunal and Ileal Microbiota of Broiler Chickens with Clinical Coccidiosis (*Eimeria maxima*). *Animals* **2024**, *14*, 2976. https://doi.org/10.3390/ani14202976

Academic Editors: Jiakui Li and Kenneth Bafundo

Received: 26 August 2024
Revised: 4 October 2024
Accepted: 9 October 2024
Published: 15 October 2024

1. Introduction

Consumption of broilers has been increasing steadily since the modern chicken industry began selecting birds for rapid weight gain in the 1960s. In 2024, it is estimated that the per capita consumption of meat derived from broilers will reach 100 pounds in the USA (Per Capita Consumption of Poultry and Livestock, 1960 to Forecast 2024, in Pounds—National Chicken Council). Chicken meat is a good source of protein, and its price is lower compared to other options such as pork and beef (Wholesale and Retail USDA Prices for Chicken, Beef, and Pork—National Chicken Council), making it an affordable alternative. Though broiler chicken meat is in demand and its growth is expanding, there are challenges to maintaining high production while maintaining affordable cost. Modern broilers are predominantly grown in intensive farming operations where they are exposed to many types of pathogens. One of the most prevalent diseases is coccidiosis, which is caused by protozoan parasites belonging to genus *Eimeria* (Mesa-Pineda et al., 2021) [1]. There are at least seven described species; however, other cryptic species have been recently described (Blake et al., 2021) [2]. Coccidiosis in broiler chickens results in reduction of growth parameters since *Eimeria* infects the intestinal tract and causes diarrhea, inappetence, lethargy, and in some cases mortality (Mesa-Pineda et al., 2021) [1]. Currently coccidiosis is controlled largely by vaccines and chemotherapeutic agents, but new chemotherapeutic agents have not been developed recently (Ahmad et al., 2024) [3].

Eimeria maxima is one of the established species that causes coccidiosis in chickens. It infects the small intestine (jejunum, ileum, and can appear in the duodenum) (Noack et al., 2019) [4]. The *Eimeria* life cycle is initiated when a chicken ingests sporulated oocysts, which are excysted releasing sporozoites into the intestine. The excysted sporozoites infect epithelial cells (other cell types can be infected as well) and undergo several stages of asexual reproduction until they develop sexual stages (micro- and macrogametocytes) that undergo fertilization followed by meiosis to produce oocysts that are excreted in the feces. To become infectious, the oocysts must mature by sporulating in the litter. The entire life cycle is completed in 6–7 days (depending on the species), with clinical symptoms appearing as early as four days post-infection (PI) (Martins et al., 20222) [5]. Much of the research concerning *Eimeria* has been focused on gathering data at the height of infection; however, changes in the intestine occur prior, beginning with the invasion of sporozoites, and persist past the average of seven days needed to complete the life cycle (Cloft et al., 2023; Elsasser et al., 2018; Rothwell et al., 1995) [6–8], but it is unclear when during the infection the microbiota becomes altered. It has previously been shown that changes to the microbiota can take place in the lumen and mucosa of the small and large intestine at the height of infection and remain up to 14 days PI (Campos et al., 2023, 2024) [9,10]. The aim of the current study was to establish and measure a clinical *E. maxima* infection and investigate the changes in bird growth parameters, gut morphology, and microbiota through 16S rRNA sequencing in the small intestine. Because the primary target of *E. maxima* infection is the small intestine, changes in microbiota were measured that are most likely caused directly by the pathogen. Many studies measuring microbiota populations sample intestinal contents, even though it has been observed that the populations of bacteria that reside in the lumen of the gut are distinct from those that adhere to the gut epithelium (mucosal population) (Borda-Molina et al. 2018; Campos et al., 2022; Proszkowiec-Weglarz, 2022) [11–13]. It was hypothesized that *E. maxima* infection would disrupt the microbiota most severely during the height of infection, and that the changes of the luminal and mucosal microbiota would differ.

2. Materials and Methods

2.1. Animal Husbandry and Tissue Sampling

All animal care procedures were approved by the Institutional Animal Care and Use Committee (IACUC, protocol #18-025) of the Beltsville Agricultural Research Center (BARC). Two hundred and eighty-eight Ross 708 male broiler hatchlings were obtained from Longnecker's Hatchery (Elizabethtown, PA, USA) and placed into 1.00 m^2 open-top wire brooder pens (approximately 25 chicks per pen). At 19 days of age, all birds were moved to 72 cages (Alternative Designs, Siloam Springs, AR, USA) with 4 birds per pen. A corn–soybean-based diet (approximately 24% crude protein, crumble) and water were provided to chicks ad libitum for the duration of the study. One-half of the birds (144 birds) were infected (IF) with 1×10^3 *E. maxima* oocysts (USDA APU1 isolate) in a volume of 1.0 mL per bird by oral gavage at 21 days of age. The remaining 144 birds were sham-infected with water (control [C]). A total of 36 pens of C birds and 36 pens of IF birds were housed in a single room, on opposite sides to prevent cross-transmission.

At six time points (0, 3, 5, 7, 10, and 14 days PI), birds and feed were weighed to calculate body weight (BW), body weight gain (BWG), feed intake (FI), and feed conversion ratio (FCR). For this study, the pen was treated as an experimental unit, with 6 replicates per time point. For the BW, BWG, FI, and FCR, the average value per cage was determined from all 4 birds/cage. Blood samples were collected via cardiac puncture from one bird per cage following cervical dislocation into an EDTA tube and centrifuged at $2000\times g$ at 4 °C to collect plasma for carotenoid measurements using previously described methods (Allen, 1987) [14]. The concentration of plasma carotenoids was used to evaluate the level of *E. maxima* infection. For tissue morphology, a 2–3 cm segment of the jejunum was dissected, rinsed with sterile PBS, and placed into neutral buffered formalin for later histological analysis.

For luminal contents and mucosal microbiota sampling, the jejunum and ileum of one bird per cage (6 replicates per treatment) were dissected, and the jejunal and ileal contents (J-C and IL-C) and scrapings (J-M and IL-M) were collected. The contents of the jejunum and ileum were collected by squeezing the digesta by hand into cryovials. The intestine was then placed in sterile ice cold 1× PBS, opened with scissors longitudinally and rinsed to remove any contents still adhering to the mucosa. The mucosa was then scraped using a microscope slide and the tissues were also placed into cryovials. Isolated specimens were snap frozen in liquid nitrogen and stored at −80 °C until bacterial DNA isolation.

2.2. Identification of Stem Cells and Morphological Measurements

Jejunum segments collected for histological analysis were transferred to 70% ethanol and kept at 4 °C until they were embedded in paraffin (StageBio, Mount Jackson, VA, USA). Formalin-fixed, paraffin-embedded jejunum tissues were utilized for in situ hybridization of olfactomedin 4 (Olfm4; NM_001040463.1) as described in Cloft et al. (2023) [6]. Olfm4 is an intestinal stem cell marker that defines the functional crypts as opposed to the visible crypt, providing a more objective measure (Cloft et al., 2023) [6]. Images were captured with a Nikon Eclipse 80i microscope with a DS-Ri1 digital camera (Nikon Instruments, Inc., Melville, NY, USA) at 40× magnification. Morphological measurements of crypt depth (CD) and villus height (VH) were taken with Image J software (version 15.0) from the National Institutes of Health (Bethesda, MD, USA). Approximately 75 crypts and 30 villi were measured on 4 birds per treatment timepoint.

2.3. DNA Isolation and 16s rRNA Library Preparation

DNA was extracted from J-C, J-M, IL-C, and IL-M and was evaluated as described previously (Campos et al., 2022) [12]. The 16S rRNA gene amplicon libraries were generated using the workflow and chemistry supplied by Illumina (Illumina, Inc., San Diego, CA, USA), and PCR primers (Forward: 5′-TCGTCGGCAGCGTCAGATGTGTATAAGAGACAGCCTACGGGNGGCWGCAG-3′ and Reverse: 5′-GTCTCGTGGGCTCGGAGATGTGTATAAGAGACAGGACTACHV GGGTATCTAATCC-3′) targeted the V3-V4 variable region of the 16S gene (Darwish et al., 2021) [15]. Amplicon PCR was followed by index PCR followed by amplicon cleaning, as described previously (Campos et al., 2022) [12]. The concentration and quality of the amplicons were determined using QIAxcel DNA Hi-Resolution cartridge, proprietary QIAxcel ScreenGel software (version 1.6.0, and QIAxcel Advanced System (Qiagen, Germantown, MD, USA) per manufacturing instructions. The pooled (96 barcoded amplicons) DNA library (4 nM) and PhiX control v3 (Illumina, Inc., San Diego, CA, USA)

(4 nM) were denatured with 0.2 N NaOH (Sigma-Aldrich, Corp., St. Louis, MO, USA) and diluted to a final concentration of 4 pM. The library was mixed with PhiX control (20% v/v), and pair-end 2 × 300 bp sequencing was performed using the Illumina MiSeq platform and a MiSeq Reagent Kit v3 (Illumina, Inc.).

2.4. Bioinformatics and Data Processing

Quantitative Insight Into Microbial Ecology (QIIME) software package 2 (version 2023.2, http://qiime2.org accessed on 3 October 2024) (Bolyen et al., 2019) [11] was used to perform quality control and analysis of the sequence reads as described before (Campos et al., 2022) [12]. Raw fastq files were demultiplexed using q2-demux and quality filtered and denoised with DADA2 via q2-dada2 (Callahan et al., 2016) [16]. Sequences were trimmed with DADA2 where remaining base positions had a median Phred quality score of 30 or above (see Table 1 for truncation parameters). Representative sequence sets containing DADA2 amplicon sequence variants (ASVs) were used for taxonomy classification. MAFFT (Katoh et al., 2002) [17] was used for multiple sequence alignment, and FastTree2 (Price et al., 2010) [18] was used to generate phylogenetic trees. Naïve Bayesian classifier was used for taxonomic classification against the SILVA version 138 database (Quast et al., 2012) [19], which was chosen over the Greengenes 2013 database for its larger size and more up-to-date taxonomy (Balvočiūtė et al., 2017; Campos et al., 2022) [12,20]. For this step, RESCRIPt pre-formatted SILVA 99% reference sequences, and taxonomy files were obtained from the QIIME 2 Data Resources page (https://docs.qiime2.org/2023.2/data-resources/ accessed on 3 October 2024), as RESCRIPt reduces inconsistencies by removing duplicate sequences that are assigned different taxonomies (Robeson et al., 2021) [21]. Data were rarefied to sequencing depths (Table 1) using alpha rarefaction plots to consider where bacterial diversity would be maintained for the calculation of alpha and beta diversities. Alpha diversity indices (observed features (ASVs), Shannon's diversity index, Pielou's Evenness (evenness), and Faith's Phylogenetic Diversity (richness)) were obtained through QIIME 2 package. Analysis of beta diversity was performed by QIIME2 employing unweighted (Lozupone et al., 2005) [22] and weighted UniFrac (Lozupone et al., 2007) [23] distance metrics. The linear discriminant analysis (LDA) effect size (LEfSe) algorithm was used to identify taxa with significant differential abundance between C and IF birds (Segata et al., 2011) [24]. Genus-level feature tables were converted to relative abundance and analyzed at the Huttenhower Lab Galaxy web server (http://huttenhower.sph.harvard.edu/galaxy accessed on 3 October 2024) using the default parameters (LDA threshold of 2.0). Phylogenetic Investigation of Communities by Reconstruction of Unobserved States 2 (PICRUSt2 version 2.4.2) software was used to predict functional abundances based on marker gene sequences (Douglas et al., 2020) [25]. The MetaCyc Metabolic Pathways Database (Caspi et al., 2020) [26] was used to produce functional abundance data, and the data were analyzed and visualized using STAMP 2.1.3 (Parks et al., 2014) [27] to determine the biological relevance of features.

Visualizations for alpha diversity comparisons and beta diversity PCoA were produced in R 4.3.2 (R Core Team, 2023) [28] using the packages QIIME2R 0.99.35 (Bisanz et al., 2020) [29] to import QIIME 2 PCoA results and tidyverse 2.0.0 (Wickham et al., 2020) [30] for data wrangling with dplyr (Wickham et al., 2023) [31] and figure production with ggplot2 (Wickham 2016) [32]. To produce taxonomic bar plot visualizations, samples were merged into groups (infection status × time point) using the R package phyloseq version 1.48.0 (McMurdie and Holmes, 2013) [33], and taxonomic data were processed at the genus level using microViz version 0.12.5 (Barnett et al., 2021) [34].

Table 1. Sequencing summary of the datasets generated from microbiota samples (JC, J-M, IL-C, IL-M) processed through QIIME 2.

	J-C	J-M	IL-C	IL-M
Number of samples	60	60	60	60
Raw reads	6,321,969	1,906,686	8,520,563	3,669,527
DADA2 --p-trunc-len-f	269	235	269	271
DADA2 --p-trunc-len-r	217	205	229	208
Reads after DADA2	5,212,955	1,312,059	6,443,444	2,663,546
Reads after filtering	5,060,831	1,158,586	6,434,195	2,652,074
Reads per sample (range)	114–256,509	81–255,261	13,847–257,570	197–503,642
Mean reads per sample	84,347	19,310	107,237	44,201
Total number of ASVs	1001	1273	984	1371
ASV read length (range)	269–465	234–427	269–473	271–466
Mean ASV read length	418	358	420	385
Sequencing depth	27,462	5658	49,158	13,067

2.5. Statistical Analysis

Performance data (BW, BWG, FI, FCR) and plasma carotenoid concentration were analyzed by 2-factor ANOVA considering time post-infection and infection status as main factors via JMP v15.0 (SAS Institute Cary, NC, USA). When significant ANOVA results were observed, Tukey's honestly significant difference test was conducted for mean separation.

Morphological data were unable to be transformed to meet conditions of normal distribution due to unequal variances; therefore, data were analyzed using the nonparametric Welch's one-way test for the effect of infection status at each timepoint.

Differences between alpha diversity indices were tested using the Kruskal–Wallis test (QIIME2). The difference in community structure based on groups (infection status × time point) were statistically tested by non-parametric multivariate ANOVA (PERMANOVA) with 999 permutations using QIIME2, and distances between samples were visualized with PCoA. Statistical significance was established at $p \leq 0.05$ for all analyses.

3. Results

3.1. Parameters Associated with Infection and Growth

The performance parameters of birds included in this study are shown in Table 2. These include body weight (BW), body weight gain (BWG), feed intake (FI), and feed conversion ratio (FCR). Statistical analysis of significant differences within, as well as between, main effects (infection status and age) were determined by analysis of variance. The BW of birds increased through time ($p < 0.0001$), and IF birds weighed less than C birds ($p < 0.0246$). The BWG of birds increased through time ($p < 0.0001$), and IF birds gained less weight than C birds ($p < 0.0001$). The FI did not differ significantly between C and IF birds ($p = 0.33$) but increased significantly through time ($p < 0.0001$). There was a significant interaction between the main effects ($p = 0.0487$) in FCR.

To further ensure that an *E. maxima* infection was present, the level of plasma carotenoids (Allen, 1987) [14] was measured at each of the six time-points and is shown in Figure 1. There was a significant decrease in plasma carotenoids in IF birds ($p < 0.001$) beginning at 7 days PI. The carotenoids increased at days 10 and 14 PI but were still significantly lower than those in C birds.

Table 2. Growth performance data of infected and uninfected broiler chickens [1].

Day Post-Infection	Infection Status	BW (kg)	BWG [2] (kg)	FI (kg)	FCR
3 PI	C	1.12	0.23 [f]	1.36	1.49 [c]
	IF	1.17	0.22 [f]	1.37	1.55 [c]
5 PI	C	1.23	0.38 [e]	2.40	1.58 [c]
	IF	1.24	0.37 [e]	2.30	1.57 [c]
7 PI	C	1.33	0.61 [d]	3.72	1.54 [c]
	IF	1.50	0.45 [e]	3.77	2.13 [ab]
10 PI	C	1.51	0.78 [c]	5.77	1.85 [abc]
	IF	2.05	0.63 [d]	5.51	2.27 [a]
14 PI	C	1.95	1.19 [a]	7.85	1.65 [bc]
	IF	2.05	1.03 [b]	7.27	1.79 [abc]
SEM [3]		0.12	0.03	0.29	0.11
Main effect day PI					
	3 PI	1.15 [b]	0.23 [e]	1.36 [e]	1.52 [b]
	5 PI	1.23 [b]	0.37 [d]	2.35 [d]	1.57 [b]
	7 PI	1.41 [b]	0.53 [c]	3.75 [c]	1.83 [ab]
	10 PI	1.78 [a]	0.70 [b]	5.64 [b]	2.06 [a]
	14 PI	2.00 [a]	1.11 [a]	7.56 [a]	1.72 [b]
	SEM [3]	0.08	0.02	0.20	0.08
Main effect infection status					
	C	1.43 [b]	0.64 [a]	4.22	1.62 [b]
	IF	1.60 [a]	0.54 [b]	4.04	1.86 [a]
	SEM [3]	0.05	0.01	0.13	0.05
Analysis of variance			Probabilities [4]		
Day post-infection × infection status		0.18	0.0053	0.82	*0.0487*
Day post-infection		*<0.0001*	*<0.0001*	*<0.0001*	*0.0001*
Infection status		*0.0246*	*<0.0001*	0.33	*0.0015*

[1] Each value represents the least-square means from 6 replicate cages with 4 male chickens/cage. Chickens at 21 d of age were infected with 1000 *Eimeria maxima* oocysts (IF) or sham infected with sterile water (C) and evaluated at 14 d post-infection (PI). [2] Body weight gain is calculated starting at 0 PI. On 0 PI, average body weight = 0.89 kg. [3] SEM = pooled standard error. [4] Italicized p-values are statistically significant ($p < 0.05$). [a–f] Means in the same column with different superscripts are significantly different ($p < 0.05$) based on Tukey HSD mean separation.

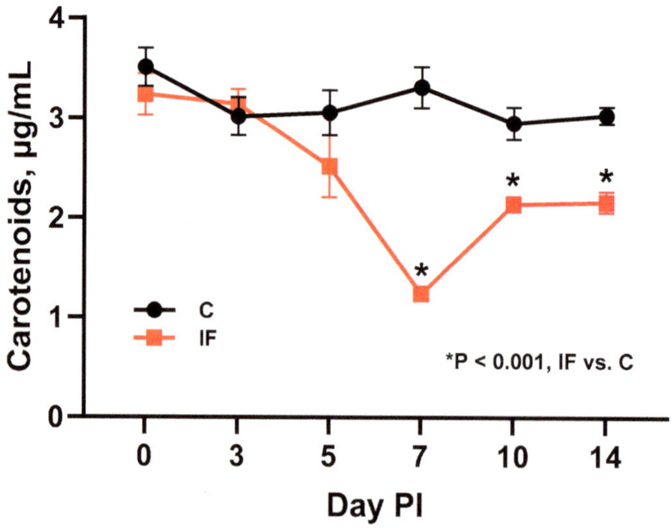

Figure 1. Concentration of plasma carotenoids in control (C) and *Eimeria maxima* (IF)-infected chickens (day PI, day post-infection; mean ± SE).

3.2. Gut Morphology

The crypt depth of the jejunum of IF birds increased significantly beginning at 7 days PI and remained significantly elevated through the remainder of the study ($p \leq 0.0051$) compared to C birds (Figure 2A). At days 5 and 7 PI, the villus height of IF birds was significantly decreased ($p = 0.0074$) compared to C birds (Figure 2B).

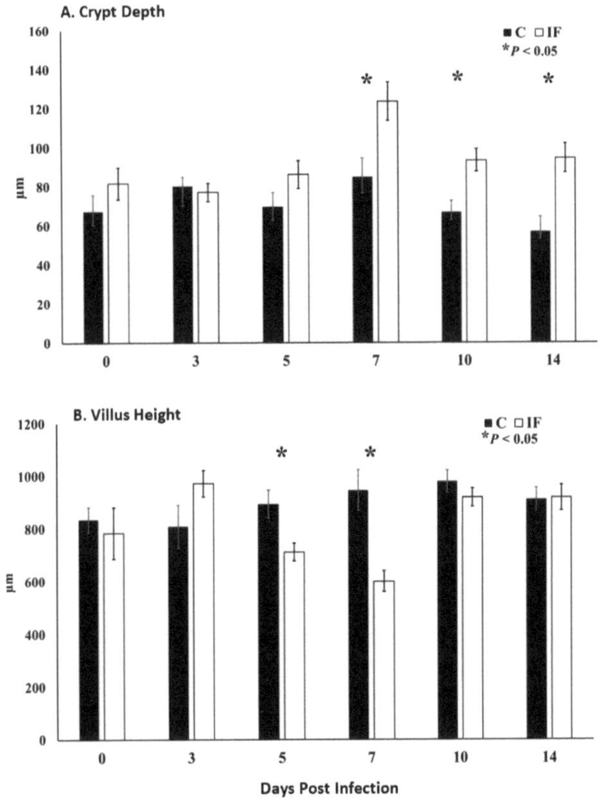

Figure 2. Crypt elongation following infection with *Eimeria maxima*. Expression of *Olfm4* mRNA by in situ hybridization in the jejunum of broiler chickens that were infected at 21 d of age with either 1000 *E. maxima* oocysts (IF) or sham infected with sterile water (C) and sampled at 0, 3, 5, 7, 10, and 14 d post-infection (PI). All tissues were counterstained with 50% hematoxylin. Images were captured at 40× magnification (n = 4). (**A**) Crypt depth and (**B**) villus height was measured on jejunal sections stained for *Olfm4* by in situ hybridization. Measures were analyzed by infection status using the nonparametric Welch's one-way test. Significances ($p < 0.05$) are indicated by asterisks (*).

3.3. Sequencing Summary

The summary of the sequences produced during this study is shown in Table 1. The sequence depth for rarefaction was lower in the mucosa samples (5658 J-M and 13,067 IL-M) compared to the luminal contents (J-C 27,462 and IL-C 49,158).

3.4. Alpha and Beta Diversity of the Microbiota of the Small Intestine

There were no significant differences in alpha diversity in either the J-C or IL-M samples, and the significant differences in J-M are shown in Figure 3. Shannon's entropy was significantly lower in the IF samples at days 3 ($p = 0.02$) and 7 PI ($p = 0.04$) (Figure 3A), and Faith's PD was significantly lower in the IF samples at day 10 PI ($p = 0.03$). The significant differences in alpha diversity of IL-C are shown in Figure 4. Shannon's entropy and Evenness were significantly lower between C and IF samples and at days 3 ($p = 0.04$) and 7 ($p = 0.02$) (Figure 4A and 4D, respectively). Observed features ($p = 0.04$) and Faith's PD ($p = 0.047$) were significantly lower between C and IF samples on day 7 PI (Figure 4B and 4C, respectively).

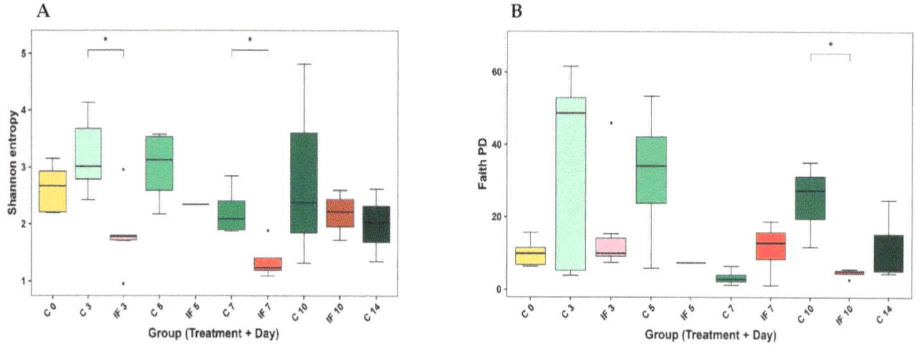

Figure 3. Effect of *Eimeria maxima* infection at days 0, 3, 5, 7, 10, and 14 post-infection (PI) on the alpha diversity indices (**A**) Shannon's entropy and (**B**) Faith's PD of the mucosal bacterial populations of the jejunum (J-M). Non-infected birds = C, infected birds = IF. Significant ($p < 0.05$) differences are indicated by asterisks (*).

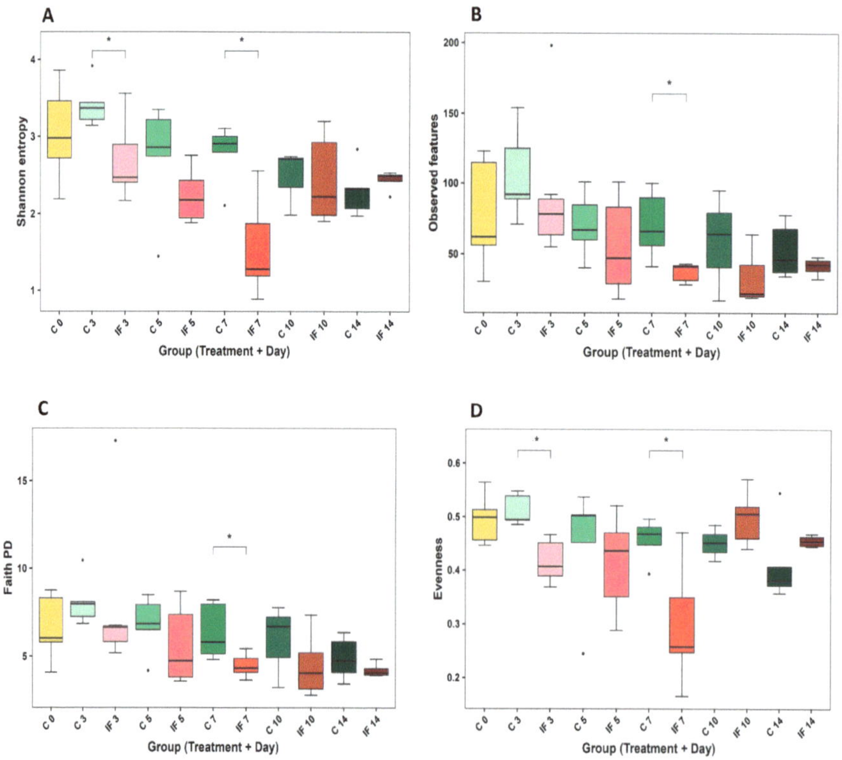

Figure 4. Effect of *Eimeria maxima* infection at days 0, 3, 5, 7, 10, and 14 post-infection (PI) on the alpha diversity indices (**A**) Shannon entropy, (**B**) observed features, Faith's PD (**C**), and Evenness (**D**) of the luminal bacterial populations of the ileum (IL-C). Non-infected birds = C, infected birds = IF. Significant ($p < 0.05$) differences are indicated by asterisks (*).

Beta diversity measures the amount of variability between bacterial communities (Figures 5 and 6) and was measured using phylogenetic-based unweighted (UWT) or weighted Uni-Frac (WT) methods. The UWT Uni-Frac analysis of J-C was not significant, but WT Uni-Frac of J-C was significantly different (p = 0.04) on day 7 PI between C and IF (Figure 5A). Conversely, the WT UniFrac analysis of J-M was not significant, but UWT UniFrac of J-M was significantly different (p = 0.03) on day 10 PI between C and IF (Figure 5B).

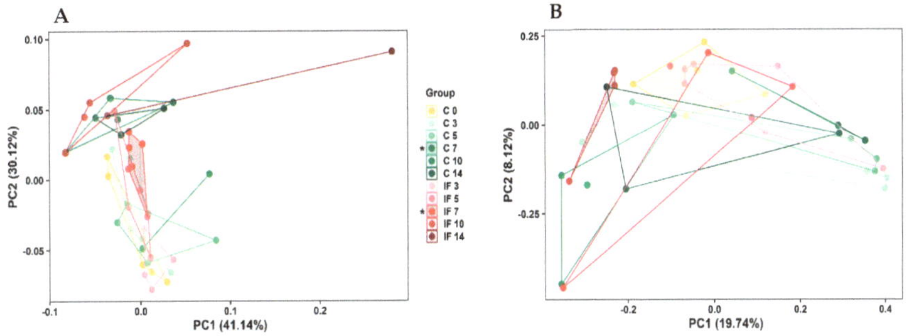

Figure 5. Effect of *Eimeria maxima* at days 0, 3, 5, 7, 10, and 14 post-infection (PI) on the beta diversity of jejunal luminal (**A**) (J-C) and jejunal mucosa (**B**) (J-M) bacterial populations using the principal coordinate analysis (PcoA) based on the weighted (**A**) and unweighted UniFrac (**B**) distances between groups. Non-infected birds = C, infected birds = IF. Significant (p < 0.05) differences are indicated by asterisks (*).

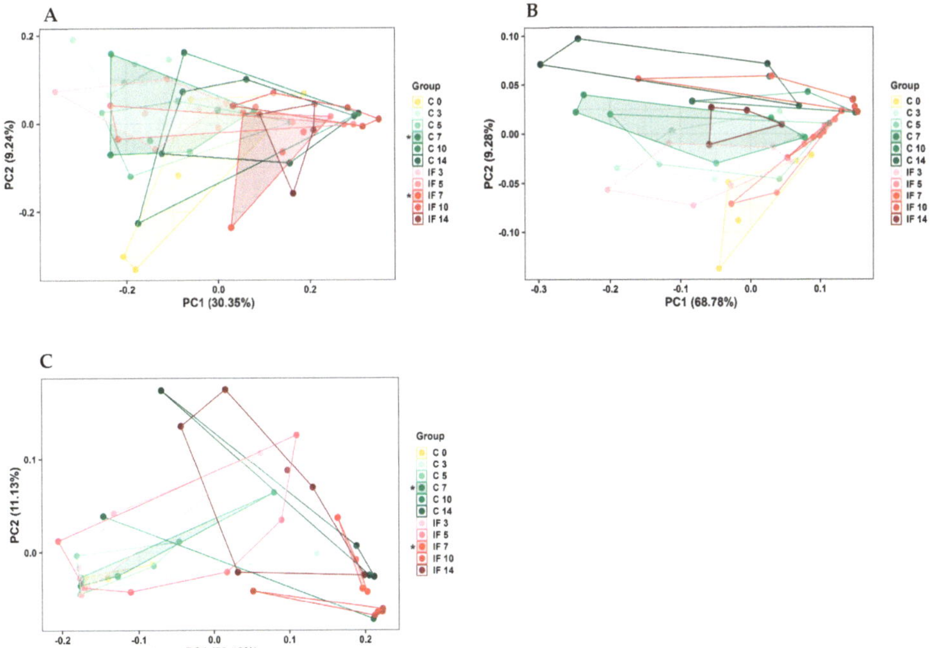

Figure 6. Effect of *Eimeria maxima* at days 0, 3, 5, 7, 10, and 14 post-infection (PI) on the beta diversity of ileal luminal (**A**,**B**) (IL-C) and ileal mucosa (**C**) (IL-M) bacterial populations using the principal coordinate analysis (PcoA) based on the weighted (**A**,**C**) and unweighted UniFrac (**B**) distances between groups. Non-infected birds = C, infected birds = IF. Significant (p < 0.05) differences are indicated by asterisks (*).

The beta diversity in both IL-C and IL-M samples was significantly different on day 7 PI ($p = 0.02$ and 0.01, respectively) (Figure 6A,B). For IL-M, the UWT Uni-Frac was not significant, but significant differences were measured using WT Uni-Frac ($p = 0.02$) between C and IF at day 7 PI (Figure 6C). In general, the results of beta diversity analysis showed significant differences between bacterial communities at day 7 PI that corresponded to height of *E. maxima* infection.

3.5. Taxonomy

The 11 most abundant taxa (% of the genera present) are shown for J-C and J-M (Figure 7) and IL-C and IL-M (Figure 8). The taxa present between C and IF in the J (both C and M) (Figure 7A,B) at corresponding days were very similar, except for IF at D14, where the abundance of *Lactobacillus* was lower. In the jejunum (both C and M), the most abundant genus was *Lactobacillus*. In all samples with the exception of those from IF birds at day 14, it made up more than 75% of the genera present. The taxa that were present in J-M but not J-C were *Escherichia–Shigella*, *Staphylococcus*, *Streptococcus*, and the *Lachnospiraceae*. On the other hand, *Candidatus Arthromitus*, *Enterococcus*, Bacteria kingdom (which includes unclassified sequences that could not be classified beyond the kingdom), and *Ralstonia* were present in J-M but not J-C. Of the taxa present in IL-C (Figure 8A), *Lactobacillus* was more abundant in IF samples at 5, 7, and 10 days PI. *Streptococcus* and *Staphyloccocus* were present in IL-C but not IL-M. On the other hand, unclassified bacteria and *Subdoligranulum* were found in IL-M but not IL-C. In the IL-M, the level of *Lactobacillus* was higher in samples from IF7, C10, IF10, and C14.

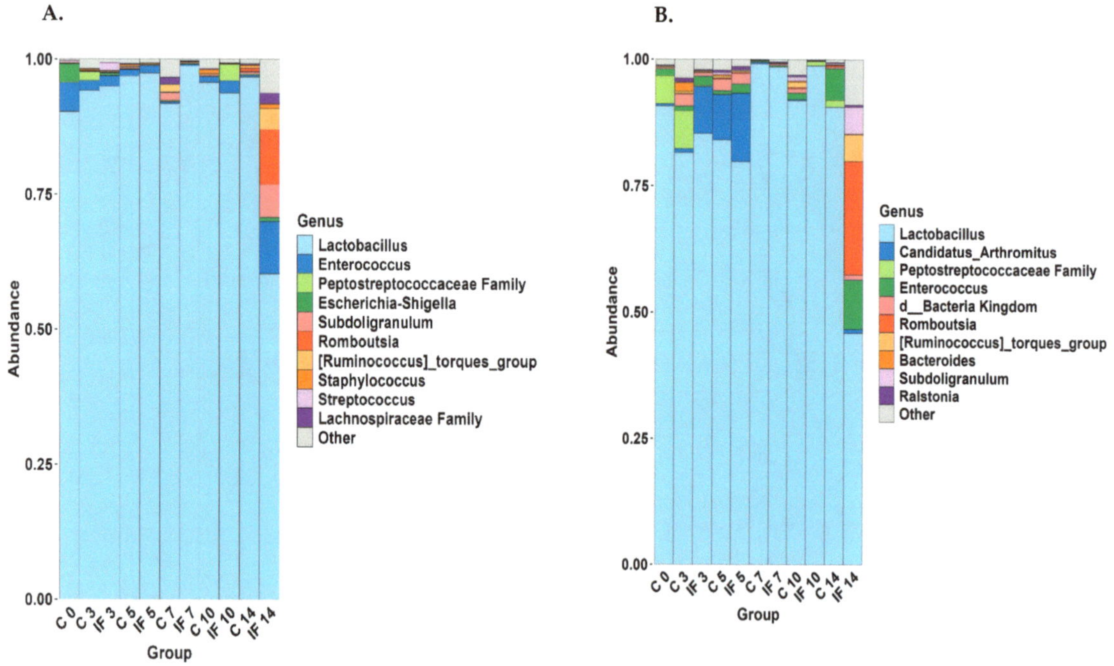

Figure 7. Effect of *Eimeria maxima* at days 0, 3, 5, 7, 10, and 14 post-infection (PI) on relative bacterial abundance (%) at the genus level in the (**A**) jejunal lumen (J-C) and (**B**) jejunal mucosa (J-M). Non-infected birds = C, infected birds = IF.

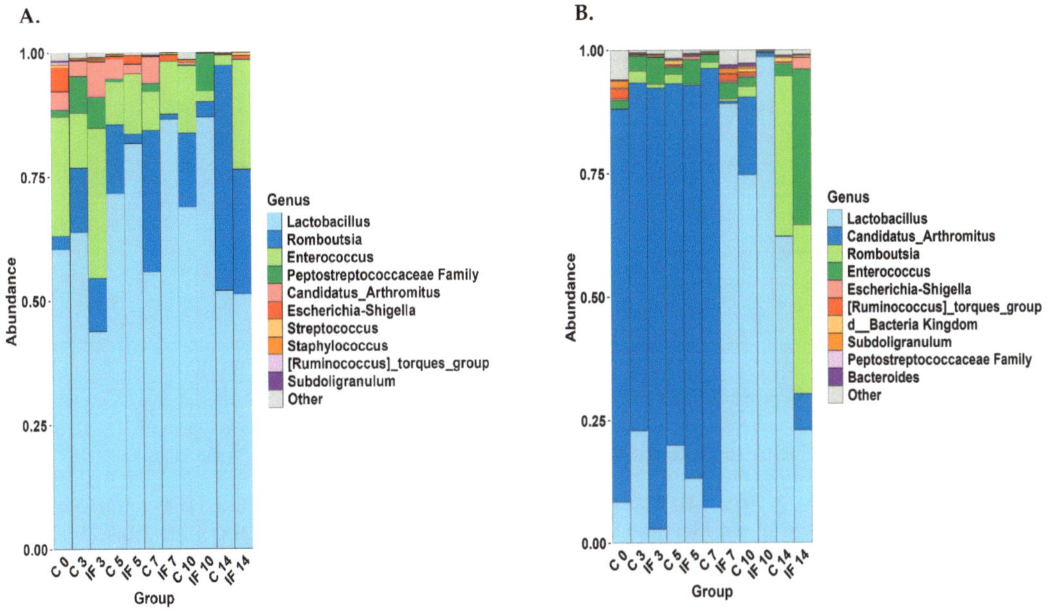

Figure 8. Effect of *Eimeria maxima* at days 0, 3, 5, 7, 10, and 14 post-infection (PI) on relative bacterial abundance (%) at the genus level in the (**A**) ileal lumen (IL-C) and (**B**) ileal mucosa (IL-M). Non-infected birds = C, infected birds = IF.

3.6. Differentially Abundant Taxa

LEfSe analysis was performed to determine differentially abundant taxa in the J-C and J-M (Figure 9) and in the IL-C and IL-M (Figure 10). In the J-C, bacteria belonging to the genus unclassified Peptostreptococcaceae and family Aerococcaceae were more abundant in samples from IF animals, while family Micrococcaceae and genus *Oceanobacillus* were more abundant in C samples (Figure 9A). In the J-M, members of the unclassified genus *Staphyloccocus* were more abundant in IF samples, and eight genera (CHKCI001, *Ochrobactrum*, [*Ruminococcus*] *gauvreauii* group, uncultured Ruminococcaceae, *Eisenbergiella*, *Pseudomonas*, unclassified Peptostreptococcaceae) and three families (Aerococcaceae, Rhizobiaceae, and Micrococcaceae) were more abundant in C samples (Figure 9B).

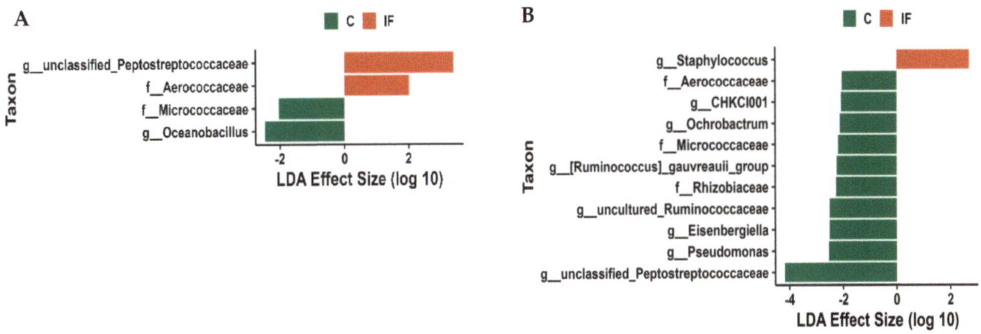

Figure 9. Effect of *Eimeria maxima* on differentially abundant bacterial genera as determined by linear discriminant analysis (LDA) effect size (LEfSe) analysis in jejunal lumen (J-C, (**A**)) and mucosal (J-M, (**B**)) bacterial populations. C—uninfected chickens, IF—infected chickens.

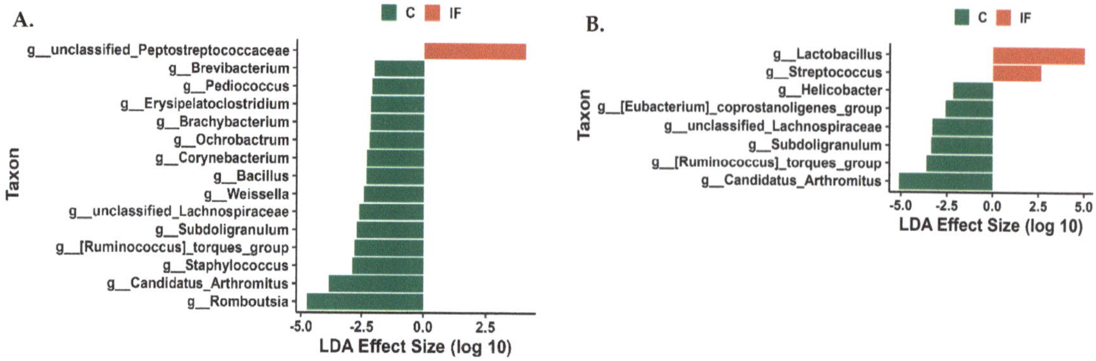

Figure 10. Effect of *Eimeria maxima* on differentially abundant bacterial genera as determined by linear discriminant analysis (LDA) effect size (LEfSe) analysis in ileal lumen (IL-C, (**A**)) and mucosal (IL-M, (**B**)) bacterial populations. C—uninfected chickens, IF—infected chickens.

In the IL-C, members of the unclassified genus *Staphyloccocus* were more abundant in IF samples, and 12 different genera (shown in Figure 10A) were more abundant in C samples. In the IL-M, the genus *Lactobacillus* and *Streptococcus* were more abundant in IF samples, and the genera *Helicobacter*, [*Eubacterium*] *coprostanoligenes* group, unclassified *Lachnospiraceae*, *Subdilogranulum*, [*Ruminoccocus*] *torques* group, and *Candidatus Arthromitus* were more abundant in C samples (Figure 10B).

3.7. Analysis of Predicted Functional Processes

The metabolic pathways that were predicted to be significantly different between IF and C samples collected from J-C and J-M are shown in Figure 11A and 11B, respectively. In C samples from J-C, the glycolysis and Entner–Doudoroff superpathway, photorespiration, and lactose and galactose degradation were predicted to be in significantly greater abundance (Figure 11A). In IF samples from J-M, the L-arabinose, androtestosterone, L-rhamnose II, gallate I and III, L-valine I, syringate, methylgallate, and protocatechuate I degradation pathways, as well as the superpathway of lipopolysaccharide biosynthesis, were in significantly greater abundance compared to C samples (Figure 11A).

In C samples from J-M, the methyl ketone, mono-trans, poly-cis decaprenyl phosphate, mycothiol, cob(II)yrinate a,c-diamide II, adenosylcabalamin I and II, and superpathway of heme from glycine biosynthesis, as well as the superpathway of salicylate degradation, toluene degradation III, 4-methylcatechol degradation, and methanol oxidation to carbon dioxide, were in greater abundance compared to IF samples (Figure 11B). Only the pentose phosphate pathway was predicted to be in significantly greater abundance in IF samples (Figure 11B).

The metabolic pathways that were predicted to be significantly different between IF and C samples collected from IL-C and IL-M are shown in Figure 12A and 12B, respectively. In the IL-C, 21 pathways were predicted to be in greater abundance in C samples, including nitrate reduction, L-leucine I, L-arabinose IV, L-tyrosine I, toluene IV, superpathway of toluene degradation, chlorosalicylate, protocatechuate II, L-tryptophan, and catechol degradation, as well as photorespiration, ectoine, NAD II, ergothioneine, cob(II)yrinate a,c-diamide II, adenosylcobalamin I and II, isoprene II and norspermidine biosynthesis, fatty acid salvage, and the meta cleavage pathway of aromatic compounds, compared to IF samples (Figure 12A). No metabolic pathways were predicted to be in greater abundance in IF samples.

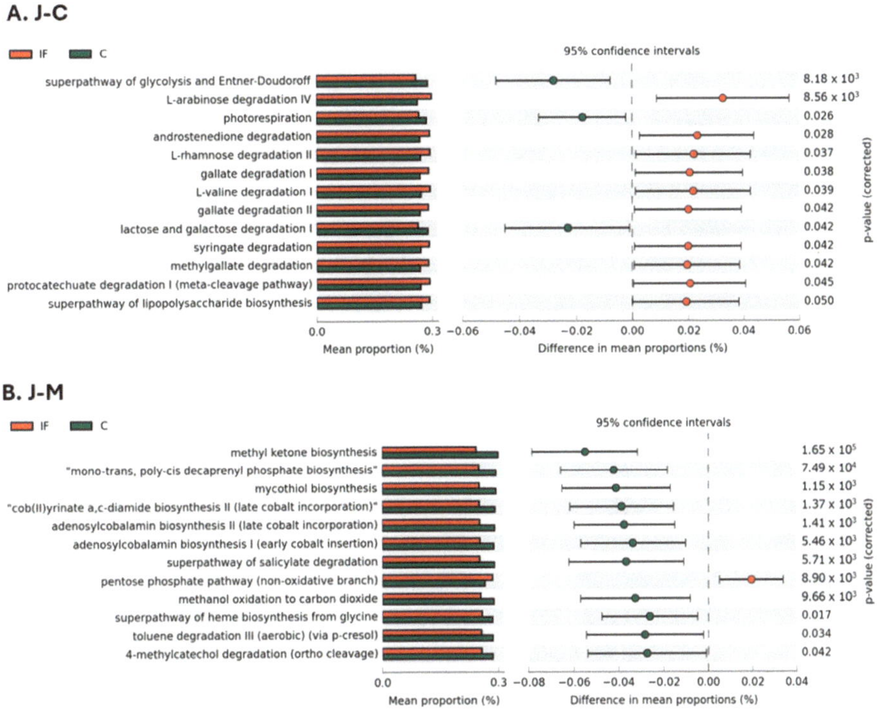

Figure 11. Effect of *Eimeria maxima* infection on mean proportion (%) of predicted MetaCyc pathways (up to top 21 pathways shown) in the jejunal luminal (J-C) (**A**) and jejunal mucosal (J-M) (**B**) populations.

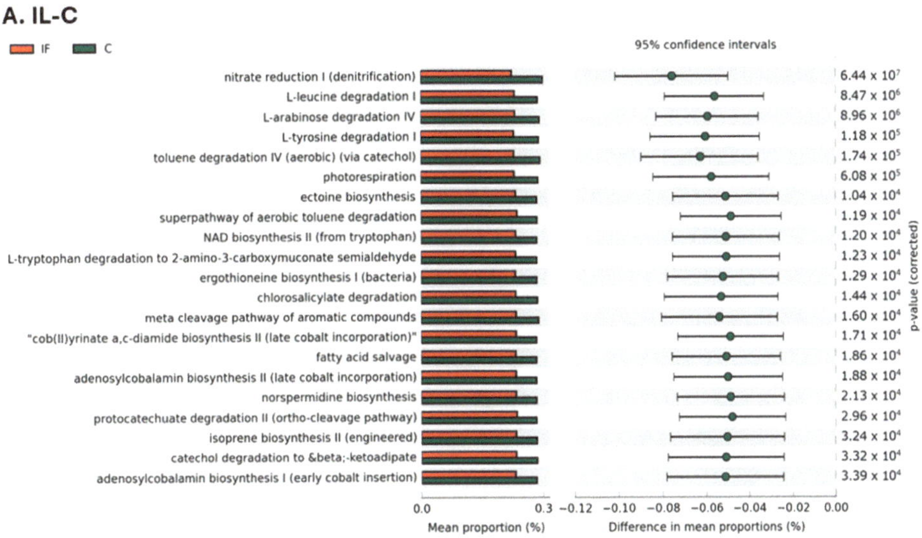

Figure 12. *Cont.*

B. IL-M

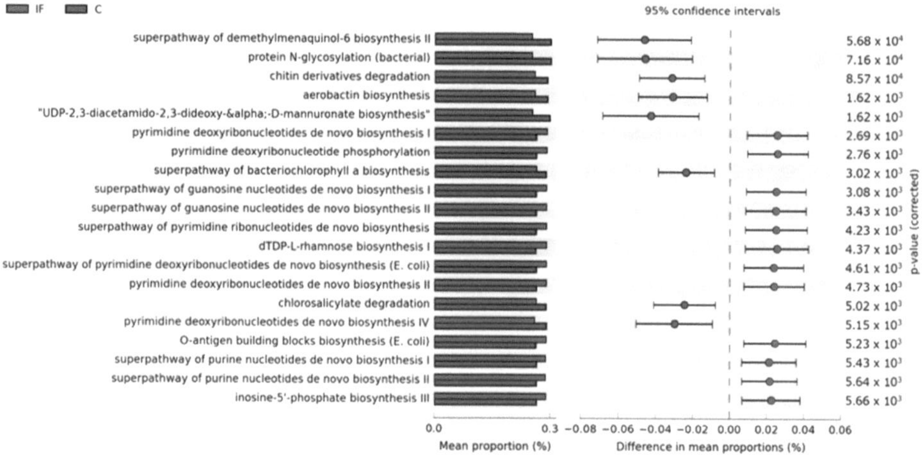

Figure 12. Effect of *Eimeria maxima* infection on the mean proportion (%) of predicted MetaCyc pathways (up to top 21 pathways shown) in the ileal luminal (IL-C) (**A**) and ileal mucosal (IL-M) (**B**) populations.

In the IL-M, eight metabolic pathways were predicated to be in greater abundance in C samples, namely, chitin derivatives and chlorosalicylate degradation, protein N-glycosylation, aerobactin, D-mannuronate biosynthesis, superpathway of bacteriochlorophyll a biosynthesis, superpathway of demethylmenaquinol-6 biosynthesis II, and pyrimidine deoxyribonucleotides de novo biosynthesis IV (Figure 12B). In the IL-M, 12 metabolic pathways were predicted to be in greater abundance in IF samples in comparison to C samples: pyrimidine deoxyribonucleotide phosphorylation, pyrimidine deoxyribonucleotides de novo biosynthesis I and II, superpathway of guanosine nucleotides de novo biosynthesis I, superpathways of guanosine nucleotides de novo synthesis I and II, superpathway of pyrimidine deoxyribonucleotides de novo synthesis, superpathways of purine nucleotides de novo synthesis I and II, O-antigen building blocks biosynthesis, dTDP-L-rhamnose biosynthesis I, and inosine-5′-phosphate biosynthesis III (Figure 12B).

4. Discussion

In the current study, a clinical *E. maxima* infection was produced that affected the production parameters (BWG and FCR), plasma carotenoid levels, and gut morphology of the jejunum. The course of the infection was tracked, beginning on the day of infection and continuing to 14 days PI. The biggest effects on the performance parameters were observed on day 7 PI, which is standard with the previously obtained results of this *E. maxima* strain (Hansen et al., 2021; Jenkins et al., 2017) [35,36]. Measuring the levels of carotenoids in the plasma has been a useful tool in determining the level of *Eimeria* infection in the species that invade the small intestine. The levels of carotenoids in the plasma decrease following infection presumably due to decreased gut integrity where these compounds "leak" from the plasma via the compromised gut (Conway et al., 1993; Fetterer et al., 2015; Sakkas et al., 2018) [37–39]. Measuring the villus height and crypt depth is another way of determining presence of infection, since *Eimeria* causes a decrease in the height of the villi and an increase in the depth of the crypts (Fernando and McCraw, 1973) [40]. This was also observed at the height and recovery phase (10–14 days PI) of this experimental infection.

The major goal of this study was to determine the effects of a clinical *E. maxima* infection on the luminal and mucosal populations of the microbiota in the small intestine (the target infection site of *E. maxima*) over the two-week period. Most studies concentrate on measuring infection effects at the height of infection around days 5–7 PI (depending on species) and predominantly measure these populations in the gut contents or the excreta (Choi and Kim, 2022; Zhou et al., 2020) [41–43]. The small intestine is the site of nutrient absorption, and it also houses a robust population of microbiota that can be affected by *Eimeria* infection. Sufficient sequencing depth was obtained for the analysis; however, the mucosal scrapings generated fewer sequences compared to the luminal contents. This is not surprising since much less material is obtained during scraping of the mucosa, and this can also reflect that less bacteria occupy the mucosa since these are the populations that adhere to the gut epithelium (Borda-Molina et al., 2018) [13].

Significant differences in alpha diversity (within a population) were detected in the population of the jejunum scrapings and ileum contents. These differences were detected as early as day 3 PI and as late as day 10 PI. In samples from infected birds, the alpha diversity was lower compared to controls, and by day 14 PI, the populations were once again similar. By day 14 PI, the birds are in the recovery phase from coccidiosis, even though oocysts can still be shed in the feces by day 11 PI (Cha et al., 2018) [44]. It has previously been reported that coccidiosis can decrease different metrics of alpha diversity and results in a decrease in low-abundance species in the ceca (Stanley et al., 2014; Wu et al., 2014) [45,46]. However, in a study where low doses of mixed *Eimeria* species were used for infection, no significant changes in alpha diversity were observed (Leung et al., 2019) [47]. In a previous study, investigating the supplementation of feed with short-chain fatty acid, we used the same strain of *E. maxima* (APU1) and dose (1×10^3 oocysts) and found that the alpha diversity of the ileum contents was the most affected (Proszkowiec-Weglarz et al., 2020) [48], which was corroborated in the current study.

Beta diversity measures the differences between microbial communities, and in the current study differences were observed in both the ileum and jejunum, primarily at the height of infection, when the effect of parasite invasion and development/replication is the highest. The effects of *E. maxima* APU1 strain infection on beta diversity in the ileum and ceca (luminal and mucosal populations) was previously reported at days 7 and 10 PI, and it was found that interactions among the main effects were observed (age, infection, tributyrin supplementation); however, in the current study, the effect of infection was the most prominent. Even though the same strain of *E. maxima* was used in Ross 308 chickens using the same poultry facilities, differences were present. This highlights the concept that bacterial populations of the gut are highly dynamic and affected by many factors, underlining the necessity for independent reproduction of experiments to truly define these populations.

The taxonomic analysis showed that most of the taxa present in the jejunum (both L and M) are made up of *Lactobacillus*. Previously, this was also observed in birds of the same age and breed but infected with *E. acervulina* (Campos et al., 2023) [9] and that *Lactobacillus* dominated both the lumen of the jejunum as well as the duodenum; however, in the mucosa, *Lactobacillus* was not the major component. It is possible that sampling methodology affected the current results; however, further investigation of the make-up of the jejunal microbiota is warranted. While the taxa of the IL-L were dominated by *Lactobacillus*, in the IL-M, a changeover from majority *Candidatus Arthromitus* to *Lactobacillus* was observed over the course of the study. This changeover was more rapid in IF animals, and the presence of both these taxa have been associated with the growth performance of chickens (Markova et al., 2024) [49].

LEfSe analysis was conducted to determine the differential abundance of bacteria between C and IF samples. Surprisingly, only a total of members of six genera and families were differentially abundant in IF samples. *Lactobacillus,* which is a beneficial probiotic genus, was more differentially abundant in the mucosal population of the IL. We have previously found that *Lactobacillus* was more differentially abundant in the ceca (both mucosal and luminal populations) and ileum (mucosa) of birds infected with *E. acervulina*, *E. tenella*, and necrotic enteritis (infected with *E. maxima* and *Clostridium perfringens*) (Campos et al., 2022; Campos et al., 2024; Latorre et al., 2018; Miska et al., 2023) [10,12,50,51]. An in vitro infection model (Tierney et al., 2004) [52] showed that indigenous *Lactobacillus* species inhibited the invasion of *E. tenella* sporozoites, and therefore it is perplexing that species that may be inhibitory to *Eimeria* invasion are more abundant in infected birds. Markova et al. (2024) [49] recently noted that the presence of many genera of lactic-acid-producing bacteria are often contradictory as to expectation, and in many cases, *Lactobacillus* is associated with poor performance. The members *Staphylococcus* and *Streptococcus* can be opportunistic pathogens of the intestinal tract, and these were also more abundant in IF animals; however, *Streptococcus thermophilus* has been used in probiotic supplements, and therefore these bacteria can have different effects on intestinal health depending on the species present (Mirsalami and Mirsalami, 2023) [53].

Unclassified members of the family *Lachnospiraceae* and [*Ruminococcus*] *torques group*, another member of *Lachnospiraceae*, were more abundant in the ileum (C and M) of C birds. These bacteria are known to be important producers of short-chain fatty acids like butyrate (Pryde et al., 2002) [54] that are beneficial to chickens in having anti-inflammatory properties and maintaining the integrity of the gut (Onrust et al., 2015) [55]. It is therefore not surprising that bacteria involved in short-chain fatty acid production would be more abundant in non-infected birds.

The family *Micrococcaceae* was found to be more abundant in the J of control birds. These bacteria are Gram-positive cocci that are frequently found in milk and cheeses and could be associated with fermentative processes (Basbas et al., 2022) [56] but are also associated with epithelial and mucosal tissues in humans (skin, small intestine, respiratory tract) (Igartua et al., 2017) [57].

Candidatus Arthromitus are spore-forming, Gram-positive bacteria of the phylum Firmicutes, also known as segmented filamentous bacteria, and they were found to be more abundant in the IL (C and M) of control healthy chickens. This taxon has shown a pattern of decreasing in microbiota of *Eimeria*-infected chickens across multiple studies, such as in the cecal mucosa, ileal lumen, and duodenal lumen of *E. acervulina*-infected chickens (Campos et al., 2023; Campos et al., 2024) [9,10], the cecal lumen of *E. tenella*-infected chickens (Campos et al., 2022) [12], and the ileal digesta of chickens infected with a mixture of *E. maxima* and *E. brunetti* (Feng et al., 2022) [58]. The presence of this genus is often associated with better performance in chickens (Markova et al., 2024) [49], although it has also been associated with chickens with "runting and stunting" syndrome (Lages de Silva at al., 2024) [59]. Its function is unknown at this point. Abundance of bacterial taxa in the intestine of poultry can be affected by other parameters besides infection, such as age, environmental conditions, feed and is therefore highly variable; however, additional studies should be able to distill the true effect of coccidia infection on the intestinal microbiota.

To connect the relative abundance of bacteria to metabolic processes they may be involved in, we carried out a predictive analysis. Although the accuracy of these analyses could be limited since the database is primarily composed of metabolic processes that occur in mammalian hosts (Sun et al., 2020) [60], the information gathered could be useful when planning future physiological and biochemical studies on the effect of microbiota on gut health. There was some but not a lot of overlap between the metabolic processes and sections of the small intestine (J or IL) and their C or M. It was previously found that metabolic pathways that could be involved in short-chain fatty acid production in the duodenum, jejunum, and ceca were reduced in birds infected with *E. acervulina* (Campos et al., 2024) [10]. This was not found to be the case in this study in which birds were infected with *E. maxima*, which are related but cause somewhat different symptomology. In the J-C of IF birds, metabolic pathways related to sugar and amino acid degradation were more abundant, and in the IL-M, the predominant pathways in IF birds were related to nucleotide biosynthesis. At this point, the significance of these findings is unknown but should be investigated in future physiological studies on the effects of *Eimeria* infection on the intestinal tract.

5. Conclusions

In conclusion, a symptomatic *E. maxima* infection was induced that lowered performance parameters such as BWG and FCR, as well as the morphology of the small intestine. The infection resulted in significantly decreased levels of carotenoids in the plasma, indicating loss of gut integrity. The effect of the infection was followed for 14 days PI, and its effect on the microbiota in the small intestine (J and IL) was analyzed. The infection affected alpha and beta diversity, particularly at the height of infection. The analysis of relative abundance of bacterial taxa suggested that butyrate-producing bacteria may be greater in C animals. *Lactobacillus* species were more abundant in IF animals, which is a paradoxical finding. It was predicted that pathways involved in sugar and amino acid degradation and nucleotide synthesis were more abundant in the microbiota of IF birds. It was also clear that the luminal and mucosal populations of bacteria differed, and therefore including a separate analysis of gut contents and scrapings produces a more complete picture of the ecology of the gut microbiota.

Author Contributions: Conceptualization, K.B.M.; methodology, K.B.M., P.M.C., S.E.C., M.C.J., M.P.-W.; software, P.M.C.; validation, P.M.C., S.E.C.; formal analysis, K.B.M., P.M.C., S.E.C.; resources, K.B.M., M.P.-W.; data curation, P.M.C.; writing—original draft preparation, K.B.M.; writing—review and editing, K.B.M., P.M.C., S.E.C., M.C.J., M.P.-W.; visualization, K.B.M., S.E.C., P.M.C.; supervision, K.B.M., M.P.-W.; project administration, M.P.-W.; funding acquisition, K.B.M., M.P.-W. All authors have read and agreed to the published version of the manuscript.

Funding: This work was supported by an internal CRIS project number 8042-31000-114-00D funded by USDA-ARS.

Institutional Review Board Statement: All animal care procedures were approved by the Institutional Animal Care and Use Committee (IACUC, protocol #18-025) of the Beltsville Agricultural Research Center (BARC).

Informed Consent Statement: Not applicable.

Data Availability Statement: The 16S rRNA gene sequences determined in this study were deposited in the NCBI Sequence Read Archive (SRA) database (SRA accession # PRJNA1122195).

Acknowledgments: The authors would like to acknowledge the assistance of veterinary and animal services at BARC.

Conflicts of Interest: The authors declare no conflicts of interest.

References

1. Mesa-Pineda, C.; Navarro-Ruíz, J.L.; López-Osorio, S.; Chaparro-Gutiérrez, J.J.; Gómez-Osorio, L.M. Chicken coccidiosis: From the parasite lifecycle to control of the disease. *Front. Vet. Sci.* **2021**, *8*, 787653. [CrossRef] [PubMed]
2. Blake, D.P.; Vrba, V.; Xia, D.; Jatau, I.D.; Spiro, S.; Nolan, M.J.; Underwood, G.; Tomley, F.M. Genetic and biological characterisation of three cryptic Eimeria operational taxonomic units that infect chickens (*Gallus gallus domesticus*). *Int. J. Parasitol.* **2021**, *51*, 621–634. [CrossRef] [PubMed]
3. Ahmad, R.; Yu, Y.H.; Hua, K.F.; Chen, W.J.; Zaborski, D.; Dybus, A.; Hsiao, F.S.; Cheng, Y.H. Management and control of coccidiosis in poultry—A review. *Anim. Biosci.* **2024**, *37*, 1–15. [CrossRef] [PubMed]
4. Noack, S.; Chapman, H.D.; Selzer, P.M. Anticoccidial drugs of the livestock industry. *Parasitol. Res.* **2019**, *118*, 2009–2026. [CrossRef]
5. Martins, R.R.; Silva, L.J.G.; Pereira, A.M.P.T.; Esteves, A.; Duarte, S.C.; Pena, A. Coccidiostats and poultry: A comprehensive review and current legislation. *Foods* **2022**, *11*, 2738. [CrossRef]
6. Cloft, S.E.; Miska, K.B.; Jenkins, M.; Proszkowiec-Weglarz, M.; Kahl, S.; Wong, E.A. Temporal changes of genes associated with intestinal homeostasis in broiler chickens following a single infection with *Eimeria acervulina*. *Poult. Sci.* **2023**, *102*, 102537. [CrossRef]
7. Elsasser, T.H.; Miska, K.; Kahl, S.; Fetterer, R.H.; Martínez Ramirez, A. Temporal pattern changes in duodenal protein tyrosine nitration events in response to *Eimeria acervulina* infection in chickens. *J. Anim. Sci.* **2018**, *96*, 2125–2138. [CrossRef]
8. Rothwell, L.; Gramzinski, R.A.; Rose, M.E.; Kaiser, P. Avian coccidiosis: Changes in intestinal lymphocyte populations associated with the development of immunity to *Eimeria maxima*. *Parasite Immunol.* **1995**, *17*, 525–533. [CrossRef]

9. Campos, P.M.; Miska, K.B.; Jenkins, M.C.; Yan, X.; Proszkowiec-Weglarz, M. Effects of *Eimeria acervulina* infection on the luminal and mucosal microbiota of the duodenum and jejunum in broiler chickens. *Front. Microbiol.* **2023**, *14*, 1147579. [CrossRef]
10. Campos, P.M.; Miska, K.B.; Jenkins, M.C.; Yan, X.; Proszkowiec-Weglarz, M. Effects of Eimeria acervulina infection on the luminal and mucosal microbiota of the cecum and ileum in broiler chickens. *Sci. Rep.* **2024**, *14*, 10702. [CrossRef]
11. Bolyen, E.; Rideout, J.R.; Dillon, M.R.; Bokulich, N.A.; Abnet, C.C.; Al-Ghalith, G.A.; Alexander, H.; Alm, E.J.; Arumugam, M.; Asnicar, F.; et al. Reproducible, interactive, scalable and extensible microbiome data science using QIIME 2. *Nat. Biotechnol.* **2019**, *37*, 852–857. [CrossRef] [PubMed]
12. Campos, P.M.; Darwish, N.; Shao, J.; Proszkowiec-Weglarz, M. Research Note: Choice of microbiota database affects data analysis and interpretation in chicken cecal microbiota. *Poult. Sci.* **2022**, *101*, 101971. [CrossRef] [PubMed]
13. Borda-Molina, D.; Seifert, J.; Camarinha-Silva, A. Current Perspectives of the Chicken Gastrointestinal Tract and Its Microbiome. *Comput. Struct. Biotechnol. J.* **2018**, *16*, 131–139. [CrossRef] [PubMed]
14. Allen, P.C. Physiological responses of chicken gut tissue to coccidial infection: Comparative effects of *Eimeria acervulina* and *Eimeria mitis* on mucosal mass, carotenoid content, and brush border enzyme activity. *Poult. Sci.* **1987**, *66*, 1306–1315. [CrossRef]
15. Darwish, N.; Shao, J.; Schreier, L.L.; Proszkowiec-Weglarz, M. Choice of 16S ribosomal RNA primers affects the microbiome analysis in chicken ceca. *Sci. Rep.* **2021**, *11*, 11848. [CrossRef]
16. Callahan, B.J.; McMurdie, P.J.; Rosen, M.J.; Han, A.W.; Johnson, A.J.; Holmes, S.P. DADA2: High-resolution sample inference from Illumina amplicon data. *Nat. Methods* **2016**, *13*, 581–583. [CrossRef]
17. Katoh, K.; Misawa, K.; Kuma, K.; Miyata, T. MAFFT: A novel method for rapid multiple sequence alignment based on fast Fourier transform. *Nucleic Acids Res.* **2002**, *30*, 3059–3066. [CrossRef]
18. Price, M.N.; Dehal, P.S.; Arkin, A.P. FastTree 2—Approximately maximum-likelihood trees for large alignments. *PLoS ONE* **2010**, *5*, e9490. [CrossRef]
19. Quast, C.; Pruesse, E.; Yilmaz, P.; Gerken, J.; Schweer, T.; Yarza, P.; Peplies, J.; Glöckner, F.O. The SILVA ribosomal RNA gene database project: Improved data processing and web-based tools. *Nucleic Acids Res.* **2012**, *41*, D590–D596. [CrossRef]
20. Balvočiūtė, M.; Huson, D.H. SILVA, RDP, Greengenes, NCBI and OTT—How do these taxonomies compare? *BMC Genom.* **2017**, *18* (Suppl. 2), 114. [CrossRef]
21. Robeson, M.S., 2nd; O'Rourke, D.R.; Kaehler, B.D.; Ziemski, M.; Dillon, M.R.; Foster, J.T.; Bokulich, N.A. RESCRIPt: Reproducible sequence taxonomy reference database management. *PLoS Comput. Biol.* **2021**, *17*, e1009581. [CrossRef] [PubMed]
22. Lozupone, C.; Knight, R. UniFrac: A new phylogenetic method for comparing microbial communities. *Appl. Environ. Microbiol.* **2005**, *71*, 8228–8235. [CrossRef] [PubMed]
23. Lozupone, C.A.; Hamady, M.; Kelley, S.T.; Knight, R. Quantitative and qualitative beta diversity measures lead to different insights into factors that structure microbial communities. *Appl. Environ. Microbiol.* **2007**, *73*, 1576–1585. [CrossRef] [PubMed]
24. Segata, N.; Izard, J.; Waldron, L.; Gevers, D.; Miropolsky, L.; Garrett, W.S.; Huttenhower, C. Metagenomic biomarker discovery and explanation. *Genome Biol.* **2011**, *12*, R60. [CrossRef]
25. Douglas, G.M.; Maffei, V.J.; Zaneveld, J.R.; Yurgel, S.N.; Brown, J.R.; Taylor, C.M.; Huttenhower, C.; Langille, M.G.I. PICRUSt2 for prediction of metagenome functions. *Nat. Biotechnol.* **2020**, *38*, 685–688. [CrossRef]
26. Caspi, R.; Billington, R.; Keseler, I.M.; Kothari, A.; Krummenacker, M.; Midford, P.E.; Ong, W.K.; Paley, S.; Subhraveti, P.; Karp, P.D. The MetaCyc database of metabolic pathways and enzymes—A 2019 update. *Nucleic Acids Res.* **2020**, *48*, D445–D453. [CrossRef]
27. Parks, D.H.; Tyson, G.W.; Hugenholtz, P.; Beiko, R.G. STAMP: Statistical analysis of taxonomic and functional profiles. *Bioinformatics* **2014**, *30*, 3123–3124. [CrossRef]
28. R Core Team. *R: A Language and Environment for Statistical Computing, Version 4.3.2*; R Foundation for Statistical Computing: Vienna, Austria, 2023. Available online: https://www.R-project.org/ (accessed on 4 March 2024).
29. Bisanz, J. *qiime2R: Importing QIIME2 Artifacts and Associated Data into R Sessions, R Package Version 0.99*; R Foundation for Statistical Computing: Vienna, Austria, 2018; Volume 13.
30. Wickham, H.; Averick, M.; Bryan, J.; Chang, W.; McGowan, L.D.A.; François, R.; Grolemund, G.; Hayes, A.; Henry, L.; Hester, J.; et al. Welcome to the Tidyverse. *J. Open Source Softw.* **2020**, *4*, 1686. [CrossRef]
31. Wickham, H.; François, R.; Henry, L.; Müller, K.; Vaughan, D. *dplyr: A Grammar of Data Manipulation, R Package Version 1.1.0*; R Foundation for Statistical Computing: Vienna, Austria, 2023. Available online: https://dplyr.tidyverse.org (accessed on 4 March 2024).
32. Wickham, H. *ggplot2: Elegant Graphics for Data Analysis*; Springer: New York, NY, USA, 2016; ISBN 978-3-319-24277-4. Available online: https://ggplot2.tidyverse.org (accessed on 4 March 2024).
33. McMurdie, P.J.; Holmes, S. phyloseq: An R package for reproducible interactive analysis and graphics of microbiome census data. *PLoS ONE* **2013**, *8*, e61217. [CrossRef]
34. Barnett, D.J.M.; Arts, I.C.W.; Penders, J. microViz: An R package for microbiome data visualization and statistics. *J. Open Source Softw.* **2021**, *6*, 3201. [CrossRef]
35. Hansen, V.L.; Kahl, S.; Proszkowiec-Weglarz, M.; Jiménez, S.C.; Vaessen, S.F.C.; Schreier, L.L.; Jenkins, M.C.; Russell, B.; Miska, K.B. The effects of tributyrin supplementation on weight gain and intestinal gene expression in broiler chickens during *Eimeria maxima*-induced coccidiosis. *Poult. Sci.* **2021**, *100*, 100984. [CrossRef] [PubMed]

36. Jenkins, M.C.; Dubey, J.P.; Miska, K.; Fetterer, R. Differences in fecundity of *Eimeria maxima* strains exhibiting different levels of pathogenicity in its avian host. *Vet. Parasitol.* **2017**, *236*, 1–6. [CrossRef] [PubMed]
37. Conway, D.P.; Sasai, K.; Gaafar, S.M.; Smothers, C.D. Effects of different levels of oocyst inocula of *Eimeria acervulina*, *E. tenella*, and *E. maxima* on plasma constituents, packed cell volume, lesion scores, and performance in chickens. *Avian Dis.* **1993**, *37*, 118–123. [CrossRef] [PubMed]
38. Fetterer, R.H.; Barfield, R.C.; Jenkins, M.C. Protection of broiler chicks housed with immunized cohorts against infection with *Eimeria maxima* and *E. acervulina*. *Avian Dis.* **2015**, *59*, 98–105. [CrossRef]
39. Sakkas, P.; Oikeh, I.; Blake, D.P.; Nolan, M.J.; Bailey, R.A.; Oxley, A.; Rychlik, I.; Lietz, G.; Kyriazakis, I. Does selection for growth rate in broilers affect their resistance and tolerance to *Eimeria maxima*? *Vet. Parasitol.* **2018**, *258*, 88–98. [CrossRef]
40. Fernando, M.A.; McCraw, B.M. Mucosal morphology and cellular renewal in the intestine of chickens following a single infection of *Eimeria acervulina*. *J. Parasitol.* **1973**, *59*, 493–501. [CrossRef]
41. Choi, J.; Tompkins, Y.H.; Teng, P.-Y.; Gogal, R.M., Jr.; Kim, W.K. Effects of Tannic Acid Supplementation on Growth Performance, Oocyst Shedding, and Gut Health of in Broilers Infected with *Eimeria maxima*. *Animals* **2022**, *12*, 1378. [CrossRef]
42. Choi, J.; Kim, W. Interactions of Microbiota and Mucosal Immunity in the Ceca of Broiler Chickens Infected with *Eimeria tenella*. *Vaccines* **2022**, *10*, 1941. [CrossRef]
43. Zhou, B.H.; Jia, L.S.; Wei, S.S.; Ding, H.Y.; Yang, J.Y.; Wang, H.W. Effects of *Eimeria tenella* infection on the barrier damage and microbiota diversity of chicken cecum. *Poult. Sci.* **2020**, *99*, 1297–1305. [CrossRef]
44. Cha, J.O.; Zhao, J.; Yang, M.S.; Kim, W.I.; Cho, H.S.; Lim, C.W.; Kim, B. Oocyst-Shedding Patterns of Three Eimeria Species in Chickens and Shedding Pattern Variation Depending on the Storage Period of Eimeria tenella Oocysts. *J. Parasitol.* **2018**, *104*, 18–22. [CrossRef]
45. Stanley, D.; Wu, S.B.; Rodgers, N.; Swick, R.A.; Moore, R.J. Differential responses of cecal microbiota to fishmeal, *Eimeria* and *Clostridium perfringens* in a necrotic enteritis challenge model in chickens. *PLoS ONE* **2014**, *9*, e104739. [CrossRef]
46. Wu, S.B.; Stanley, D.; Rodgers, N.; Swick, R.A.; Moore, R.J. Two necrotic enteritis predisposing factors, dietary fishmeal and *Eimeria* infection, induce large changes in the caecal microbiota of broiler chickens. *Vet. Microbiol.* **2014**, *169*, 188–197. [CrossRef] [PubMed]
47. Leung, H.; Yitbarek, A.; Snyder, R.; Patterson, R.; Barta, J.R.; Karrow, N.; Kiarie, E. Responses of broiler chickens to I challenge when fed a nucleotide-rich yeast extract. *Poult. Sci.* **2019**, *98*, 1622–1633. [CrossRef] [PubMed]
48. Proszkowiec-Weglarz, M.; Miska, K.B.; Schreier, L.L.; Grim, C.J.; Jarvis, K.G.; Shao, J.; Vaessen, S.; Sygall, R.; Jenkins, M.C.; Kahl, S.; et al. Research Note: Effect of butyric acid glycerol esters on ileal and cecal mucosal and luminal microbiota in chickens challenged with *Eimeria maxima*. *Poult. Sci.* **2020**, *99*, 5143–5148. [CrossRef] [PubMed]
49. Marková, K.; Kreisinger, J.; Vinkler, M. Are there consistent effects of gut microbiota composition on performance, productivity and condition in poultry? *Poult. Sci.* **2024**, *103*, 103752. [CrossRef]
50. Latorre, J.D.; Adhikari, B.; Park, S.H.; Teague, K.D.; Graham, L.E.; Mahaffey, B.D.; Baxter, M.F.A.; Hernandez-Velasco, X.; Kwon, Y.M.; Ricke, S.C.; et al. Evaluation of the epithelial barrier function and ileal microbiome in an established necrotic enteritis challenge model in broiler chickens. *Front. Vet. Sci.* **2018**, *5*, 199. [CrossRef]
51. Miska, K.B.; Proszkowiec-Weglarz, M.; Buiatte, V.; Mahmoud, M.; Lesko, T.; Jenkins, M.C.; Chopra, S.; Lorenzoni, G. Comparative analysis of microbiota in the ceca of broiler chickens with necrotic enteritis fed a commercial corn diet or with corn high in flavonoids (PennHFD1). *Front. Microbiomes* **2023**, *2*, 1212130. [CrossRef]
52. Tierney, J.; Gowing, H.; Van Sinderen, D.; Flynn, S.; Stanley, L.; McHardy, N.; Hallahan, S.; Mulcahy, G. In vitro inhibition of *Eimeria tenella* invasion by indigenous chicken *Lactobacillus* species. *Vet. Parasitol.* **2004**, *122*, 171–182. [CrossRef]
53. Mirsalami, S.M.; Mirsalami, M. Effects of duo-strain probiotics on growth, digestion, and gut health in broiler chickens. *Vet. Anim. Sci.* **2023**, *24*, 100343. [CrossRef]
54. Pryde, S.E.; Duncan, S.H.; Hold, G.L.; Stewart, C.S.; Flint, H.J. The microbiology of butyrate formation in the human colon. *FEMS Microbiol. Lett.* **2002**, *217*, 133–139. [CrossRef]
55. Onrust, L.; Ducatelle, R.; Van Driessche, K.; De Maesschalck, C.; Vermeulen, K.; Haesebrouck, F.; Eeckhaut, V.; Van Immerseel, F. Steering Endogenous Butyrate Production in the Intestinal Tract of Broilers as a Tool to Improve Gut Health. *Front. Vet. Sci.* **2015**, *2*, 75. [CrossRef] [PubMed]
56. Basbas, C.; Aly, S.; Okello, E.; Karle, B.M.; Lehenbauer, T.; Williams, D.; Ganda, E.; Wiedmann, M.; Pereira, R.V. Effect of Intramammary Dry Cow Antimicrobial Treatment on Fresh Cow's Milk Microbiota in California Commercial Dairies. *Antibiotics* **2022**, *11*, 963. [CrossRef] [PubMed]
57. Igartua, C.; Davenport, E.R.; Gilad, Y.; Nicolae, D.L.; Pinto, J.; Ober, C. Host genetic variation in mucosal immunity pathways influences the upper airway microbiome. *Microbiome* **2017**, *5*, 16. [CrossRef]
58. Feng, X.; Li, T.; Zhu, H.; Liu, L.; Bi, S.; Chen, X.; Zhang, H. Effects of challenge with *Clostridium perfringens*, *Eimeria* and both on ileal microbiota of yellow feather broilers. *Front. Microbiol.* **2022**, *13*, 1063578. [CrossRef]

59. Lages da Silva, D.H.; Marques da Silva, R.L.; Rios, D.L.; de Souza, D.D.G.; Aburjaile, F.; de Freitas Neto, O.C.; Camargos Lara, L.J.; Dias Araújo, M.; Ecco, R. Intestinal microbiota diversity from broilers with runting and stunting syndrome performed by metagenomics. *Avian Pathol.* **2024**, *53*, 408–418. [CrossRef]
60. Sun, S.; Jones, R.B.; Fodor, A.A. Inference-based accuracy of metagenome prediction tools varies across sample types and functional categories. *Microbiome* **2020**, *8*, 46. [CrossRef]

animals MDPI

Article

The Immunoprotective Effect of ROP27 Protein of *Eimeria tenella*

Menggang Li [1], Xiaoling Lv [2], Mingxue Zheng [2] and Yingyi Wei [1,3,*]

1 Guangxi Key Laboratory of Animal Breeding, Disease Control and Prevention, College of Animal Science and Technology, Nanning 530004, China; limenggang2023@163.com
2 College of Veterinary Medicine, Shanxi Agricultural University, Jinzhong 030801, China; lvxiaoling19@163.com (X.L.); zhengmingxue288@163.com (M.Z.)
3 Guangxi Zhuang Autonomous Region Engineering Research Center of Veterinary Biologics, Nanning 530004, China
* Correspondence: weiyingyi@gxu.edu.cn; Tel.: +86-13978862113

Simple Summary: *Eimeria tenella* is the most pathogenic and common coccidia that causes chicken coccidiosis. At present, the control of this disease mainly relies on anticoccidial drugs, but the use of these drugs is greatly limited due to issues such as drug-resistant strains and drug residues. This study aims to investigate the immunoprotective effect of recombinant *Et*ROP27 virulence protein, which can provide theoretical support for the development of chicken coccidiosis vaccines. This study successfully constructed a prokaryotic expression vector of *Et*ROP27 and purified the recombinant protein, identified the expression pattern of the protein, and tested its immune protection. *Et*ROP27 has been shown to have a certain immunoprotective effect.

Abstract: *Eimeria tenella* rhoptry protein has the properties of a protective antigen. *Et*ROP27 is a pathogenic gene that is detected via a transcriptome, but its expression pattern, immunogenicity, and potency are unknown. Therefore, a gene segment of *Et*ROP27 was amplified and transplanted into the pET28a prokaryotic vector for the expression of the recombinant protein, and it subsequently purified for the generation of a polyclonal antibody. Then, RT-PCR and Western blotting were performed to understand the expression pattern of *Et*ROP27. Subsequently, animal experiments were conducted to evaluate the immunoprotective effect of the recombinant protein with different immunizing doses (50, 100, and 150 μg). The results showed that the expression of *Et*ROP27 gradually increased with the prolongation of infection time, reaching the highest level at 96 h and then decreasing. Additionally, *Et*ROP27 is a natural antigen of coccidia that can stimulate the body to produce high levels of IgY. As with recombinant protein vaccines, the results of immune protection evaluation tests showed that the average weight gain rates of the immune challenge groups were significantly higher than that of the challenged control group, and their average lesion scores were significantly lower than that of the challenged control group. Furthermore, the oocyst excretion decreased by 81.25%, 86.21%, and 80.01%, and the anticoccidial index was 159.45, 171.47, and 166.75, respectively, for these groups. *Et*ROP27 is a promising antigen gene candidate for the development of a coccidiosis vaccine.

Keywords: *Eimeria tenella*; recombinant *Et*ROP27 protein; expression changes; immunoprotective

Citation: Li, M.; Lv, X.; Zheng, M.; Wei, Y. The Immunoprotective Effect of ROP27 Protein of *Eimeria tenella*. *Animals* **2023**, *13*, 3500. https://doi.org/10.3390/ani13223500

Academic Editor: Kenneth Bafundo

Received: 12 October 2023
Revised: 4 November 2023
Accepted: 8 November 2023
Published: 13 November 2023

1. Introduction

Chicken coccidiosis is a parasitic protozoonosis caused by one or more of the seven species of *Eimeria* that are parasitic within the epithelial cells of the chicken intestinal tract [1,2]. The economic losses caused by chicken coccidiosis infection worldwide amount to GBP 10.4 billion annually [3]. China spends hundreds of millions of CNY annually on medicines to prevent and control coccidiosis, accounting for one-third of the cost of chicken disease prevention and control, and it is the second largest chicken production and consumption country in the world [4]. Coccidiosis is characterized by intestinal injury, diarrhea, or bloody stools [5]. *Eimeria tenella* is the most virulent, widely distributed, and harmful of the seven chicken coccidia species, mainly attacking the cecum [6]. The highest incidence rate was found in 3~6-week-old chicks. The symptoms of excreting bloody stool or even blood only appear after 4~5 days of infection. Large numbers of deaths begin one to two days after the appearance of bloody stools, with a mortality rate of up to 80% in severe cases [7–9].

Drug control has been the main method for chicken coccidiosis for a long time. However, many problems such as the emergence of drug-resistant strains, drug residues, and increased treatment costs during the prevention and treatment process have greatly limited drug control [3]. Therefore, the key strategy and methods for effective prevention and control of this disease is to develop a safe, effective, low-toxicity, and environmentally friendly vaccine against coccidiosis in chickens. Currently, live coccidian oocyst vaccines, with virulent and precocious strains, are widely used. However, traditional coccidiosis vaccines have shortcomings that limit their wide application such as high production costs, cumbersome production processes, environmental pollution caused by the spread of pathogens to the outside world, and the improper use of wild virus live worm vaccines that can cause outbreaks of coccidiosis [10]. Recent studies have demonstrated that recombinant protein vaccines can be used as effective control measures against coccidiosis [11].

Rhoptry proteins (ROPs) are conserved and immunoprotective in nature, and some members of this protein family are among the main virulence factors of *Toxoplasma gondii* [12]. Therefore, this protein family is considered as a potential target in the development of anti-*toxoplasmosis* vaccines. As one of the key virulence factors of the *T. gondii* type I strain, ROP18 participates in multiple mechanisms to escape host immune response. ROP5 controls the activity of ROP18 and affects IFN-γ and other effectors dependent on IRG protein, Irgm3, to regulate the acute toxicity response of *T. gondii* [13]. Similarly, ROPs play an important role in the process of *E. tenella* invading host cells. They exhibit the property of a protective antigen, and their corresponding antibodies can inhibit the infection of the host by coccidia [14]. Research has proven that *Et*ROP17 can inhibit host cell apoptosis [15]. In ROPs such as ROP14, ROP30, ROP52, and ROP53, *E. tenella* was detected in the proteome at different developmental stages; they have some degree of homology with the ROPs of *T. gondii* [15], but their functions are unclear. Zhang Li used RNA-Seq to detect and compare the transcriptome of the precocious strain and the virulent strain at the mitotic stage. qPCR was used to verify and screen the pseudopathogenic gene ROP27 of *E. tenella*, which is presumed to have a good immunogenicity [16,17]. However, it is not yet known whether the recombinant protein can effectively prevent the *E. tenella* disease.

Therefore, this study aims to investigate whether chicken *Et*ROP27 can reduce the effects of *E. tenella* infection by expressing the recombinant protein ROP27 and detecting its immunoprotective effect. Additionally, theses results may influence the development of an effective coccidiosis vaccine that can be used on a commercial scale.

2. Materials and Methods

2.1. Ethics Statement

All experiments involving animals were carried out in accordance with national regulations for the protection of animal welfare, and these guidelines were strictly followed. Additionally, this study was approved by the Animal Experimental Ethical Inspection Form of Guangxi University, China (Approval Code: Gxu-2023-0197; Approval Date: 16 February 2023). This study did not involve any human research.

2.2. Experimental Animals and Parasite

Firstly, 15-day-old SPF chicken embryos were obtained from Beijing Meri Avigon Laboratory Animal Technology Co., Ltd. (Beijing, China). Secondly, the *E. tenella* Shanxi virulent and precocious strains were used in this experiment, which were donated by the Veterinary Pathology Laboratory of the College of Veterinary Medicine, Shanxi Agricultural University.

2.3. Reagents

PrimeSTAR® (Ōsaka shi, Japan) Max DNA Polymerase, DL2000 DNA Marker, DL5000 DNA Marker, pMD™ (Ōsaka shi, Japan) 18-T Vector Cloning Kit, *E.coli* BL21(DE3) Competent Cells, *E.coli* DH5α Competent Cells, QuickCut™ (Ōsaka shi, Japan) EcoR I, QuickCut™ Xho I, T4 DNA Ligase, TritonX-100, Tris, Glycine, SDS, Acrylamide, and SYBR Green qPCR Kit were purchased from TaKaRa (Ōsaka shi, Japan); pET-28a (+) DNA was purchased from Sangon Biotech (Shanghai, China); isopropyl β-D-thiogalactoside (IPTG), LB broth powder, LB agar powder, Ampicillin, and Kanamycin were purchased from Solarbio (Beijing, China); 6*His, His-Tag Monoclonal antibody, HRP-conjugated Affinipure Goat Anti-Rabbit IgG(H+L), HRP-conjugated Affinipure Goat Anti-Mouse IgG(H+L), Fluorescein (FITC)-conjugated Affinipure Goat Anti-Rabbit IgG(H+L), and ECL detection kit were purchased from Proteintech (Chicago, IL, USA).

2.4. Plasmid Construction

Specific primers were designed to measure the ROP27 sequence (Cluster-10347.5943) based on the transcriptome, and they were sent to a company for synthesis. Primer sequences were as follows: ROP27 sense primer, 5′-GAATTCAGGTAACGAGTCTCTGC-3′; and anti-sense primer, 5′-TTGCCAGAATTGGCTCTACTACG-3′. The PCR product was purified and cloned into the pMD18-T vector, according to the operation steps of the gel extraction kit. A single bacterial colony was selected for amplification, culture, and sequencing, and the bacterial solution with the most correct sequencing was selected for plasmid extraction. The plasmid and pET-28a (+) DNA were digested using EcoR I and Xho I endonucleases, and the digested products were purified. The target fragment was cloned into the eukaryotic expression vector, and then, a single colony was selected for amplification, culture, and sequencing. The plasmid with the correct sequencing was named pET-*Et*ROP27.

2.5. Protein Expression and Purification

The recombinant plasmid pET-*Et*ROP27 was transformed into BL21 (DE3) Competitive Cells. Subsequently, positive colonies were selected for enrichment. IPTG with a final concentration of 1 mmol/L was added to induce expression for 8 h when the OD_{600} value of the bacterial solution was about 0.5. The heavy suspension liquid was placed on ice for ultrasonic crushing treatment, and then, the sediment and supernatant were collected. We followed the method of protein purification as described by Barkhordari F. Protein purity and concentration were determined using 12% (w/v) sodium dodecyl sulfate polyacrylamide gel electrophoresis (SDS-PAGE) and the BCA protein determination kit. The purified protein was stored at −80 °C and used for subsequent experiments.

2.6. Preparation of Anti-E. tenella and Anti-rEtROP27 Positive Serum

In this experiment, 1×10^4 sporulated *E. tenella* oocysts were intragastrically administered to 2-week-old SPF chickens. After 3 days, each chicken was infected with 5000 sporulated oocysts every 3 days for a total of four times. Blood was collected with cardiac puncture and centrifuged to prepare anti-*E. tenella* positive serum; this was then stored at −80 °C for standby.

Three 2-month-old New Zealand white rabbits were immunized with r*Et*ROP27, which was emulsified with Freund's adjuvant at a dose of 200 µg per rabbit three times, once every two weeks. Before each immunization, 30 µL of rabbit blood was taken from the ear vein, and the serum was isolated to detect the antibody titer. Blood was collected from the heart and centrifuged to collect serum after the third immunization. The collected serum was deactivated in a water bath at 56 °C for 30 min and stored at −80 °C for standby.

2.7. Indirect ELISA Determination of Antibody Titer

Rabbit anti-*Et*ROP27 serum is the sample to be tested, and the non-immunized serum is the negative control. The purified *Et*ROP27 recombinant protein was used as the detection antibody. PBST was used to test the serum according to gradient dilutions of 1:500, 1:1000, 1:2000, 1:4000, 1:8000, 1:16,000, and 1:32,000. Serum antibody titer was tested according to the method reported by Dong-chao Zhang [18].

2.8. Western Blotting Analysis

The protein samples were separated with 12% polyacrylamide gels. After separation, the target band was transferred to 0.22 μm polyvinylidene fluoride membranes using the Bio-Rad wet transfer system. After blocking with 5% BSA in PBS at 37 °C for 1 h, the membranes were probed with anti-*E. tenella* (1:400), anti-r*Et*ROP27(1:500), and anti-HIS (1:2000), respectively, with an overnight incubation at 4 °C. Subsequently, the membranes were washed four times with PBS containing tween-20 (PBST) for 10 min each time. Next, the membranes were incubated with horseradish peroxidase-linked secondary anti-chicken, anti-rabbit, or anti-mouse IgG antibodies for 1 h at 37 °C. After washing four times with PBST for 10 min each time, the signal was visualized using an ECL hypersensitivity detection kit.

2.9. Primary Culture of Chicken Embryo Caecal Epithelial Cells

The 15-day-old SPF chicken embryo was placed in a bioclean bench. The cecum was isolated and thoroughly washed with the PBS buffer. Then, it was cut into a tissue of approximately 1 mm^3 size. After washing, the tissue was resuspended and mixed with thermolysin (50 mg/L) and digested under shaking at 41 °C for 2 h; the PBS buffer was used to stop the digestion, and the supernatant was pipetted and centrifuged at 1200 r/min for 5 min. The PBS and enzyme solution were discarded; the cell pellet was resuspended in 10% fetal bovine serum (FBS) low-glycemic DMEM cell culture medium. The supernatant was collected after 70 min of adherent culture. After centrifugation at 1200 r/min for 5 min, it was resuspended and precipitated with DMEM/F12 culture medium containing 2.5% FBS. The cells were inoculated into a cell culture plate for cultivation. This culture can be used for subsequent experiments when the cell adhesion rate reaches 85%.

2.10. *E. tenella* Sporozoite Preparation

An appropriate amount of *E. tenella* sporulation oocysts was removed and centrifuged at 2000 r/min for 5 min, and the supernatant was discarded. The sediments were washed thoroughly with PBS. The oocysts were suspended in 2 mL PBS. This suspension was grinded with a homogenizer until its shelling rate reached 80%. The solution was centrifuged at 1800 r/min for 5 min. An appropriate amount of sporocyst digestion solution containing 0.75% trypsin and 10% chicken bile was added to the precipitate and was shaken for digestion at 41 °C (150 r/min) until 80% of the spores were released. After filtration, the solution was centrifuged at 3000 r/min for 10 min, and then suspended and precipitated in the DMEM culture medium. The amount of sporozoites used in this experiment was 1×10^4.

2.11. RT-PCR Analysis

Total RNA from chicken cecum or primary cecal epithelial cells of chicken embryo cells infected with *E. tenella* precocious strains was extracted using TRIzol Reagent. The cDNA was generated with reverse transcription using PrimeScript reverse transcriptase. ROP27 mRNA was assessed using quantitative RT-PCR with a TaKaRa SYBR Green reagent. RT-PCR Kit was used with a CFX96 Touch real-time PCR detection system. The primer sequences were as follows: ROP27 sense primer, 5′-AGCTACGACACTCCTGTTGC-3′; and anti-sense primer, 5′-ACTCAAGACGGAGTTGCTGG-3′. The PCR product was purified with a gel extraction kit and then cloned into the pMD18-T vector. The extracted plasmid was continuously diluted and used as a standard for quantitative analysis. The initial copy number of ROP27 gene in each group was calculated using the following formula: $X = -K \log Ct + b$, where X is the initial copy number, and K, Ct, and b refer to the slope rate, the cycle threshold, and a constant, respectively.

2.12. Animal Experiment

Firstly, 14-day-old chickens were randomly divided into 5 groups, with 10 chickens in each group. In three experimental groups, legs of the chicken were subcutaneously injected with recombinant proteins emulsified using Freund's complete adjuvant (50 μg, 100 μg, and 150 μg, respectively). After 7 days, the recombinant protein was mixed with Freund's incomplete adjuvant in equal volume and fully emulsified for enhanced immunity. The infection control group (challenged control) and the uninfected control group (unchallenged control) were injected with PBS and Freund's complete adjuvant or Freund's incomplete adjuvant at the same time. After 7 days, each chicken in the experimental group and the infection control group received an oral administration of 5×10^4 sporulation oocysts of *E. tenella*, whereas PBS was administered to the uninfected control group (Table 1).

Table 1. Grouping of experimental animals and immune challenge procedures.

Groups	Immunization	Dose	Challenge
Unchallenged control	PBS + adjuvant	/	/
Challenged control	PBS + adjuvant	/	*E. tenella* sporulated oocysts (5×10^4)
*Et*ROP27 (50 μg)	r*Et*ROP27 protein + adjuvant	50 μg	*E. tenella* sporulated oocysts (5×10^4)
*Et*ROP27 (100 μg)	r*Et*ROP27 protein + adjuvant	100 μg	*E. tenella* sporulated oocysts (5×10^4)
*Et*ROP27 (150 μg)	r*Et*ROP27 protein + adjuvant	150 μg	*E. tenella* sporulated oocysts (5×10^4)

2.13. Concentration of Serum Antibody

The concentration of IgY antibody level in serum were detected using ELISA commercial kits, which were named as "chick cytokine ELISA Quantitation Kits" (catalog number: CSB-E11635Ch, for IgY CUSABIO, Wuhan, China), according to manufacturer's instructions.

2.14. In Vivo Immunoprotective Parameters

In vivo immunoprotective parameters of the r*Et*ROP27 protein include clinical, pathological, and parasitological factors [19]. Clinical factors include weight gain, survival, and mortality rate, which can be directly obtained and calculated. The pathological factor is the cecal injury score, which is a continuous number from 0 (none) to 4 (severe), representing varying degrees of cecal lesions, evaluated by three independent observers, according to Johnson and Reid [20]. Parasitological factors include oocyst output and ACI. The output of oocysts mainly detects the content of cecum per gram of oocysts (OPG), calculated using McMaster counting technique [21]. ACI is a synthetic criterion that can measure the effectiveness of anticoccidial activity. The calculation formula for ACI = (relative weight gain rate + survival rate) × 100 − Mean lesion score × 10 − OPG value). ACI > 180 indicates good protection, 160 < ACI < 179 indicates moderate protection, 120 < ACI < 159 indicates limited protection, and ACI < 120 indicates no protection [22].

2.15. Image and Statistical Analyses

All data were analyzed using SPSS 17.0 statistical software (Chicago, IL, USA) and expressed as arithmetic mean ± standard deviation. Histograms were prepared via GraphPad Prism 5.0 software (San Diego, CA, USA). Each experiment was repeated at least 3 times. The results are considered statistically significant when the p-value is less than 0.05.

3. Results

3.1. Gene Clone and Plasmid Construction

*Et*ROP27 showed a specific band at 1068 bp after PCR amplification. Gene sequence and bioinformatics were analyzed, details in Supplementary Data S1 and S2. The PCR product was then cloned into the pMD-18T vector, and positive colonies were selected for bacterial liquid PCR and sequencing analysis (Figure 1A,B). The target fragment was cut out from the T vector and cloned into the pET-28a vector. Then, the positive colonies were selected with bacterial liquid PCR for double digestion identification and sequencing analysis (Figure 1C,D). The prokaryotic expression vector pET-*Et*ROP27 was constructed successfully.

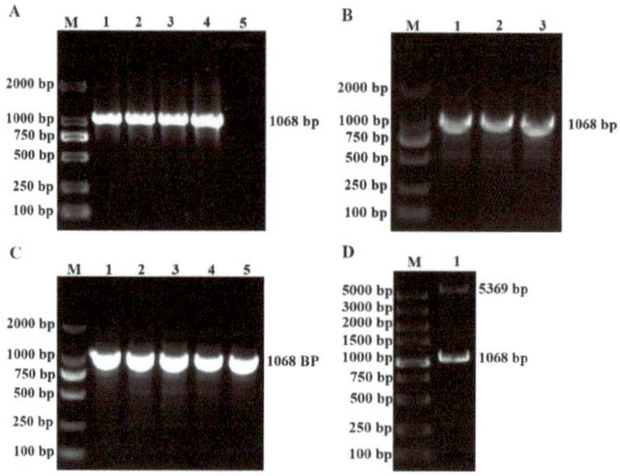

Figure 1. The construction of pET-*Et*ROP27 Expression Vector. (**A**) *EtROP27* gene amplification. 1–4: annealing temperatures are 52 °C, 54 °C, 56 °C, and 58 °C, respectively. 5: negative control. (**B**) The bacterial solution PCR results of pMD-*Et*ROP27. 1–3: monoclonal colony amplified bacterial liquid. (**C**) The bacterial solution PCR results of pET-*Et*ROP27. (**D**) The results of double enzyme digestion of pET-*Et*ROP27. M: DNA marker.

3.2. Expression and Purification of rEtROP27 Protein

Coomassie brilliant blue staining showed that the highest expression of *Et*ROP27 was observed after 6 h of IPTG induction (Figure 2A). After purification, the target band appears at about 48 kDa, and the elution buffer with a concentration of 150 mmol/L imidazole has the best elution effect (Figure 2B). Western blotting results showed that His tag anti body and chicken serum infected with coccidia were primary antibodies that could produce clear and reasonably sized target bands, while healthy chicken serum as primary antibody did not show any bands (Figure 2C). Chicken naturally infected with coccidia produced antibodies against *Et*ROP27. The results indicated that *Et*ROP27 was a natural antigen of *E. tenella* and could be a potential vaccine protein candidate.

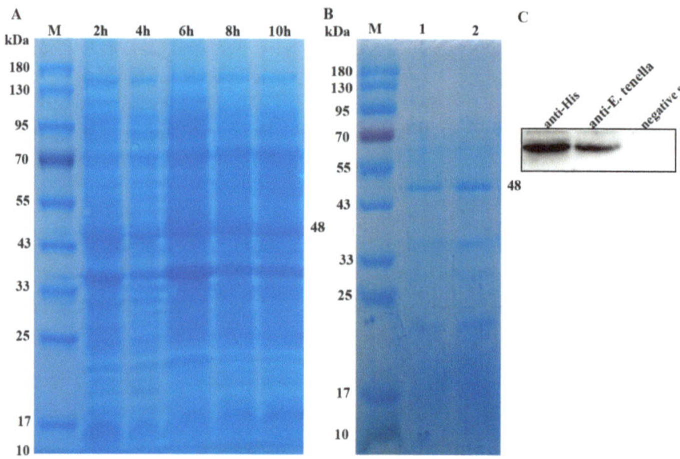

Figure 2. The expression, purification, and identification of r*Et*ROP27. (**A**) Coomassie bright blue staining of rEtROP27 protein. (**B**) Coomassie brilliant blue staining of purified r*Et*ROP27 protein. 1: 100 mmol/L imidazole. 2: 150 mmol/L imidazole. (**C**) Western blotting identification results.

3.3. Determination of rEtROP27 Polyclonal Antibody Titer

The rabbit anti-*Et*ROP27 serum was the sample chosen to be tested; the non-immunized serum was the negative control. The purified *Et*ROP27 recombinant protein was used as the detection antibody. After three immunizations with *Et*ROP27 fusion protein, blood and serum were collected to determine the antibody titer. From Table 2, it can be seen that the rabbit serum titer reaches 1:2.56 × 10^4 after three immunizations. This indicates that the immunogen injection achieves a good immune effect, and the positive serum can be used for subsequent experiments. Western blotting results showed that r*Et*ROP27 protein could be detected with *Et*ROP27 polyclonal antibody (Figure 3).

Table 2. Titer determination of *Et*ROP27 polyclonal antibody (OD_{450}).

Antibody Dilution Ratio									Negative Serum
1:100	**1:200**	**1:400**	**1:800**	**1:1600**	**1:3200**	**1:6400**	**1:12,800**	**1:25,600**	
*	*	*	3.442	2.532	2.015	1.503	1.022	0.516	0.36

Note: * indicates that the OD_{450} value exceeds 3.5.

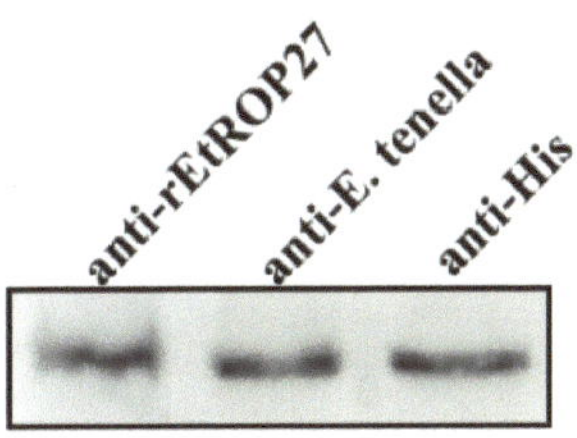

Figure 3. Specific detection of *Et*ROP27 polyclonal antibody.

3.4. Changes in EtROP27 Expression

The expression of *Et*ROP27 mRNA and protein was detected at different time points in primary chicken embryo cecal epithelial cells or 14-day-old chickens infected with *E. tenella*. The result showed that the expression of *Et*ROP27 mRNA in the primary cecal epithelial cells of chicken embryos infected with *E. tenella* gradually increased with the prolongation of infection time, reaching a peak at 96 h, and then decreasing. The protein expression pattern was also consistent with the one described above (Figure 4A–D). The results showed that *Et*ROP27 had the highest expression level during the first generation of schizogenesis.

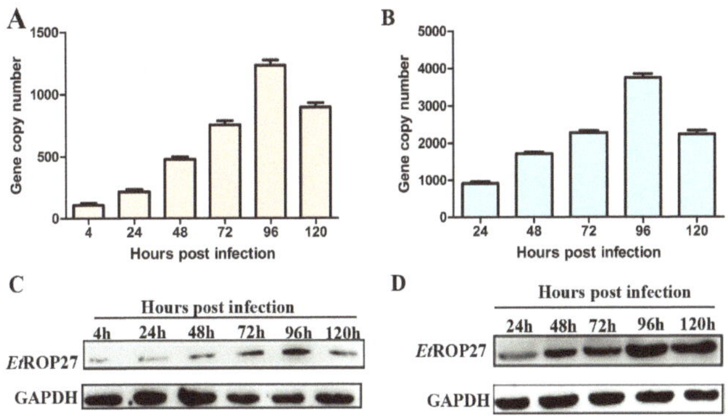

Figure 4. The expression of *Et*ROP27 in cells and tissues. (**A**,**B**) The expression of *Et*ROP27 mRNA was detected using qRT-PCR. A: chicken embryo cecal epithelial cells. (**B**) cecal tissue. (**C**,**D**) The expression of *Et*ROP27 protein was detected using Western blotting. (**C**) chicken embryo cecal epithelial cells. (**D**) cecal tissue. All of the data are representative of at least three independent experiments.

3.5. *In Vivo Immunoprotective Effect of rEtROP27 Protein*

Four chickens died in the challenged control group, while no deaths were observed in the other groups. Average and relative body weight gains of chickens that were immunized with the r*Et*ROP27 protein were significantly higher than in the challenged control groups (Figure 5A,B). Similarly, chickens vaccinated with the r*Et*ROP27 protein showed a reduction in the oocyst output compared with the challenged control group (Figure 5C,D), and lower lesion scores and bloody stool (refer Supplementary Materials) compared with the infection control group. The ACI value of the low-dose group (50 µg) of the r*Et*ROP27 protein was more than 155, and that of the medium-dose group (100 µg) and the high-dose group (150 µg) was more than 160 (Table 3). The results indicated that the optimal immune dose for protein was 100 µg.

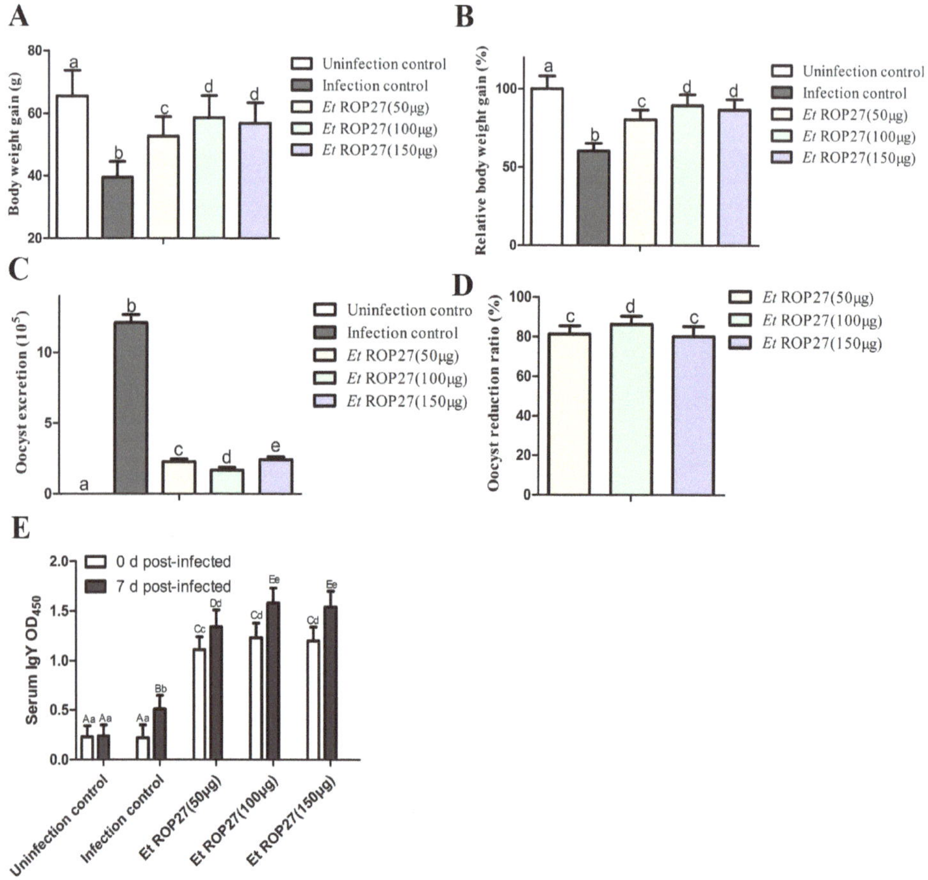

Figure 5. Effects of recombinant *Et*ROP27 protein against the *E. tenella* challenge. (**A**) Body weight gain. (**B**) Relative body weight gain. (**C**) OPG. (**D**) Oocyst reduction ratio. (**E**) IgY concentration. All of the data are representative of at least three independent experiments. The same column of shoulder markers with the same lowercase letters indicates no significant difference ($p > 0.05$), shoulder markers with different lowercase letters indicate significant differences ($p < 0.05$), shoulder markers with different uppercase letters indicate extremely significant differences ($p < 0.01$).

Table 3. Experimental ACI value for chicken.

Group	Relative Body Weight Gain (%)	Oocyst Index	Cecum Mean Lesion Score	Anticoccidial Index
Unchallenged control	100	0	0	200.00
Challenged control	60.25	5	2.9	87.25
*Et*ROP27 (50 μg)	80.45	1	2.0	159.45
*Et*ROP27 (100 μg)	89.47	1	1.7	171.47
*Et*ROP27 (150 μg)	86.75	1	1.9	166.75

The IgY antibody levels of the serum with different doses of *Et*ROP27 in immune groups were significantly higher than those of the control group at 7 days after the second immunization (0 day post infection), with the group administered 100 μg of *Et*ROP27 having the highest antibody level (Figure 5E). The serum IgY antibody levels in the *Et*ROP27 immune group with different doses and in the challenged group at 7 day post infection were significantly higher than those at 0 day post infection (Figure 5E). These results indicated that *Et*ROP27 was an antigen of *E. tenella* and could induce a higher level of humoral immune response for resisting an *E. tenella* infection.

4. Discussion

Chicken coccidiosis is one of the most important diseases threatening the health and welfare of poultry [23–25]. There are poultry farming enterprises in various provinces of China, and the overall infection rate of chicken coccidia in China is over 50%, which seriously hinders the development of China's poultry industry [3]. The life cycle of *E. tenella* is relatively complex and can be generally divided into two developmental stages: the exogenous stage (spore-stage reproduction) and the endogenous stage (schizogeny and gametogenesis). Coccidial infection impairs the nutrient uptake by its host, and the formation and release of merozoites and gametophytes destroy cell structures and functions, causing hemorrhagic inflammation, emaciation, and death in the cecum during *E. tenella* schizogeny and gametogenesis [26]. At present, the control of coccidiosis mainly relies on anticoccidial drugs, but they have many shortcomings. Therefore, there is an urgent need to develop a safe and effective vaccine against chicken coccidia to effectively prevent and control this disease.

E. tenella is a member of the apicomplexan protozoa [27]. The protozoa of the *apicomplexan* have unique secretory organelles, including rod-shaped bodies, dense particles, and microlines, which allow invasion mechanisms against host cells [28]. The secretion of rhoptry neck protein (RON) related to parasite invasion in the early stage is crucial for the formation and function of mobile connections between parasites and host cell membranes, such as RON1, RON2, and RON3 [29]. The release of ROPs from the base of the rod-shaped body into the PV changes the environment of the PV and enters the host cell [30,31]. They interact with the host cell through signal transduction, causing a series of downstream effects and becoming a key determinant of protozoan virulence. After entering the cell, ROPs can disrupt host cell signaling and defense mechanisms and assist in recruiting host organelles [32].

*Et*ROP plays an important role in the parasitic invasion of host cells, the modification of host vacuoles, and the regulation of host cells, and it has been proven to be a key virulence factor. It also has the properties of a protective antigen, and its corresponding antibodies can inhibit the infection of the host by coccidia [32]. Studies have confirmed the expression of *Et*ROP17 in *E. tenella* merozoites. Western blotting analysis showed that *Et*ROP17 can be recognized by the chicken immune system and induce antibody responses. The vaccination of animals with r*Et*ROP17 can significantly reduce cecal lesions [33]. This indicates that *Et*ROP17 can be used as an effective vaccine candidate for *E. tenella*. Similarly, studies have shown that *Et*ROP30 and *Et*ROP35 are natural antigens of coccidia, which have a good immunogenicity and are potential vaccine protein candidates [34,35].

*Et*ROP27 is a differential gene detected by the transcriptome, which is presumed to be a pathogenic gene [16,17]. However, its expression and function are still unclear. The different expression levels of secreted proteins at different stages of the parasite's life cycle are often closely related to their function. A high expression of proteins in the mitotic stage of invading cells is associated with parasite development and host–parasite interactions [36]. The extracellular stages (sporozoite and merozoite) of *Eimeria* are fragile, but it is crucial for the invasion and development of the parasite [37]. The high expression of proteins in the sporozoite and/or merozoite stage makes *Et*ROP27 an ideal vaccine candidate. In this study, we first conducted a bioinformatics analysis on *Et*ROP27. The analysis results indicate that the protein is a secreted protein with potential antigenic properties. Subsequently, a prokaryotic expression recombinant plasmid for *Et*ROP27 was successfully constructed and purified, and polyclonal antibodies against *Et*ROP27 were successfully obtained. RT-PCR and Western blotting results indicate that *Et*ROP27 protein is naturally present and expressed at all stages of *E. tenella* development, with the highest expression level in the first generation schizonts. Therefore, it may be a good protein candidate for vaccines.

Subunit vaccines are expected to become an alternative control strategy for avoiding shortcomings of anticoccidial medication and live vaccines, and the screening of effective immunoprotective proteins is a key focus of the subunit vaccine research. Animal experiments showed that the r*Et*ROP27 protein could produce increased average body weight gain, increased IgY titer in serum, decreased oocyst output and bloody stool, and lower mean lesion scores compared with the infection control. The ACI value of r*Et*ROP27 protein for a dose of 100 μg or 150 μg reached more than 160. These results indicate that *Et*ROP27 can effectively induce immune protection against *E. tenella* in chickens. In this study, other doses, immune methods, immune ages, immune intervals, and other factors that may affect the final immune protection effect were not considered. Therefore, the immune program used in this study may not be optimal. A large number of scientific studies are still needed before the successful clinical application of r*Et*ROP27.

This study provides a basis for determining the key genes that dominate or regulate the pathogenicity of coccidia, for deciphering drug action targets that control the pathogenicity of coccidia. This study also provides new ideas and scientific basis for further revealing the pathogenic mechanism of coccidia against its host.

5. Conclusions

This study successfully constructed the eukaryotic expression vector of *E. tenella* ROP27 and purified its recombinant protein. Our findings suggest that ROP27 is a natural protein of *E. tenella* and is highly expressed during the first generation of schizogenesis. The results of animal experiments showed that medium-to-high doses of r*Et*ROP27 could have a moderate effect against coccidia, indicating that this gene can reduce the effects of *E. tenella* infection and providing new ideas for the prevention of *E. tenella* disease and the development of vaccines for the disease.

Supplementary Materials: The following supporting information can be downloaded at: https://www.mdpi.com/article/10.3390/ani13223500/s1, Data S1: EtROP27 Gene sequence; Data S2: Bioinformatics analysis of EtROP27; Figure S1: The original image of Figure 3C; Figure S2: The original image of Figure 4; Figure S3: The original images of EtROP27 and GAPDH in Figure 5C; Figure S4: The original images of EtROP27 and GAPDH in Figure 5D; Figure S5: Bloody stool picture; Figure S6: Lesion scores picture.

Author Contributions: M.L. carried out most of the experiments, wrote the manuscript, and should be considered as the primary author. Y.W. critically revised the manuscript and the experimental design. X.L. and M.Z. helped with the experiment. All authors have read and agreed to the published version of the manuscript.

Funding: This study was funded by Guangxi Innovation Team Construction Project of National Modern Agricultural Industry Technology System (Grant No. nycytxgxcxtd-2021-09), Jiangnan key research and development Program (Grant No. 20220620-2), the Science and Technology Innovation Fund Project of Shanxi Agricultural University (Grant No. 2020BQ08), the Scientific research project of Shanxi Province outstanding doctoral work award fund (Grant No. SXYBKY2019023), and Innovation Projects of College of Veterinary Medicine, Shanxi Agricultural University (Grant No. J202111305).

Institutional Review Board Statement: The animal study protocol was approved by the Animal Experimental Ethical Committee of Guangxi University (Approval Code: Gxu-2023-0197; Approval Date: 16 February 2023).

Informed Consent Statement: Not applicable.

Data Availability Statement: Data are contained within the article and Supplementary Materials.

Acknowledgments: We specially thank Li Zhang, Rui Bai, Yong-bin Li, and Kui-hao Liu for their help during this study.

Conflicts of Interest: The authors declare no conflict of interest.

References

1. Lv, X.-L.; Wang, Y.-Y.; Zheng, M.-X.; Bai, R.; Zhang, L.; Duan, B.-T.; Lei, X.; Zhang, X.-S.; Zhao, Y.-J.; Cui, K.-L.; et al. The role of Ca^{2+} in the injury of host cells during the schizogenic stage of *E. tenella*. *Poult. Sci.* **2022**, *101*, 101916. [CrossRef] [PubMed]
2. Lv, X.; Chen, Z.; Zheng, M.; Bai, R.; Zhang, L.; Zhang, X.; Duan, B.; Zhao, Y.; Yin, L.; Fan, B.; et al. The interaction between free Ca^{2+} in host cells and invasion of *E. tenella*. *Parasitol. Res.* **2022**, *121*, 965–972. [CrossRef]
3. Blake, D.P.; Knox, J.; Dehaeck, B.; Huntington, B.; Rathinam, T.; Ravipati, V.; Ayoade, S.; Gilbert, W.; Adebambo, A.O.; Jatau, I.D.; et al. Re-calculating the cost of coccidiosis in chickens. *Vet. Res.* **2020**, *51*, 115. [CrossRef]
4. Li, C.; Gu, X.L.; Lu, C.X.; Liao, Q.; Liu, X.Y.; Suo, X. Progress in epidemiological research on chicken coccidiosis in China. *Acta Parasitol. Med. Entomol. Sin.* **2018**, *25*, 230–241. [CrossRef]
5. Zhang, X.; Li, S.; Zheng, M.; Zhang, L.; Bai, R.; Li, R.; Hao, S.; Bai, B.; Kang, H. Effects of the PI3K/Akt signaling pathway on the apoptosis of early host cells infected with *Eimeria tenella*. *Parasitol. Res.* **2020**, *119*, 2549–2561. [CrossRef]
6. Zhang, Y.; Duan, B.-T.; Zhao, Y.-J.; Cui, K.-L.; Xu, T.; Zhang, X.-S.; Lv, X.-L.; Guo, L.-L.; Zheng, M.-X.; Bai, R. Pathogenic mechanism of *Eimeria tenella* autophagy activation of chicken embryo cecal epithelial cells induced by *Eimeria tenella*. *Poult. Sci.* **2023**, *102*, 102535. [CrossRef]
7. Xu, Z.; Zheng, M.; Zhang, L.; Zhang, X.; Zhang, Y.; Cui, X.; Gong, X.; Xi, R.; Bai, R. Effect of ATP and Bax on the apoptosis of *Eimeria tenella* host cells. *BMC Vet. Res.* **2017**, *13*, 399. [CrossRef]
8. Xu, Z.-Y.; Zheng, M.-X.; Zhang, L.; Gong, X.; Xi, R.; Cui, X.-Z.; Bai, R. Dynamic expression of death receptor adapter proteins tradd and fadd in *Eimeria tenella*-induced host cell apoptosis. *Poult. Sci.* **2017**, *96*, 1438–1444. [CrossRef]
9. Li, S.; Zheng, M.-X.; Xu, H.-C.; Cui, X.-Z.; Zhang, Y.; Zhang, L.; Yang, S.-S.; Xu, Z.-Y.; Bai, R.; Sun, X.-G. Mitochondrial pathways are involved in *Eimeria tenella*-induced apoptosis of chick embryo cecal epithelial cells. *Parasitol. Res.* **2017**, *116*, 225–235. [CrossRef]
10. Jenkins, M.C.; Parker, C.; Ritter, D. *Eimeria* Oocyst Concentrations and Species Composition in Litter from Commercial Broiler Farms During Anticoccidial Drug or Live *Eimeria* Oocyst Vaccine Control Programs. *Avian Dis.* **2017**, *61*, 214–220. [CrossRef]
11. Lee, Y.; Lu, M.; Lillehoj, H.S. Coccidiosis: Recent Progress in Host Immunity and Alternatives to Antibiotic Strategies. *Vaccines* **2022**, *10*, 215. [CrossRef] [PubMed]
12. Possenti, A.; Di Cristina, M.; Nicastro, C.; Lunghi, M.; Messina, V.; Piro, F.; Tramontana, L.; Cherchi, S.; Falchi, M.; Bertuccini, L.; et al. Functional Characterization of the Thrombospondin-Related Paralogous Proteins Rhoptry Discharge Factors 1 and 2 Unveils Phenotypic Plasticity in Toxoplasma gondii Rhoptry Exocytosis. *Front. Microbiol.* **2022**, *13*, 899243. [CrossRef] [PubMed]
13. Sparvoli, D.; Delabre, J.; Penarete-Vargas, D.M.; Mageswaran, S.K.; Tsypin, L.M.; Heckendorn, J.; Theveny, L.; Maynadier, M.; Cova, M.M.; Berry-Sterkers, L.; et al. An apical membrane complex for triggering rhoptry exocytosis and invasion in *Toxoplasma*. *EMBO J.* **2022**, *41*, e111158. [CrossRef] [PubMed]
14. Bermúdez, M.; Arévalo-Pinzón, G.; Rubio, L.; Chaloin, O.; Muller, S.; Curtidor, H.; Patarroyo, M.A. Receptor-ligand and parasite protein-protein interactions in *Plasmodium vivax*: Analysing rhoptry neck proteins 2 and 4. *Cell. Microbiol.* **2018**, *20*, e12835. [CrossRef]
15. Oakes, R.D.; Kurian, D.; Bromley, E.; Ward, C.; Lal, K.; Blake, D.P.; Reid, A.J.; Pain, A.; Sinden, R.E.; Wastling, J.M.; et al. The rhoptry proteome of *Eimeria tenella* sporozoites. *Int. J. Parasitol.* **2013**, *43*, 181–188. [CrossRef]
16. Zhang, L.; Zheng, M.-X.; Xi, R.; Xu, Z.-Y.; Zhang, X.-S.; Zheng, L.-L.; Bai, R.; Mi, C.-L.; Hao, F.-F.; Feng, Y.-P. Comparison of the host cells apoptosis induced by precocious strains and virulent strains of *Eimeria tenella*. *Poult. Sci.* **2019**, *98*, 4384–4390. [CrossRef]
17. Zhang, L. Study on Pathogenicity Differences and Pathogenic Genes of E. tenella Precocious Strains and Virulent Strains. 2020. Available online: https://doi.org/10.27285/d.cnki.gsxnu.2020.000042 (accessed on 16 March 2023).

18. Zhang, D.; Jiang, N.; Chen, Q. Vaccination with recombinant adenoviruses expressing *Toxoplasma gondii* MIC3, ROP9, and SAG2 provide protective immunity against acute toxoplasmosis in mice. *Vaccine* **2019**, *37*, 1118–1125. [CrossRef]
19. Morehouse, N.F.; Baron, R.R. Coccidiosis: Evaluation of coccidiostats by mortality, weight gains, and fecal scores. *Exp. Parasitol.* **1970**, *28*, 25–29. [CrossRef]
20. Johnson, J.; Reid, W. Anticoccidial drugs: Lesion scoring techniques in battery and floor-pen experiments with chickens. *Exp. Parasitol.* **1970**, *28*, 30–36. [CrossRef]
21. Long, P.L.; Millard, B.J.; Joyner, L.P.; Norton, C.C. A guide to laboratory techniques used in the study and diagnosis of avian coccidia. *Folia Vet. Lat.* **1976**, *6*, 201–217.
22. Chapman, H. Evaluation of the efficacy of anticoccidial drugs against *Eimeria* species in the fowl. *Int. J. Parasitol.* **1998**, *28*, 1141–1144. [CrossRef] [PubMed]
23. Nguyen, B.T.; Flores, R.A.; Cammayo, P.L.T.; Kim, S.; Kim, W.H.; Min, W. Anticoccidial activity of berberine against Eimeria-infected chickens. *Korean J. Parasitol.* **2021**, *59*, 403–408. [CrossRef] [PubMed]
24. Macdonald, S.E.; Nolan, M.J.; Harman, K.; Boulton, K.; Hume, D.A.; Tomley, F.M.; Stabler, R.A.; Blake, D.P. Effects of *Eimeria tenella* infection on chicken caecal microbiome diversity, exploring variation associated with severity of pathology. *PLoS ONE* **2017**, *12*, e0184890. [CrossRef] [PubMed]
25. Bafundo, K.; Gomez, L.; Lumpkins, B.; Mathis, G.; McNaughton, J.; Duerr, I. Concurrent use of saponins and live coccidiosis vaccines: The influence of a quillaja and yucca combination on anticoccidial effects and performance results of coccidia-vaccinated broilers. *Poult. Sci.* **2021**, *100*, 100905. [CrossRef] [PubMed]
26. Fatoba, A.J.; Adeleke, M.A. Transgenic *Eimeria* parasite: A potential control strategy for chicken coccidiosis. *Acta Trop.* **2020**, *205*, 105417. [CrossRef]
27. Huang, J.; Chen, S.; Zhou, Z.; Sun, X.; Haseeb, M.; Lakho, S.A.; Zhang, Y.; Liu, J.; Shah, M.A.A.; Song, X.; et al. Poly (D, L-lactide-co-glycolide) delivery system improve the protective efficacy of recombinant antigen TA4 against *Eimeria tenella* infection. *Poult. Sci.* **2021**, *100*, 101083. [CrossRef]
28. Patarroyo, M.A.; Molina-Franky, J.; Gómez, M.; Arévalo-Pinzón, G.; Patarroyo, M.E. Hotspots in *Plasmodium* and RBC Receptor-Ligand Interactions: Key Pieces for Inhibiting Malarial Parasite Invasion. *Int. J. Mol. Sci.* **2020**, *21*, 4729. [CrossRef]
29. Lim, D.C.; Cooke, B.M.; Doerig, C.; Saeij, J.P. Toxoplasma and Plasmodium protein kinases: Roles in invasion and host cell remodelling. *Int. J. Parasitol.* **2012**, *42*, 21–32. [CrossRef]
30. Pan, M.; Zhou, Y.; Wang, Y.; Li, L.; Song, Y.; Hou, L.; Zhao, J. Screening and Identification of the Host Proteins Interacting with Toxoplasma gondii Rhoptry Protein ROP16. *Front. Microbiol.* **2017**, *8*, 2408. [CrossRef]
31. Ihara, F.; Nishikawa, Y. Toxoplasma gondii manipulates host cell signaling pathways via its secreted effector molecules. *Parasitol. Int.* **2021**, *83*, 102368. [CrossRef]
32. Meng, Y.-J.; Mu, B.-J.; Liu, X.-X.; Yu, L.-M.; Zheng, W.-B.; Xie, S.-C.; Gao, W.-W.; Zhu, X.-Q.; Liu, Q. Transcriptional changes in LMH cells induced by *Eimeria tenella* rhoptry kinase family protein 17. *Front. Vet. Sci.* **2022**, *9*, 956040. [CrossRef] [PubMed]
33. Liu, X.; Mu, B.; Zheng, W.; Meng, Y.; Yu, L.; Gao, W.; Zhu, X.; Liu, Q. Identification and Protective Efficacy of *Eimeria tenella* Rhoptry Kinase Family Protein 17. *Animals* **2022**, *12*, 556. [CrossRef] [PubMed]
34. Wang, B.; Zhao, N.; Sun, J.; Sun, L.; Li, H.; Wu, Z.; Li, H.; Zhang, X.; Zhao, X. Identification and Characterization of *Eimeria tenella* Rhoptry Protein 35 (EtROP35). *Vet. Sci.* **2022**, *9*, 465. [CrossRef] [PubMed]
35. Bingxiang, W.; Ningning, Z.; Yakun, W.; Lingyu, S.; Hongmei, L.; Zhang, X.; Zhao, X. Characterization of the *Eimeria tenella* rhoptry protein with a nuclear localization sequence (EtROP30). *Parasitol. Res.* **2022**, *121*, 1507–1516. [CrossRef]
36. Fernandez-Becerra, C.; Bernabeu, M.; Castellanos, A.; Correa, B.R.; Obadia, T.; Ramirez, M.; Rui, E.; Hentzschel, F.; López-Montañés, M.; Ayllon-Hermida, A.; et al. *Plasmodium vivax* spleen-dependent genes encode antigens associated with cytoadhesion and clinical protection. *Proc. Natl. Acad. Sci. USA* **2020**, *117*, 13056–13065. [CrossRef]
37. Rezaei, F.; Sarvi, S.; Sharif, M.; Hejazi, S.H.; Pagheh, A.S.; Aghayan, S.A.; Daryani, A. A systematic review of *Toxoplasma gondii* antigens to find the best vaccine candidates for immunization. *Microb. Pathog.* **2019**, *126*, 172–184. [CrossRef]

animals

Article

Automated Image Analysis for Detection of Coccidia in Poultry

Isaac Kellogg [1], David L. Roberts [1] and Rocio Crespo [2,*]

[1] Department of Computer Science, College of Engineering, North Carolina State University, Raleigh, NC 27695, USA; dlrober4@ncsu.edu (D.L.R.)
[2] Department of Population Health and Pathobiology, College of Veterinary Medicine, North Carolina State University, Raleigh, NC 27606, USA
* Correspondence: rcrespo@ncsu.edu

Simple Summary: Coccidiosis is one of the most common and costly diseases faced by commercial poultry. To establish effective control measures, it is essential to identify the infective species and the numbers of oocysts. Standard methods for analysis require highly skilled technicians or veterinarians to manually identify and enumerate these protozoal parasites. This process is labor intensive, time-consuming, and susceptible to human error. None of the current methods available report the infectivity status of these protozoal parasites. Therefore, an automated, reproducible protocol for counting, speciation, and determination of infectivity of these protozoa using Artificial Intelligence capable of enumerating, speciating, and determining the infectivity status of the coccidia has the potential to improve diagnostics and refine control strategies to mitigate the impacts of coccidiosis on farms.

Abstract: Coccidiosis, caused by the protozoan *Eimeria* sp., is one of the most common and costly diseases impacting the poultry industry. To establish effective control measures, it is essential to identify these protozoa. Typical methods for identifying and determining the severity of the protozoal infection include intestinal lesion scoring or enumeration of the protozoal oocysts in fecal samples. Standard analysis methods require highly skilled technicians or veterinarians to manually identify and manually enumerate these protozoal parasites. This process is labor intensive, time-consuming, and susceptible to human error. None of the current methods available, including molecular flow cytometry or even digital image analysis, can determine if an oocyst is sporulated or not. Oocysts are not infectious until they sporulate. The goal of this study was to design an automated model using Artificial Intelligence (AI) to expedite the process of enumeration, improve the efficiency and accuracy of the species identification, and determine the ability of the oocysts to infect. To this end, we trained and evaluated computer vision models based on the Mask RCNN neural network architecture. A model was trained to detect and differentiate three species and to determine sporulation for each (totaling six detection groups). This model achieved a mean relative percentage difference (RPD) of 5.64%, representing a slight overcount compared to manual counts, averaging across all groups. The mean RPD for each group individually fell within a range from −33.37% to 52.72%. These results demonstrate that these models were speedy and had high agreement with manual counts, with minimal processing of field-quality samples. These models also could differentiate the sporulation status of the oocysts, providing critical diagnostic information for potential field applications.

Keywords: Artificial Intelligence; coccidiosis; computer vision; oocyst

Citation: Kellogg, I.; Roberts, D.L.; Crespo, R. Automated Image Analysis for Detection of Coccidia in Poultry. *Animals* **2024**, *14*, 212. https://doi.org/10.3390/ani14020212

Academic Editor: Kenneth Bafundo

Received: 10 December 2023
Revised: 5 January 2024
Accepted: 6 January 2024
Published: 9 January 2024

1. Introduction

Coccidiosis, a disease caused by *Eimeria* sp. parasites, is one of the most common diseases faced by the poultry industry worldwide [1,2]. Coccidia parasites can cause decreased growth rates, diarrhea, or even death. In addition, *Eimeria* sp. harm gut health and enable other enteric conditions, including clostridial enteritis. It is estimated that the annual economic impact of coccidiosis, including prevention, treatment, and production loss, is almost USD 2 billion in the United States alone, and over USD 15.5 billion worldwide [3].

The basis for successful control measures against coccidiosis relies on identification of these protozoal parasites. Typical methods for identifying and determining the severity of the protozoal infection include intestinal lesion scoring or enumeration of the protozoal oocysts in fecal samples [4]. While many poultry producers give little consideration to the species of coccidia, knowledge of the species can be important for developing vaccines or improving management strategies [5,6]. Standard methods to manually identify and enumerate these protozoal parasites require highly skilled technicians or veterinarians. This process is labor intensive, time-consuming, and susceptible to human error.

Automatic and semi-automatic protocols, including molecular tests, cytometry, and digital image analysis, for the enumeration and identification of coccidia are available [7–10]. Most PCR reactions target the intergenic transcribed spacer region 1 (ITS1) of the ribosomal RNA (rRNA) gene operon [7], a multi-copy gene, which makes these protocols unsuitable for estimation of the relative abundance of species in mixed infections of *Eimeria* sp. Vrba et al. [11] validated a quantitative PCR (qPCR) protocol to quantify samples with mixed *Eimeria* populations. However, to improve the sensitivity, the oocysts need to be sporulated beforehand, which can generate a delay in the results [12]. On the other hand, while there is no delay for protocols based on cytometry and digital image analysis, sample preparation can be cumbersome. Furthermore, if the samples contain debris of similar size to coccidia parasites, the number of oocysts can be overestimated. While improved preparation protocols may help to minimize these errors, the accuracy of the machine to differentiate between oocysts and debris cannot improve over time.

Poultry management and diagnostics is undergoing a significant transition with the introduction of machine learning or Artificial Intelligence models. These models have the potential to be an excellent diagnostic tool for coccidia identification, by exploiting *Eimerian* oocyst morphology. The human process of scanning for oocysts shapes and making decisions based on observation of morphological features (size, shape, and internal and external features) is simultaneous. This process is best modeled using a deep-learning-based approach, such as the Region-based Convolutional Neural Network-based (RCNN) model. This project aimed to demonstrate a computer vision model approach that would enable fast and accurate detection, speciation, and determination of sporulation status of coccidia oocysts without the need for extensive personnel training or subjectivity of traditional microscopic methods.

2. Materials and Methods

2.1. Coccidia Preparation

We prepared 30 samples of species-specific oocysts (*Eimeria acervulina*, *E. maxima*, and *E. tenella*, kindly donated by Dr. Mark Jenkins, USDA), for a total of 90 samples, by combining 100 μL of a pure isolate of coccidia species oocysts with 1 mL flotation solution (Feca-Med Sodium Nitrate, Vedco, Inc., St. Joseph, MO, USA). Additionally, we mixed 20 samples of sterile 100 μL of peptone buffer solution (Sigma-Aldrich, St. Louis, MO, USA) with 1 mL flotation solution.

To mimic field samples, we spiked 0.5 teaspoon (2.5 g approximately) fecal samples from poultry free of coccidia with 0.05 mL of a commercial coccidia vaccine (i.e., Advent® by Huvepharma®, Peachtree City, GA, USA; Coccivac®-B52 by Merck & Co, Inc., Rahway, NJ, USA; Hatchpak® Cocci III by Boehringer Ingelheim, Duluth, GA, USA; and Immucox® 5 by Ceva Animal Health, Lenaxa, KS, USA). A minimum of 20 samples for each vaccine type will be prepared. The feces were then deposited in one side of a 7 oz (207 mL) whirl-pak filter bag (Nasco, Pleasant Prairie, WI, USA) that contained 22.5 mL Fecasol, to achieve a 1:10 dilution (feces to salt solution).

Each sample, including the negative mixture, was vortexed, centrifuged at 280 G, and allowed to sit for 10 min. Next, the samples were transferred from the top of each tube to a cell counter slide chamber (Countess™ Cell Counting Chambe Slide, Thermo Scientific,

Waltham, MA, USA). We used an automated cell counter (Invitrogen Countess 3, Thermo Scientific, MA) to acquire images. The JPG photographs of the slides measured 2592 pixels (2257.63 µm) in width and 1944 pixels (1693.22 µm) length. We labeled the oocysts in the sample images using the LabelMe version 5.0.1 application (https://github.com/wkentaro/labelme, accessed on 10 August 2022). These manually labelled oocysts were used for training of the model. For each oocyst, labels indicated the *Eimeria* species and whether the oocyst was sporulated or not. Sporulated oocysts were identified by the presence of sporocysts within. On the other hand, non-sporulated oocysts were characterized by a single central mass (sporont) occupying most of the oocyst [13]. Manual enumeration, speciation, and sporulation status of all the samples were done by the same person. Counts were expressed as a total of units observed in each image, without transformations. Once labeling was complete, the LabelMe application saved the data as JPG and JSON file pairs, where each JSON file contained label information for its associated JPG file. These files formed the basis of the model training dataset.

2.2. Data Augmentation

To expand the training dataset, we performed multiple rotation and crop operations on all images and associated label data. First, we rotated each image at 45-degree intervals between 0 and 90 degrees (inclusive). Rotated images were cropped in-place to remove border artifacts from the rotation process. Next, we took multiple crops of each rotated image at fixed dimensions (1280 pixels horizontally, 960 pixels vertically). Consecutive crops used horizontal and vertical increments (425 pixel horizontal increment, 320 pixel vertical increment). These cropped images comprised the final dataset.

To preserve ground-truth labels for training, the rotation and cropping parameters used on the images were reused to apply an equivalent transformation to the manual label data. This process enables a single large image with manual segmentation labels to be converted into a multi-image dataset made of many smaller training images with associated segmentation labels. To create the dataset used to train our multi-species models, we applied this process to an initial dataset of 110 images (Figure 1). This dataset consisted of 30 unchanged images of each *E. acervulina*, *E. maxima*, and *E. tenella* oocysts, for a total of 90 images. In addition, 20 images with no oocysts (negative) were included in the initial dataset. Each image was paired with its corresponding labeled file. Then we cropped and rotated the images, as described above, to expand the number of images to an augmented dataset of 2928 file pairs in the training dataset. Of this dataset, 732 file pairs were used in the validation dataset.

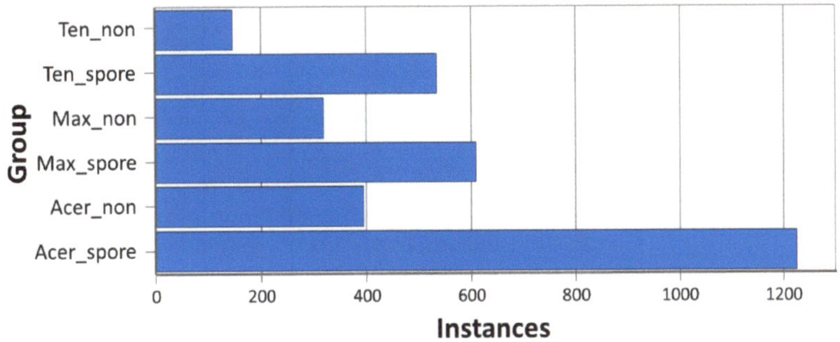

Figure 1. Instance counts for initial multi-species dataset. Oocyst counts for each group in the dataset used to train the multi-species models, across all images (n = 110), prior to any augmentation. Where Ten_non = *E. tenella* non-sporulated, Ten_spore = *E. tenella* sporulated, Max_non = *E. maxima*, non-sporulated, Max_spore = *E. maxima* sporulated, Acer_non = *E. acervulina* non-sporulated, and Acer_spore = *E. acervulina* sporulated.

The dataset created using the process described here was further augmented during the training process by the Mask-RCNN library used for training.

2.3. Model Training and Performance

To train the model, we used PixelLib version 3.7 (https://github.com/ayoolaolafenwa/PixelLib accessed on 1 September 2022), a sophisticated open-source software library built for training Mask-RCNN models to perform instance segmentation. For the instance segmentation process, the software analyzed the group of pixels that made the labelled oocysts, rather than the individual pixels. As a starting point for training, we used Matterport's Mask-RCNN 2.0 model (https://github.com/matterport/Mask_RCNN/releases/tag/v2.0 accessed on 1 September 2022) with MS COCO trained weights. For our implementation, we used the library's default Learning Rate of 0.01, and adjusted the Learning Momentum to 0.95. For the models shown here, we used a batch size of 3 over 250 epochs. In other words, the system analyzed 3 images (batch size) concurrently before updating the internal model parameters. When it completes analyzing all the training images in the dataset, the model repeats the same process 250 times, also known as epochs. For each epoch, the system used updated model parameters based on the previous epoch. Our increased Learning Momentum (up from the library's default value of 0.9) meant that the system considered the overall history of parameter changes to a greater degree for each subsequent epoch, making the system accelerate towards its final parameters more quickly. For feature extraction, we used ResNet101 as a network backbone.

To determine how similar the manually labelled and the model's automatic counts were, we used mean relative percent difference (mRPD), according to the formula

$$mRPD = \frac{\sum_{i=1}^{n} \frac{a_i - m_i}{\frac{1}{2}(a_i + m_i)}}{n} \times 100\%$$

where n is the number of images in the dataset, a_i is the AI model's automatic count for the relevant group for image i, and m_i is the ground-truth manual count for the relevant group for image i.

Note: To allow true-negative cases (where $a_i = m_i = 0$) to impact mRPD,

$$\frac{a_i - m_i}{\frac{1}{2}(a_i + m_i)}$$

was defined as 0 for these cases, corresponding to a mRPD of 0%. These cases indicate situations where the AI model correctly counted 0 instances of a group, and thus did not disagree with manual counts.

For an intuitive performance visualization, we also plotted ground-truth manual counts (X axis) against counts automatically generated from the model output (Y axis). Each point represents the counts for a specific group in each input image. Each chart has a regression line, with a reference line representing linear correlation (automatic count = manual count) at y = x. A regression line closer to y = x generally reflects better correlation between model counts and ground-truth counts for the input set. To measure the linear correlation of manual and automated oocyst counts, we used the Pearson correlation coefficient. We created Bland–Altman plots to visualize and quantify the agreement of the results between manual and AI-based methods. For correlation charts in Figure 2, and for mRPD calculations, we used the Pandas and Vega-Altair libraries for Python. We used JMP-Pro v.16 to calculate Pearson correlation coefficients and to create Bland–Altman plots.

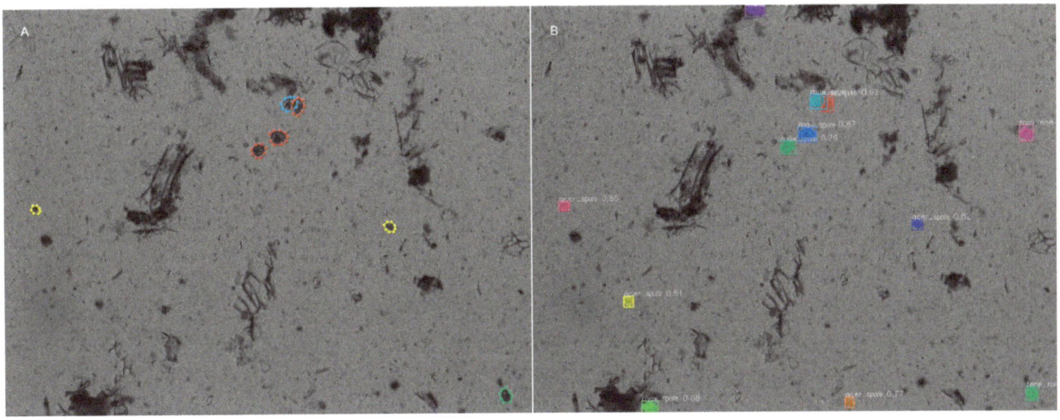

Figure 2. Comparison between an image of a multi-species sample that was manually labeled (**A**) and the automated output (**B**). Photo (**A**) is a manually labeled image used to train the model. For manual labels in (**A**), red is *E. maxima* sporulated, blue is *E. maxima* non-sporulated, yellow is *E. acervulina* sporulated), green is *E. tenella* sporulated. Image (**B**) shows the output from the multi-species model with confidence 0.5. Where label max_spore = *E. maxima* sporulated, max_non = *E. maxima* non-sporulated, acer_spore = *E. acervulina* sporulated, and tene_spore = *E. tenella* sporulated. The number adjacent to the label represents the model's confidence of that classification (from 0 to 1, with 1 representing 100% certainty). The colors in this image are random to help differentiate individual oocysts, and do not relate to the species or sporulation status.

3. Results

To evaluate the performance of the trained model we used mRPD, which assessed the relative difference between the counts from the manually labelled images and the model's automatic counts (Table 1). Models trained on one species generally exhibited higher accuracy than models trained on multiple species. In a scale of 0–1, when the confidence threshold was 0.7, all the models for individual groups were over 90% accurate. In general, models recognized oocysts and ignored debris in images accurately (Figure 2). While there were a few false positive instances and misclassifications of oocysts, the model indicated a lower likelihood that these predictions were accurate by giving them a lower confidence score (below 0.7).

Unsurprisingly, models produced the most accurate results when detecting only *E. maxima* oocysts. The multi-species model, when set to a confidence threshold of 0.5, correctly identified all the instances marked in our sample ground-truth reference for *E. maxima* (Figure 3). In a few instances, in samples that only had *E. maxima*, there was a mismatch between manual and automated identification. These mismatches generally occurred when the oocyst laid on the border of the image or two oocysts overlapped almost entirely. In the latter case, the automated system did not identify one of the instances. Very few of the visibly labeled instances were misclassified for sporulation.

On the other hand, when set to a confidence level of 0.5, the same multi-species model revealed discrepancies when identifying *E. acervulina* (Figure 4) and *E. tenella* oocysts. Most notably, this model over-identified sporulated *E. acervulina* oocysts, and under-identified non-sporulated *E. tenella* instances. Despite these tendencies, reducing the confidence threshold for the multi-species model from 0.7 to 0.5 resulted in a measurable improvement in overall RPD metrics. Considering all groups, the multi-species model with a confidence threshold of 0.5 had mean RPD values that were generally closer to 0, meaning that the manual and automatic counts are more similar.

Table 1. Relative Percent Difference (RPD) between the manual and the model's automatic counts. This value was used to assess the performance for single- and multi-species models. Each model was trained on a dataset containing oocyst examples of each group listed for that model. The model counted as oocysts the units that were above the set confidence threshold. Mean RPD values closer to 0 indicate better consensus between both manual and automatic counts.

Model Type	Model Confidence Threshold	No. of Images in Input Dataset	Group [1]	Mean % RPD
Single-Species				
E. acervulina	0.7	140	Acer_spore	1.26
			Acer_non	−7.01
E. maxima	0.7	141	Max_spore	4.49
			Max_non	8.17
E. tenella	0.7	129	Ten_spore	4.03
			Ten_non	−6.20
Multi-Species (*E. acervulina, E. maxima, E. tenella*)	0.7	732	Acer_spore	25.32
			Acer_non	−47.40
			Max_spore	−1.18
			Max_non	−9.21
			Ten_spore	−10.89
			Ten_non	−37.61
	0.5	732	Acer_spore	52.72
			Acer_non	−4.40
			Max_spore	7.16
			Max_non	1.58
			Ten_spore	10.13
			Ten_non	−33.37

[1] Acer_spore = *E. acervulina* sporulated, Acer_non = *E. acervulina* non-sporulated Max_spore = *E. maxima* sporulated, Max_non = *E. maxima* non-sporulated, Ten_spore = *E. tenella* sporulated, and Ten_non = *E. tenella* non-sporulated.

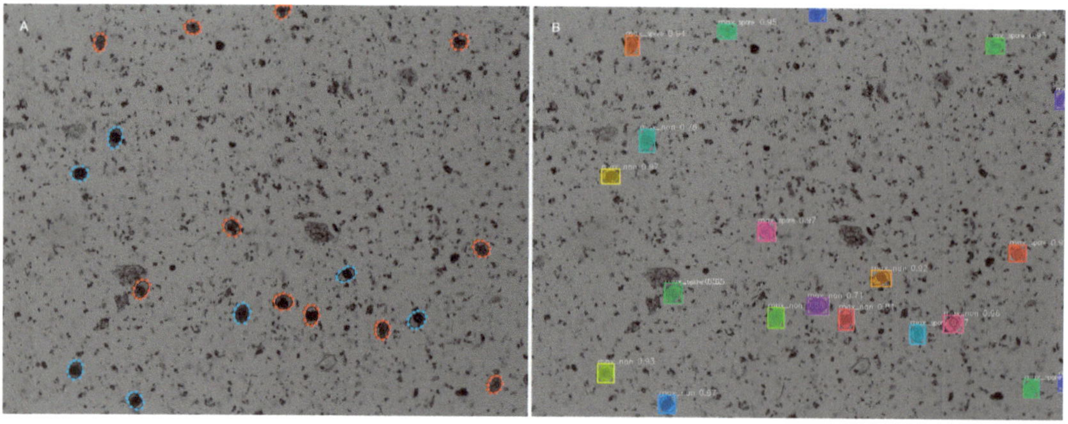

Figure 3. Comparison of manual labels and automated output for *E. maxima*. (**A**) Manually labelled oocysts, used to train the model. Red represents sporulated oocysts of *E. maxima*, blue indicates non-sporulated *E. maxima* oocysts. (**B**) Output from the multi-species model with confidence threshold set to 0.5. Where label max_spore = *E. maxima* sporulated and max_non = *E. maxima* non-sporulated. The number adjacent to the label represents the model's confidence of that classification (from 0 to 1, with 1 representing 100% certainty). The colors in this image are random to help differentiate individual oocysts, and do not relate to the species or sporulation status.

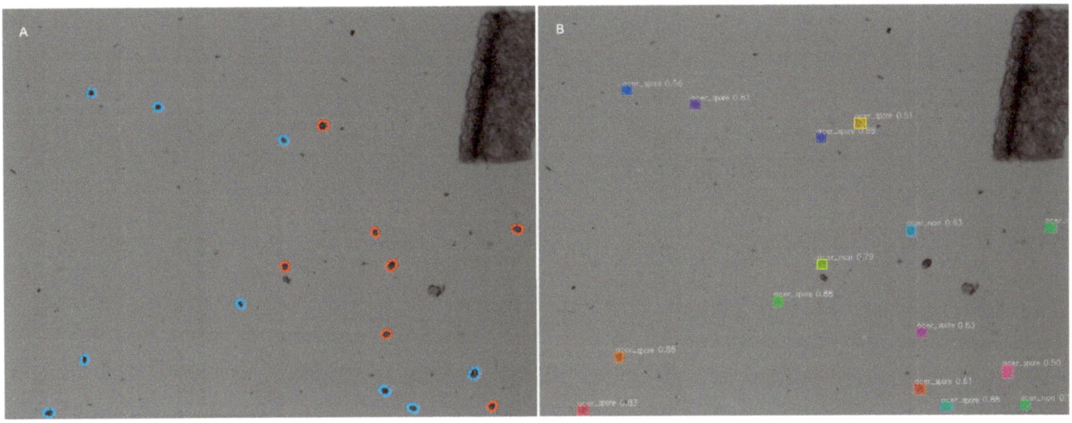

Figure 4. Comparison of manual labels and automated output for *E. acervulina*. Image (**A**) shows manually labeled oocyst, used to train the model. Blue represents *E. acervulina* sporulated oocysts, red indicates *E. acervulina* non-sporulated oocysts. Image (**B**) shows output from the multi-species model with confidence threshold set to 0.5. Where label acer_spore = *E. acervulina* sporulated, acer_non = *E. acervulina* non-sporulated oocysts. The number adjacent to the label represents the model's confidence of that classification (from 0 to 1, with 1 representing 100% certainty). The colors in this image are random to help differentiate individual oocysts, and do not relate to the species or sporulation status. Here, the model manages to correctly identify 3 non-sporulated instances. Most other instances were misclassified as sporulated, though one non-sporulated instance remains unidentified.

Visualizations in Figure 5 show per-group correlation between manual and automated counting of oocysts for the most accurate multi-species model. The sporulated *E. acervulina* (acer_spore) group dominates the upper region of the chart (Figure 5b). This suggests that the models over-counted this group. On the other hand, the non-sporulated *E. acervulina* (acer_non) and the non-sporulated *E. tenella* (ten_non) groups dominate the lower region of this chart (Figure 5c,g), which suggests that the model under-counted these groups. Furthermore, the undercounts were worse as manual counts increased along the x-axis. Notably, the correlations between the manual and automated counts, for both the sporulated and non-sporulated, *E. maxima* (Figure 5d,e), as well as the sporulated *E. tenella* (tene_spore) (Figure 5f), were better and did not show any obvious trend towards under- or over-counting for these groups.

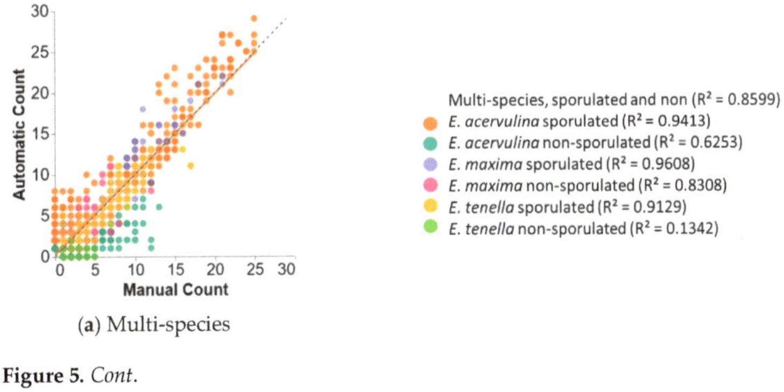

(**a**) Multi-species

Figure 5. *Cont.*

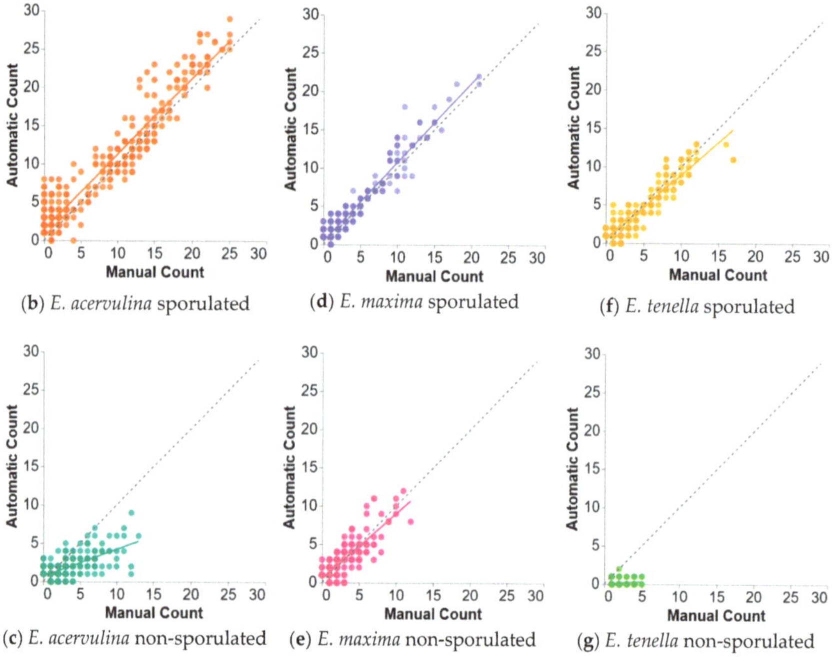

(**b**) *E. acervulina* sporulated (**d**) *E. maxima* sporulated (**f**) *E. tenella* sporulated

(**c**) *E. acervulina* non-sporulated (**e**) *E. maxima* non-sporulated (**g**) *E. tenella* non-sporulated

Figure 5. Pearson correlation between manual and automatic counts. Charts: Multi-species model (**a**) and all individual groups (**b–g**), at 0.5 confidence. For these charts, the multi-species model analyzed an input set that contained instances of all 3 *Eimeria* species, with both sporulated and non-sporulated instances of each species. The combined chart and most single group charts show good linear correlation. Chart (**a**) shows all groups in one chart, with groups separated by color. The number in parenthesis next to each group in the legend represents the linear regression R^2 for sporulated and non-sporulated individual oocyst species. Counts are expressed as a total of units observed in each image.

The Bland–Altman plots (Figure 6) show the agreement between manual and automatic counts. The solid line indicates the average difference between both counts. Under ideal conditions, the average between both counting methods is expected to be =0; dotted lines are the upper and lower 95% confidence interval for the mean. The closer the data points are to the mean line or are within the 95% confidence interval for the mean represent the agreement between the automated and manual (or ground-truth) counts.

Our model showed a high agreement (>95%) between the automated and manual counts of individual species. For *E. acerulina* oocysts, the agreement was 99.5% (or 729/732 samples within the 95% confidence internal), 99.79% (or 731/732 samples within the 95% confidence internal) for *E. maxima*, and 96.18% (or 728/732 samples within the 95% confidence internal) for *E. tenella*. For the *Eimeria* multi-species plot, the agreement was moderate (50–94%) between automated and manual counts, at 92.32% (or 128/1667 samples outside the 95% confidence internal). Furthermore, while each species showed high agreement, regardless of the sporulation status, the model showed poor agreement for the non-sporulated *E. acervulina* and *E. tenella*. This lack of agreement is in accordance with the lower correlation shown for these two groups.

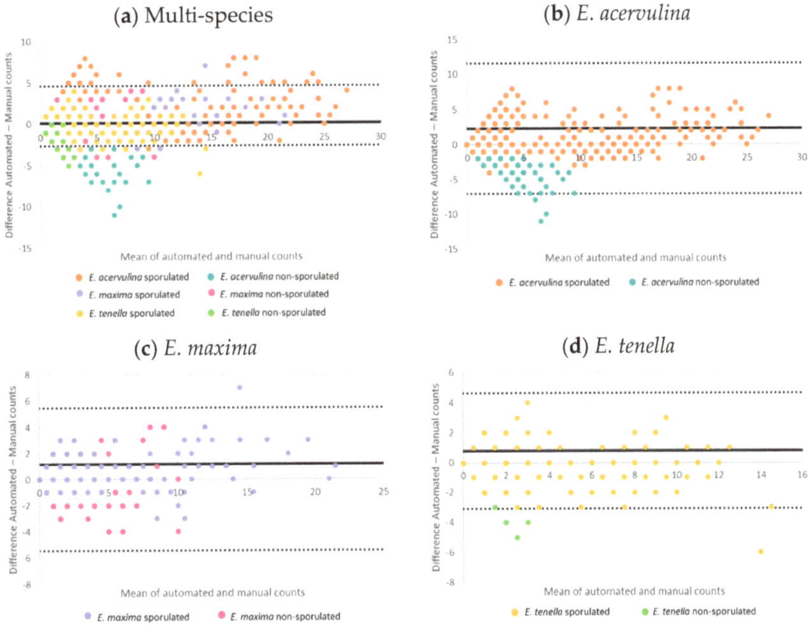

Figure 6. Bland–Altman plot comparing oocysts counts done automatically and manually: difference between both counts (Y axis) plotted against the mean difference between the counts (X axis). Counts were expressed as a total of units observed in each image, without transformations.

4. Discussion

The purpose of this study was to compare manual and automated analyses, using a custom Mask-RCNN model, for the enumeration, speciation, and determination of sporulation status of three species of coccidia that infect chickens. We used digitalized microscopic images of floatation samples as a validation technique.

Enumeration and speciation of coccidia oocysts are commonly performed in research and even in some clinical investigations. Enumeration and speciation can be laborious because they are routinely performed manually. Consequently, this problem represents a substantial bottleneck for research projects and vaccine evaluations, requiring well-trained and experienced parasitologists. Validated qPCR protocols to quantify samples with mixed *Eimeria* populations are available [11]. However, the practical utility of PCR assays for routine diagnostics is questionable, because many PCR reactions described in the literature target the multi-copy ITS1 gene [7], which makes these protocols unsuitable for the quantification of oocysts in mixed infections. Results for protocols capable of quantifying samples with mixed *Eimeria* populations may be delayed because sporulation is recommended prior to performing the assay [11].

To date, there are a few automated protocols for oocyst enumeration and speciation using flow cytometry [9] and digital image analysis [10]. However, we found that the protocols used for sample preparation for these applications are complex, and the oocyst counts may be overestimated. A recent publication describes a fast and automated method for the enumeration of *Eimeria* oocysts [8]. None of the available protocols differentiates sporulated and non-sporulated oocysts. Information about the sporulation status can be important for clinicians and vaccine producers, as only sporulated oocysts can cause infection in poultry [13].

Recent advances in computer technology have enabled software-based automation of standard laboratory data analysis, including analysis of digital images. Artificial Intelligence represents a promising technology for microscopic parasite examination [14]. The

Convolutional Neural Network (CNN) is the most commonly used artificial neural network to examine visual images [15]. Initial versions of our pipeline used OpenCV version 4.6 and TensorFlow CNN version 2.13 as foundational software packages to construct a complete working pipeline prototype. These packages were primarily used for image classification, object detection, and image segmentation tasks. We were able to prepare coccidia data by first labeling it, then using OpenCV to augment and expand it. Then we developed custom-trained models using TensorFlow. These models provided enumeration and speciation information for new input images in a process (inference) that again used OpenCV.

Once we had the prototype, we used the Mask-RCNN pipeline for the image analysis. Mask-RCNN is a newer CNN algorithm that makes object detection and its classification more accurate and faster [16]. Our machine learning pipeline for the identification and enumeration of coccidia parasites constituted three steps: data preparation, training, and inference. As with any machine learning pipeline, dataset preparation was the key to successfully achieving a high-accuracy model [17]. When creating such datasets, labeling images is often time-consuming. In this study, data labeling means to identify the species of each oocyst and whether it is sporulated or not. For the model to extract the proper information, accurate labeling is essential. For example, the model must learn to discriminate debris from oocysts, determine the species of *Eimeria*, or whether the oocyst is sporulated or not. Also, we needed to label as many oocysts as possible, including partial ones at the margin of the slide and overlapping oocysts, to prevent the algorithm from misclassifying them. Irrelevant features, such as the orientation of the oocysts, can also interfere with data extraction during training, and affect the results. Data enhancement, including horizontal flipping, aims to teach the model to learn the invariance of the data [18]. A weakness of our model was the inability to include images with no oocysts, as PixelLib, our training library, rejected these images. This difficulty should be further investigated, and solutions further developed.

Training the application by species was the most time intensive. This required a thoughtful approach to (1) obtaining examples of manually labeled oocyst and (2) devising a protocol that would provide adequate data to the learning algorithms without becoming overly burdensome for humans. In other words, our efforts focused on making each manually labeled image as useful as possible in building a generalizable vision model that would perform accurately and reliably on novel samples. For this step, we used 80% of file pairs from the augmented dataset. The other 20% of the file pairs were used for validation and fine tuning of the model. We used novel ways of manipulating the labeled images to improve the accuracy of the computer vision model without needing a large number of labeled images. The original dataset contained 110 file pairs that, after manipulation, were expanded to 2928 file pairs.

The Learning Rate and Learning Momentum parameters are important in identifying over-learning and when to stop training [19]. In our model, a small Learning Rate of 0.01 and a Learning Momentum of 0.95 provided the appropriate combination for the performance prediction. The batch size determines the training time, memory usage, and accuracy. There is no single "best" batch size. A small batch size may be slower to train but consumes less memory and provides more accurate results. For this reason, we selected a batch size of three images. For the number of epochs, we started with 18 epochs, and we increased the number until the model no longer improved.

Finally, we tested our model with samples prepared from three commercial coccidia vaccines that contained the three *Eimeria* spp. Our approach was consistently able to achieve good performance for single species samples. The automated oocyst counts showed high agreement with the counts obtained by the manual method for the individual species models. On the other hand, for the multi-species model, the automated oocyst counts showed moderate agreement compared to the manual method. Between all models for each of the three species, the average deviation from the correct count ranged from an average underestimate of 0.18% to an average overestimate of 0.39%. The poorest agreement was in the non-sporulated *E. acervulina* and *E. tenella* oocysts. This is not surprising, as these oocysts are smaller than those of *E. maxima*, and it may be difficult to identify the sporocysts within the oocysts. Further, our training dataset contained fewer instances of these groups compared to some other groups.

We wanted our model to be capable of differentiating sporulated and non-sporulated oocysts, because only the first are infective. This sporulation step occurs in the environment. While sporulated oocysts may survive in the environment for more than a year, non-sporulated oocysts can only survive a short time in the environment [20]. If environmental conditions prevent the oocysts from sporulating, it is possible that a vaccinated flock does not get good coverage and a coccidia outbreak occurs later in life, when losses can be more severe. An automated model for speciation of *Eimeria* sp. oocysts, with an accuracy of 96.9%, has been published recently [21]. For this validation model, these researchers used a publicly available database (http://www.coccidia.icb.usp.br/ accessed on 15 January 2023 [22], which only contains sporulated oocysts. Therefore, while this model may identify coccidia by species, it cannot differentiate sporulation status of the oocysts, which can be important for assessing the management conditions, vaccine cycling, or quality of a vaccine.

To differentiate between sporulated and non-sporulated coccidia, we fine-tuned the parameters of the neural network incorporating the sporulation status as a new group, thus resulting in six different group-labels (one for each of the three species in one of two sporulation states). We experimented by creating single-species models that could differentiate sporulated from non-sporulated oocysts, as well as creating a combined model capable of both speciating and determining sporulation simultaneously.

The single-species and multi-species models excelled in identifying *E. maxima* oocysts. Furthermore, the models were also successful in differentiating sporulated and non-sporulated oocysts of *E. maxima*. This is expected as these species are the largest [13] and easy to resolve visually and by digital computing. On the other hand, the model was least accurate for *E. acervulina*, which has the smallest oocysts of the three species investigated. Notably, most of the false positives had a low confidence value (below 0.7). A higher confidence threshold might have produced more accurate speciation results by eliminating these false positive detections. Furthermore, to develop a successful model, it is most important to have digital images that have high resolution, that is in focus. Therefore, better imaging solutions that produce higher-resolution images are another potential avenue for improvement.

Important components of successful analytical methods used in routine diagnosis and research include performance, time, and resource-efficiency. Our results show that automated image analysis is promising, and it can drastically reduce the analysis time compared to manual measurements. The model demonstrated good accuracy, with just a few digital images presented. Further validation of the Artificial Intelligence model would enable accurate and rapid analysis of a larger number of samples in a short period of time. This would decrease the risk of bias by the analyst.

This study was a proof of concept to demonstrate we can classify and enumerate coccidia using AI. We worked with the three most common coccidia, which are included in all the commercial vaccines and whose oocyst morphology is relatively easy to separate visually. In the future, we plan to include other *Eimeria* spp. in the model. It would be interesting to develop a model that is capable of discriminating oocysts from *Eimeria* protozoa that have similar dimensions and sizes, such as *E. tenella* vs. *E. brunetti*, and *E. acervulina* vs. *E. praecox*. Similarly, we would like to develop a model for turkey coccidia, whose oocysts are difficult to separate visually and biochemically [23].

5. Conclusions

In conclusion, the proposed AI model enables rapid and automated enumeration, speciation, and identification of sporulation of three species of coccidia oocysts of chickens from light-microscopic images with overall excellent precision and accuracy for *E. maxima* oocysts, as well as sporulated oocysts of *E. acervulina* and *E. tenella*. However, the model was weak identifying non-sporulated *E. acervulina* and *E. tenella*. Two factors that might have contributed to the poor correlation of the model are the image resolution due to the small size of the oocyst as well as the low numbers of non-sporulated oocysts in the samples used for training and inference. Higher-resolution images may be needed to improve the model. Currently, additional samples are being analyzed to improve the accuracy of the model. Another shortcoming of our model was the inability to analyze samples with no oocysts. The cause of rejection of these samples is being investigated and solutions will be developed. Even with these weaknesses, this method, based on open-source software, has the potential to significantly reduce the time needed for the enumeration and speciation of coccidia as well as to help with coccidia research and development of vaccines.

Although still in its infancy, AI image analysis is likely to be a key technology in the future. In the long term, this technology can lead to the development of other industry assays, such as more accurate identification of shedding patterns of specific species of coccidia in mixed infections, and evaluation of vaccines and anticoccidial drugs. This system could be expanded to identify other parasites and even be used in the diagnosis of intestinal parasites in other animal species.

Author Contributions: Conceptualization, R.C.; methodology, D.L.R. and R.C.; software, D.L.R.; validation, I.K., D.L.R. and R.C.; formal analysis, I.K.; investigation, I.K.; resources, R.C.; data curation, D.L.R.; writing—original draft preparation, I.K.; writing—review and editing, D.L.R. and R.C.; funding acquisition, R.C. All authors have read and agreed to the published version of the manuscript.

Funding: This research was funded by Bioresource International (BRI).

Institutional Review Board Statement: Not applicable.

Informed Consent Statement: Not applicable.

Data Availability Statement: The data presented in this study are available on request from the corresponding author. The data are not publicly available due to [patent application].

Acknowledgments: We are thankful to BRI for supporting this study. Also, we would like to thank the College of Engineering for partially funding Kellogg. We thank Mark Jenkins for donating the *Eimeria* isolates, as well as Merck Animal Health, Boehringer Ingelheim, and Ceva for donating the vaccines used in this study. Lastly, we want to thank V. Criollo and M. Browning for their technical support.

Conflicts of Interest: The authors declare no conflicts of interest.

References

1. Fatoba, A.J.; Adeleke, M.A. Diagnosis and control of chicken coccidiosis: A recent update. *J. Parasit. Dis.* **2018**, *42*, 483–493. [CrossRef] [PubMed]
2. Jones, P.J.; Niemi, J.; Christensen, J.-P.; Tranter, R.B.; Bennett, R.M. A review of the financial impact of production diseases in poultry production systems. *Anim. Prod. Sci.* **2019**, *59*, 1585–1597. [CrossRef]
3. Blake, D.P.; Knox, J.; Dehaeck, B.; Huntington, B.; Rathinam, T.; Ravipati, V.; Ayoade, S.; Gilbert, W.; Adebambo, A.O.; Jatau, I.D.; et al. Re-calculating the cost of coccidiosis in chickens. *Vet. Res.* **2020**, *51*, 115. [CrossRef] [PubMed]
4. Conway, D.P.; McKenzie, M.E. (Eds.) *Poultry Coccidiosis Diagnostic and Testing Procedures*, 3rd ed.; Blackwell Publishing: Ames, IA, USA, 2007.
5. McDougald, L.R.; Fuller, L.; Solis, J. Drug-sensitivity of 99 isolates of coccidia from broiler farms. *Avian Dis.* **1986**, *30*, 690–694. [CrossRef] [PubMed]
6. Lee, B.H.; Kim, W.H.; Jeong, J.; Yoo, J.; Kwon, Y.K.; Jung, B.Y.; Kwon, J.H.; Lillehoj, H.S.; Min, W. Prevalence and cross-immunity of *Eimeria* species on Korean chicken farms. *J. Vet. Med. Sci.* **2010**, *72*, 985–989. [CrossRef] [PubMed]
7. Jenkins, M.C.; Miska, K.; Klopp, S. Application of polymerase chain reaction based on ITS1 rDNA to speciate *Eimeria*. *Avian Dis.* **2006**, *50*, 110–114. [CrossRef] [PubMed]
8. Smith, M.K.; Buhr, D.L.; Dhlakama, T.A.; Dupraw, D.; Fitz-Coy, S.; Francisco, A.; Ganesan, A.; Hubbard, S.A.; Nederlof, A.; Newman, L.J.; et al. Automated enumeration of *Eimeria* oocysts in feces for rapid coccidiosis monitoring. *Poult. Sci.* **2023**, *10*, 102252. [CrossRef] [PubMed]
9. Adams, D.S.; Kulkarni, R.R.; Mohammed, J.P.; Crespo, R. A flow cytometric method for enumeration and speciation of coccidia affecting broiler chickens. *Vet. Parasit.* **2022**, *301*, 109634. [CrossRef] [PubMed]

10. Adams, D.S.; Ruiz-Jimenez, F.; Fletcher, O.J.; Gall, S.; Crespo, R. Image analysis for *Eimeria* oocyst counts and classification. *J. Appl. Poult. Res.* **2022**, *31*, 100260. [CrossRef]
11. Vrba, V.; Blake, D.P.; Poplstein, M. Quantitative real-time PCR assays for detection and quantification of all seven *Eimeria* species that infect the chicken. *Vet. Parasit.* **2010**, *174*, 183–190. [CrossRef] [PubMed]
12. Edgar, S.A. Sporulation of oocysts at specific temperatures and notes on the prepatent period of several species of avian coccidia. *J. Parasitol.* **1955**, *41*, 214–216. [CrossRef] [PubMed]
13. Cervantes, H.M.; McDougald, L.R.; Jenkins, M.C. Coccidiosis. In *Diseases of Poultry*, 14th ed.; Swayne, D.E., Boulianne, M., Logue, C.M., McDougald, L.R., Nair, V., Suarez, D.L., Eds.; Wiley Blackwell: Hoboken, NJ, USA, 2020; Volume II, pp. 1193–1217, 1242–1246.
14. Lebel, P.; Dial, R.; Vemuri, V.N.P.; Garcia, V.; DeRisi, J.; Gómez-Sjöberg, R. Label-free imaging and classification of live *P. falciparum* enables high performance parasitemia quantification without fixation or staining. *PLoS Comput. Biol.* **2021**, *17*, e1009257. [CrossRef] [PubMed]
15. Bharati, P.; Pramanik, A. Deep Learning Techniques—R-CNN to Mask R-CNN: A Survey. In Proceedings of the Advances in Intelligent Systems and Computing, West Bengal, India, 19–20 January 2019; pp. 657–668.
16. He, K.; Gkioxari, G.; Dollár, P.; Girshick, R. Mask R-CNN. In Proceedings of the IEEE International Conference on Computer Vision (ICCV), Venice, Italy, 22–29 October 2017; pp. 2980–2988.
17. Krell, M.M.; Seeland, A.; Kim, S.-K. Rotational data augmentation for electroencephalographic data. In Proceedings of the 39th Annual International Conference of the IEEE Engineering in Medicine and Biology Society (EMBC), Jeju, Republic of Korea, 11–15 July 2017; pp. 471–474.
18. Cubuk, E.D.; Zoph, B.; Mane, D.; Vasudevan, V.; Le, Q.V. AutoAugment: Learning augmentation policies from data. In Proceedings of the IEEE/CVF Conference on Computer Vision and Pattern Recognition (CVPR), Long Beach, CA, USA, 15–20 June 2019; pp. 113–123.
19. Attoh-Okine, N.O. Analysis of learning rate and momentum term in backpropagation neural network algorithm trained to predict pavement performance. *Adv. Eng. Softw.* **1999**, *30*, 291–302. [CrossRef]
20. Price, K.R.; Guerin, M.T.; Barta, J.R. Success and failure: The role of relative humidity levels and environmental management in live *Eimeria* vaccination of cage-reared replacement layer pullets. *J. Appl. Poult. Res.* **2014**, *23*, 523–535. [CrossRef]
21. He, P.; Chen, Z.; He, Y.; Chen, J.; Hayat, K.; Pan, J.; Lin, H. A reliable and low-cost deep learning model integrating convolutional neural network and transformer structure for fine-grained classification of chicken *Eimeria* species. *Poult. Sci.* **2023**, *102*, 102459. [CrossRef] [PubMed]
22. Castañón, C.A.B.; Fraga, J.S.; Fernandez, S.; Gruber, A.; Costa, L.d.F. Biological shape characterization for automatic image recognition and diagnosis of protozoan parasites of the genus *Eimeria*. *Pattern Recognit.* **2007**, *40*, 1899–1910. [CrossRef]
23. Gadde, U.D.; Rathinam, T.; Finklin, M.N.; Chapman, H.D. Pathology caused by three species of *Eimeria* that infect the turkey with a description of a scoring system for intestinal lesions. *Avian Pathol.* **2019**, *49*, 80–86. [CrossRef] [PubMed]

animals
MDPI

Brief Report

Occurrence of *Eimeria* spp. and Intestinal Helminths in Free-Range Chickens from Northwest and Central Romania

Mircea Coroian [1,2,*], Tünde-Zsuzsánna Fábián-Ravasz [2], Patricia Roxana Dobrin [2] and Adriana Györke [2]

[1] Department of Poultry Management and Pathology, University of Agricultural Sciences and Veterinary Medicine of Cluj-Napoca, 400372 Cluj-Napoca, Romania
[2] Department of Parasitology and Parasitic Diseases, University of Agricultural Sciences and Veterinary Medicine of Cluj-Napoca, 400372 Cluj-Napoca, Romania; rtundike@yahoo.com (T.-Z.F.-R.); dobrin.patricia98@gmail.com (P.R.D.); adriana.gyorke@usamvcluj.ro (A.G.)
* Correspondence: mircea.coroian@usamvcluj.ro

Simple Summary: In the context of backyard poultry raising, a notable concern arises regarding the susceptibility to parasitic infections. The poultry industry holds a crucial position in ensuring food safety and nutritional requirements, emerging as the most rapidly advancing agricultural subsector. The aim of our study was to assess the prevalence of gastrointestinal parasites in chickens raised in the backyard system within the northwestern and central regions of Romania. Fecal samples were collected and tested using flotation, McMaster, and PCR (polymerase chain reaction) methods. The overall prevalence of infection with gastrointestinal parasites was 53.1%. Intestinal parasites demonstrate a pronounced prevalence within the context of backyard poultry flocks, and the substantial burdens imposed by these parasites can deleteriously influence both avian productivity and economic considerations.

Abstract: Chickens raised in backyard free-range systems are confronted with a significant threat of parasitic infections. Among the parasitic agents, protozoa belonging to the genus *Eimeria* and helminths, including *Ascaridia galli*, *Capillaria* spp., *Heterakis gallinarum*, and *Strongyloides avium*, stand out as the most prevalent. The sampling protocol included sixteen localities in four counties within the Transylvania region of Romania. Fecal samples were collected from chickens reared in a backyard system. Fecal samples were screened for oocysts (O) and eggs (E) by flotation method, and their number per gram of feces (OPG/EPG) was calculated after counting them by McMaster method. Positive samples for *Eimeria* spp. were further analyzed by PCR (polymerase chain reaction) method to identify the *Eimeria* species. A total of 145 flocks were tested and the overall prevalence of infection was 53.1%. The most prevalent infections were with *A. galli/H. gallinarum* (25.5%), and *Eimeria* spp. (24.8%), followed by *Capillaria* spp. and strongyles. The mean OPG/EPG values were as follows: 63,577 for *Eimeria* spp., 157 for *Ascaridia/Heterakis*, 362 for *Capillaria* spp., and 1671 for *Strongyle* eggs. Identified *Eimeria* species were *E. acervulina* (41.7%), *E. tenella* (27.8%), *E. praecox* (16.7%), *E. brunetti* (16.7%), OTUy (operational taxonomic unit y) (8.3%), OTUz (operational taxonomic unit z) (8.3%) and *E. mitis* (5.6%). Intestinal parasites exhibit a high prevalence among chickens in backyard poultry flocks, and the presence of significant parasite burdens can adversely affect both productive and economic aspects. To the best of our knowledge, this is the first comprehensive study that aimed to analyze the prevalence of gastrointestinal parasites in chickens raised in a backyard free-range system in Romania, and the first report of OTUy species in Europe.

Keywords: endoparasites; chicken; free-range; *Eimeria*; *Ascaridia*; *Heterakis*; *Capillaria*; *Strongyle*

Citation: Coroian, M.; Fábián-Ravasz, T.-Z.; Dobrin, P.R.; Györke, A. Occurrence of *Eimeria* spp. and Intestinal Helminths in Free-Range Chickens from Northwest and Central Romania. *Animals* **2024**, *14*, 563. https://doi.org/10.3390/ani14040563

Academic Editor: Kenneth Bafundo

Received: 22 December 2023
Revised: 21 January 2024
Accepted: 6 February 2024
Published: 7 February 2024

1. Introduction

The rising need for poultry products in human diets has led to significant expansion in both extensively and intensively managed poultry farming over the past few decades [1].

Consequently, poultry production is increasingly becoming a noteworthy contributor to the national economies of many countries [2].

In 2022, the European Union (EU) generated approximately 13 million tons of poultry meat [3]. According to EUROSTAT, Romania produced a total of 491.22 thousand metric tons, securing the 6th position among EU member states in terms of poultry meat production [4]. Poultry meat constitutes approximately 45 percent of the aggregate animal protein production in Romania, emerging as the predominant meat category in recent years. This marks a noteworthy transition from the preceding decade, during which pork production held the leading position [5]. Furthermore, in accordance with data from the National Institute of Statistics (INS), the production of eggs totaled 6.005 million units, and is anticipated to experience an annual growth rate of 7.63% [6,7].

Even though intensive farms currently dominate the primary production of poultry meat, consumer preferences are gradually moving towards alternative rearing systems, such as free-range and organic [8]. Despite the diminished impact of parasitic diseases in industrial farms attributed to modernization and effective bio-security measures, chickens raised in backyard free-range systems are confronted with a significant threat of parasitic infections. This is particularly due to factors such as unhygienic management practices, litter contamination, and abundance of intermediate hosts [9,10]. Given that numerous farm workers practice extensive chicken raising for personal consumption, they may serve as passive vectors for various diseases, thereby posing a threat to the farm's biosecurity [11].

Among the parasitic agents, protozoa belonging to the genus *Eimeria* and helminths, including *Ascaridia galli*, *Capillaria* spp., *Heterakis gallinarum*, and *Strongyloides avium*, stand out as the most prevalent [12]. Consequences attributed to parasitic infections include diminished health, welfare, and production efficacy marked by compromised feed conversion ratios, lower growth rates or weight loss, diminished egg production and compromised egg quality, as well as intestinal damage. In severe instances, fatalities may occur [11]. Additionally, an indirect effect is expressed by an increased susceptibility to secondary infectious diseases and a reduction in the host's immune response [2].

Eimeria spp. exerts a more pronounced negative impact on the health, welfare and production. The annual estimated worldwide financial burden of coccidiosis in chickens surpassed £10 billion. This cost encompasses expenses related to prevention, treatment, and economic losses [13]. Prophylaxis is achieved through a combination of chemoprophylaxis, vaccination, and dietary supplementation with various plant extracts. However, it still poses a significant threat [14–17].

For a long time, seven species have been identified as infecting chickens, inducing enteritis lesions that result in diarrhea, malabsorption, and hemorrhages [18]. Recently, three cryptic variants, designated as Operational Taxonomic Units (OTUs) X, Y, and Z, have been suggested and assigned as new species, namely *Eimeria lata*, *E. nagambie*, and *E. zaria* [19]. Although initially, the circulation of these species was associated with Australia, Africa, and South America, but recently their presence has been reported in Europe [18].

However, despite the widespread adoption of the free-range system for chicken rearing in Romania and the frequent occurrence of parasitic diseases in this species, there has been no comprehensive study to evaluate their prevalence. Therefore, the current study aims to assess the prevalence of gastrointestinal parasites in chickens raised in the backyard system within the northwestern and central regions of Romania.

2. Materials and Methods

2.1. Animals, Samples, and Sample Analysis

A total of 290 fecal samples were collected from 145 chicken backyard farms located in Northwest and Central Romania. The number of chickens in a flock varied between households, from 6 to 45, with an average of 20.1 ± 9.5 chickens/flock. The age distribution of the chickens spanned from 1 to 3 years, encompassing both males and females, with the ratio heavily favoring females. The investigation spanned from December 2016 to May 2023. The sampling protocol included 16 localities in 4 counties within the Transylvania region of Romania, as follows: Cluj (Mera, Ceanu Mare, Coruș, Frata), Harghita (Corund, Lupeni, Valea lui Pavel, Ocna de Sus), Mureș (Sovata, Chibed, Sărățeni, Sângeorgiu de Pădure), and Satu Mare (Carei, Urziceni, Foieni, Căpleni) (Figure 1). Fecal samples were collected one time from chickens reared in a backyard agricultural system. Two samples were collected randomly, by hand, from each household. Each fecal sample consisted of 10 pooled droppings, collected from the floor. Subsequently the samples were stored at 4–8 °C until the testing procedure.

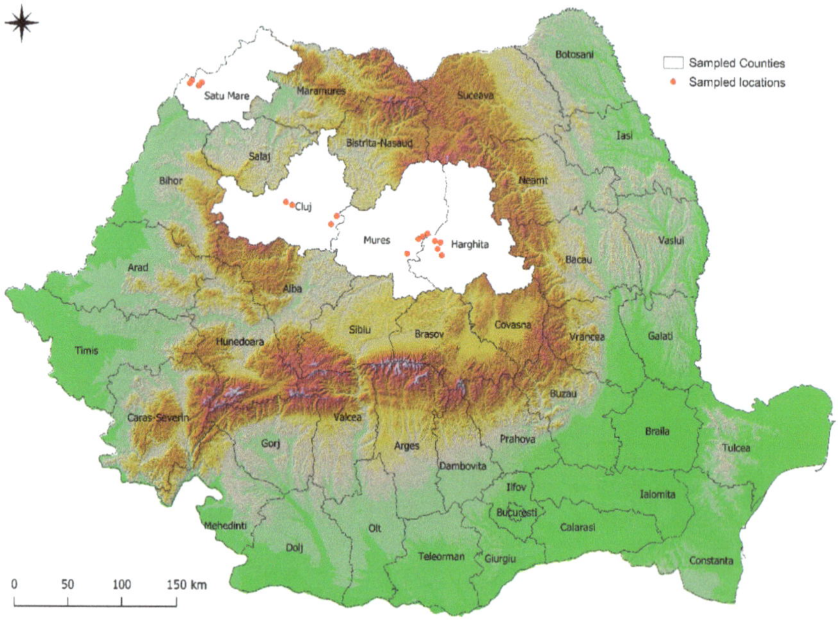

Figure 1. Sampled counties within the Transylvania region of Romania www.qgis.org (accessed on 29 November 2023).

Fecal samples were screened by the flotation method. The parasitic elements in positive samples were counted using the McMaster method [20]. Positive samples for *Eimeria* spp. identified through the flotation method were further analyzed using the polymerase chain reaction (PCR) method in order to identify the *Eimeria* species.

2.2. PCR

First, DNA extraction was performed from fecal samples (n = 18) or concentrated oocysts (n = 28) using the commercial kit Isolate Fecal DNA Kit (Bioline, London, England, United Kingdom; Cat. No. BIO-52038) [21].

Eimeria species were identified using specific primers for each species (Table 1).

2.3. Statistical Analysis

Statistical analysis of the data was conducted using EpiInfoTM 2000 software (version 7.2.0.1, Atlanta, GA, USA). The frequency, prevalence, and its 95% confidence interval (95% CI) of detected species were recorded. The differences in prevalence among identified parasites overall and by the average age of the flock, flock size, and season of sample collection were evaluated using the chi-squared test, or Pearson's chi-squared test. Based on the age of the chickens, the poultry flocks were divided into flocks with 12–23-month-old chickens and flocks with 24–36-month-old chickens. Depending on the number of chickens in a household, flocks were divided into flocks with 1–10, 11–20, 21–30, and 31–45 chickens, respectively. According to the month of sample collection, two seasons were included in the statistical analysis: winter for samples collected from December to February, and spring for samples collected from March to May. A p value of <0.05 was considered statistically significant.

Table 1. Sequence of specific primers used for identification of *Eimeria* species in backyard chickens in Romania by the PCR method [22–25].

Species	Primer Sequence 5′ 3′	Annealing Temperature (°C)	Amplicon Size (bp)
E. acervulina	F 5′-GGGCTTGGATGATGTTTGCTG-3′ R 5′-GCAATGATGCTTGCACAGTCAGG-3′	65	145
E. brunetti	F 5′-CTGGGGCTGCAGCGACAGGG-3′ R 5′-ATCGATGGCCCCATCCCGCAT-3′	58	183
E. maxima	F 5′-GTGGGACTGTGGTGATGGGG-3′ R 5′-ACCAGCATGCGCTCACAACCC-3′	65	205
E. mitis	F 5′-GTTTATTTCCTGTCGTCGTCTCGC-3′ R 5′-GTATGCAAGAGAGAATCGGGATTCC-3′	65	330
E. necatrix	F 5′-AGTATGGGCGTGAGCATGGAG-3′ R 5′-GATCAGTCTCATCATAATTCTCGCG-3′	58	160
E. praecox	F 5′-CATCGGAATGGCTTTTTGAAAGCG-3′ R 5′-GCATGCGCTAACAACTCCCCTT-3′	65	215
E. tenella	F 5′-AATTTAGTCCATCGCAACCCTTG-3′ R 5′-CGAGCGCTCTGCATACGACA-3′	65	278
OTUx	Xf2 5′-GGGTAGAGCCAGGGGTAGAG-3′ Xr2 5′-CGTAGTCCCAAGTGCCAACT-3′	58	1018
OTUy	Yf1 5′-CAAGAAGTACACTACCACAGCATG-3′ Yr1 5′-ACTGATTTCAGGTCTAAAACGAAT-3′	56	346
OTUz	Zf1 5′-TATAGTTTCTTTTGCGCGTTGC-3′ Zr1 5′-CATATCTCTTTCATGAACGAAAGG-3′	58	147

2.4. Ethical Statement

The investigation was carried out within backyard farms. The animals under consideration were neither manipulated nor subjected to constraints on their mobility or daily activities. The verbal consent of the flock owners to collect the fecal samples and to publish the results was obtained.

3. Results

The overall prevalence of infection with intestinal parasites in free-range chickens was 53.1% (77/145; 95% CI: 45.0–61.0). *Ascaridia galli/Heterakis gallinarum* (25.5%) and *Eimeria* spp. (24.8%) were the most prevalent statistically significant (p = 0.002) infections, followed by *Capillaria* spp. (23.5%) and digestive strongyles (8.3%) (Table 2). Single infection was recorded in 29.7% (43/145; 95% CI 22.8–37.5), while mixed infection was recorded in 22.8% (33/145; 95% CI 16.7–30.2) of analyzed samples (Table 2). The mean OPG/EPG values, determined through the McMaster method, were as follows: 63,577 for *Eimeria* spp., 157 for *Ascaridia/Heterakis*, 362 for *Capillaria* spp., and 1671 for *Strongyle* eggs.

No statistically significant findings were observed in relation to the age of chickens as indicated in Table 3. However, noteworthy statistical significance (p = 0.003) was observed for *A. galli/H. gallinarum*, a prevalence of 13.1% was recorded in flocks with 11–20 chickens. Also, significant results were recorded (p = 0.01) for the mixed infections in the same group of chickens (11.0%), as detailed in Table 4.

Depending on the season in which the samples were collected, namely winter and spring, statistically significant results were obtained for *Capillaria* spp., as well as between the total number of positive samples between the two seasons (Table 5).

Table 2. The frequency, prevalence, and its 95% CI of identified parasitic species by flotation technique (*n* = 145).

Species	Frequency (*n*)	Prevalence (%)	95% CI	*p*-Value
Eimeria spp.	36	24.8	18.5–32.5	0.002
A. galli/H. gallinarum	37	25.5	19.1–33.2	
Capillaria spp.	34	23.5	17.3–31.0	
Strongyle egg	12	8.3	4.8–13.9	
Single infection	43	29.7	22.8–37.5	0.251
Mixed infection	33	22.8	16.7–30.2	
Total	77	53.1	45.0–61.0	

Legend: 95% CI—95% confidence interval.

Table 3. The frequency (prevalence; 95% CI) of identified parasitic species by flotation technique according to the age of the chickens.

Species	12–23 Months (*n* = 69)	24–36 Months (*n* = 76)	*p*-Value
Eimeria spp.	14 (9.7; 5.8–15.6)	22 (15.2; 10.2–21.9)	0.182
A. galli/H. gallinarum	18 (12.4; 8.0–18.8)	19 (13.1; 8.6–19.6)	0.869
Capillaria spp.	17 (11.7; 7.5–18.0)	17 (11.7; 7.5–18.0)	1
Strongyle eggs	5 (3.5; 1.5–7.8)	7 (4.8; 2.4–9.6)	0.563
Single infection	16 (11.0; 6.9–17.2)	28 (19.3; 13.7–26.5)	0.07
Mixed infection	16 (11.0; 6.9–17.2)	17 (11.7; 7.5–18.0)	0.861
Total	32 (22.1; 16.1–29.5)	45 (31.0; 24.1–39.1)	0.138

Table 4. The frequency (prevalence; 95% CI) of identified parasitic species by flotation technique according to the size of the chicken flock.

	1–10 Chickens (*n* = 40)	11–20 Chickens (*n* = 58)	21–30 Chickens (*n* = 30)	31–45 Chickens (*n* = 17)	*p*-Value
Eimeria spp.	6 (4.1; 1.9–8.7)	10 (6.9; 3.8–12.2)	10 (6.9; 3.8–12.2)	10 (6.9; 3.8–12.2)	0.957
A. galli/H. gallinarum	6 (4.1; 1.9–8.7)	19 (13.1; 8.6–19.6)	5 (3.5; 1.5–7.8)	7 (4.8; 2.4–9.6)	**0.003**
Capillaria spp.	7 (4.8; 2.4–9.6)	12 (8.3; 4.8–13.9)	10 (6.9; 3.8–12.2)	5 (3.5; 1.5–7.8)	0.332
Strongyle egg	2 (1.4; 0.4–4.9)	5 (3.5; 1.5–7.8)	4 (2.8; 1.1–6.9)	1 (0.7; 0.1–3.8)	0.343
Single infection	13 (9.0; 5.3–14.7)	13 (9.0; 5.3–14.7)	9 (6.2; 3.3–11.4)	9 (6.2; 3.3–11.4)	0.692
Mixed infection	3 (2.1; 0.7–5.9)	16 (11.0; 6.9–17.2)	8 (5.5; 2.8–10.5)	6 (4.1; 1.9–8.7)	**0.010**
Total	16 (11.0; 6.9–17.2)	29 (20.0; 14.3–27.3)	17 (11.7; 7.5)	15 (10.3; 6.4–16.4)	0.08

Within polyspecific parasitism15.2% (22/145; 95% CI 10.2–21.9) of the flocks were positive for two parasites, while 7.6% (11/145; 95% CI 4.3–13.1) for three parasites. Statistically significant results were recorded within polyspecific parasitism ($p = 0.03$) (Table 6).

The following *Eimeria* species were identified by PCR: *E. acervulina* (41.7%), *E. tenella* (27.8%), *E. praecox* (16.7%), *E. brunetti* (16.7%), OTUy (8.3%), OTUz (8.3%) and *E. mitis* (5.6%). Coinfections involving multiple *Eimeria* spp. were also documented. Statistically significant results were recorded (Table 7). Additionally, one flock tested positive for both OTUz and OTUy species.

Table 5. The frequency (prevalence; 95% CI) of identified parasitic species by flotation technique according to the season of the samples collection.

Species	Winter (n = 66)	Spring (n = 79)	p-Value
Eimeria spp.	16 (11.0; 6.9–17.2)	20 (13.8; 9.1–20.4)	0.504
A. galli/H. gallinarum	13 (9.0; 5.3–14.7)	24 (16.6; 11.4–23.5)	0.07
Capillaria spp.	9 (6.2; 3.3–11.4)	25 (17.2; 12.0–24.2)	**0.006**
Strongyle egg	4 (2.8; 1.1–6.9)	8 (5.5; 2.8–10.5)	0.248
Single infection	17 (11.7; 7.5–18.0)	26 (17.9; 12.5–25.0)	0.169
Mixed infection	11 (7.6; 4.3–13.1)	22 (15.2; 10.2–21.9)	0.05
Total	29 (20.0; 14.3–27.3)	48 (33.1; 26.0–41.1.)	**0.03**

Table 6. Single and mixed parasitic infections. Co-occurrence between species within polyspecific parasitism.

	Frequency (n)	Prevalence (%)	95% CI	p-Value
Single infection				
Eimeria spp.	16	11.0	6.9–17.2	0.121
Ascaridia/Heterakis	11	7.6	4.3–13.1	
Capillaria spp.	10	6.9	3.8–12.2	
Strongyle	5	3.5	1.5–7.8	
Mixed infection				
E + A/H	8	5.5	2.8–10.5	**0.03**
E + C	3	2.1	0.7–5.9	
A/H + C	10	6.9	3.8–12.2	
A/H + S	2	1.4	0.4–4.9	
E + A/H + C	5	3.5	1.5–7.8	
E + C + S	4	2.8	1.1–6.9	
A/H + C + S	1	0.7	0.1–3.8	

Legend: E—*Eimeria*, A/H—*Ascaridia/Heterakis*, C—*Capillaria*, S—*Strongyle* type egg, 95% CI—95% confidence interval.

Table 7. *Eimeria* species identified by PCR and their coinfections (n = 36).

Species	Frequency (n)	Prevalence (%)	95% CI	p-Value
E. acervulina	15	41.7	27.1–57.8	0.002
E. tenella	10	27.8	15.9–44.0	
E. praecox	6	16.7	7.9–31.9	
E. brunetti	6	16.7	7.9–31.9	
OTUy	3	8.3	2.9–21.8	
OTUz	3	8.3	2.9–21.8	
E. mitis	2	5.6	1.5–18.1	
A + P	1	2.8	0.5–14.2	0.699
A + T	3	8.3	2.9–21.8	
A + B	3	8.3	2.9–21.8	
M + P	1	2.8	0.5–14.2	
M + T	1	2.8	0.5–14.2	
P + T	3	8.3	2.9–21.8	

Table 7. *Cont.*

Species	Frequency (*n*)	Prevalence (%)	95% CI	*p*-Value
A + P + T	2	5.6	1.5–18.1	
A + M + P + T	1	2.8	0.5–14.2	0.818
A + P + T + B	2	5.6	1.5–18.1	

Legend: A—*E. acervulina*, M—*E. mitis*, P—*E. praecox*, T—*E. tenella*, B—*E. brunetti*, 95% CI—95% confidence interval. Mixed infections with OTUy and OTUz are not included in the table.

4. Discussion

The poultry industry holds a crucial position in ensuring food safety and nutritional requirements, emerging as the most rapidly advancing agricultural sub-sector. Anticipated factors influencing sectoral expansion encompass ongoing urbanization trends, population increase, and rising income levels. In 2020, the poultry sector exhibited a market value of $310.7 billion, with projections indicating an ascent to surpass $400 billion by 2025 [26]. Intestinal parasites exhibit a high prevalence among chickens in backyard poultry flocks, and the presence of significant parasite burdens can adversely affect both productivity and economic aspects [27]. However, regular deworming is not commonly practiced in free-range systems, as owners often lack awareness regarding the risks posed by gastrointestinal parasites [28].

A systematic review on the prevalence of gastrointestinal nematodes in chicken published by Shifaw, encompassing nearly 200 studies published over 80 years, revealed that *A. galli*, *H. gallinarum*, and *Capillaria* spp. were the most commonly identified parasites [2]. This aligns with our findings, where these three species exhibited the highest prevalence.

The pooled prevalence reported by Shifaw for Europe (78.9%) and for the backyard production system (82.6%) surpasses the prevalence observed in our study. This variance could be attributed to ecological, environmental, and climatic factors, including seasonal dynamics, the quantity and accessibility of intermediate hosts, among others. Additionally, variations in diagnostic and sampling procedures, along with diverse host-related factors, may significantly impact the recorded prevalence values [2,29]. The complete absence of cestodes and trematodes species could be elucidated by their more intricate life cycle, in terms of intermediate hosts and environment conditions [2,30].

Polyspecific parasitism was recorded in 11.4% of the samples, strengthening the hypothesis that parasitic infestations usually co-circulate in chickens [31]. This is of significant importance, as the association within parasites with gastrointestinal predilection, such as nematodes and coccidia, may heighten their role in early chick mortality and other productivity losses among adults [31].

Similar results were documented in a study conducted in Poland, a prevalence of nearly 35% was recorded for *Eimeria* spp., with *E. acervulina* being the most frequently identified species [28].

Thus far, *Eimeria zaria* (OTUz) stands as the sole OTU species reported in Europe. However, we managed to identify two species out of three, namely, OTUy and OTUz. Further studies are imperative, particularly in the intensive sector where coccidiosis remains a significant risk, to evaluate the prevalence of the new OTU species. The heightened risk of the presence of the new species is underscored by the inadequacy of protection conferred by current anticoccidial vaccines [18]. Moreover, given that mechanical vectors constitute the most common way of *Eimeria* oocyst transmission, and considering that many employees in poultry farms own a backyard flock, preventive biosecurity measures should be implemented in order to avoid this route of contamination [32].

To the best of our knowledge, this is the first comprehensive study that aimed to analyze the prevalence of gastrointestinal parasites in chickens raised in a backyard free-range system in Romania. Moreover, this is the first report of OTUy in Europe.

5. Conclusions

The widespread distribution of gastrointestinal parasites in chickens raised in the free-range system is most likely explained by poor management practices, including sanitary deficiencies and the absence of deworming programs. Moreover, the scavenging activities of the chickens, a characteristic of this system that enhances contact with excreta, could contribute to this distribution.

Although the prevalence recorded in Romania is relatively lower compared to other European countries, we managed to identify the most prevalent gastrointestinal parasite species.

Considering the upward trend of organic growth systems, there should be an implementation of increased awareness among owners regarding the prevention and availability of treatment methods.

Additional research is necessary to obtain an optimal understanding of the epidemiological status of Romania concerning gastrointestinal parasites in chickens.

Author Contributions: Conceptualization, A.G. and M.C.; methodology, A.G.; software, M.C.; validation, A.G.; formal analysis, T.-Z.F.-R., P.R.D. and M.C.; investigation, T.-Z.F.-R. and P.R.D.; resources, A.G.; data curation, T.-Z.F.-R.; writing—original draft preparation, M.C.; writing—review and editing, A.G.; visualization, M.C.; supervision, A.G.; project administration, A.G.; funding acquisition, A.G. All authors have read and agreed to the published version of the manuscript.

Funding: This work was supported by UASVM Cluj-Napoca through an internal grant, Solution, project number 24868/5.11.2021.

Institutional Review Board Statement: The investigation was carried out within backyard farms. The fecal samples were collected from shelters in the absence of chickens. The animals under consideration were neither manipulated nor subjected to constraints on their mobility or daily activities. Consequently, formal ethical approval from a commission was deemed unnecessary.

Informed Consent Statement: Not applicable.

Data Availability Statement: All data generated or analyzed during this study are included in this published article. Other datasets used and/or analyzed can be made available by the corresponding author on reasonable request.

Conflicts of Interest: The authors declare no conflicts of interest.

References

1. Farrell, D. The role of poultry in human nutrition. In *Poultry Development Review*; Food and Agriculture Organization: Rome, Italy, 2013; pp. 2–9.
2. Shifaw, A.; Feyera, T.; Walkden-Brown, S.W.; Sharpe, B.; Elliott, T.; Ruhnke, I. Global and regional prevalence of helminth infection in chickens over time: A systematic review and meta-analysis. *Poult. Sci.* **2021**, *100*, 101082. [CrossRef] [PubMed]
3. Statista. Leading Poultry Producers in the European Union in 2022. 2023. Available online: https://www.statista.com/statistics/1407478/leading-poultry-meat-producers-in-the-eu/ (accessed on 24 November 2023).
4. Eurostat. Agricultural Production—Livestock and Meat. 2023. Available online: https://ec.europa.eu/eurostat/statistics-explained/index.php?oldid=427096 (accessed on 25 November 2023).
5. U.S. Department of Agriculture. Romania: Challenging Prospects for Romania's Poultry Industry. 2020. Available online: https://fas.usda.gov/data/romania-challenging-prospects-romanias-poultry-industry (accessed on 24 November 2023).
6. National Institute of Statistics. Efectivele de Animale şi Producţia Animală, în Anul 2022. 2023. Available online: https://insse.ro/cms/ro/content/efectivele-de-animale-%C5%9Fi-produc%C5%A3ia-animal%C4%83-%C3%AEn-anul-2022 (accessed on 24 November 2023).
7. Statista. Eggs—Romania. 2023. Available online: https://www.statista.com/outlook/cmo/food/dairy-products-eggs/eggs/romania (accessed on 24 November 2023).
8. El Jeni, R.; Dittoe, D.K.; Olson, E.G.; Lourenco, J.; Seidel, D.S.; Ricke, S.C.; Callaway, T.R. An overview of health challenges in alternative poultry production systems. *Poult. Sci.* **2021**, *100*, 101173. [CrossRef] [PubMed]
9. Javaregowda, A.K.; Kavitha Rani, B.; Revanna, S.P.; Udupa, G. Prevalence of gastro-intestinal parasites of backyard chickens (*Gallus domesticus*) in and around Shimoga. *J. Parasit. Dis.* **2016**, *40*, 986–990. [CrossRef]
10. Mujahid, T. Prevalence of Intestinal Helminth Parasites of Chicken (Gallus Gallus Domesticus Linnaeus, 1758) in Lalitpur District, Nepal. Ph.D. Thesis, Department of Zoology, Institute of Science and Technology, Kathmandu, Nepal, 2017.

11. Gržinić, G.; Piotrowicz-Cieślak, A.; Klimkowicz-Pawlas, A.; Górny, R.L.; Ławniczek-Wałczyk, A.; Piechowicz, L.; Olkowska, E.; Potrykus, M.; Tankiewicz, M.; Krupka, M.; et al. Intensive poultry farming: A review of the impact on the environment and human health. *Sci. Total Environ.* **2023**, *858*, 160014. [CrossRef] [PubMed]
12. Tomza-Marciniak, A.; Pilarczyk, B.; Tobiańska, B.; Tarasewicz, N. Gastrointestinal parasites of free-range chickens: A worldwi issue. *Bull. Univ. Agric. Sci. Vet. Med. Cluj-Napoca Vet. Med.* **2019**, *76*, 110–117. [CrossRef]
13. Blake, D.P.; Knox, J.; Dehaeck, B.; Huntington, B.; Rathinam, T.; Ravipati, V.; Ayoade, S.; Gilbert, W.; Adebambo, A.O.; Jatau, I.D., et al. Re-calculating the cost of coccidiosis in chickens. *Vet. Res.* **2020**, *51*, 115. [CrossRef] [PubMed]
14. McDonald, V.; Shirley, M.W. Past and future: Vaccination against Eimeria. *Parasitology* **2009**, *136*, 1477–1489. [CrossRef]
15. Peek, H.W.; Landman, W.J.M. Coccidiosis in poultry: Anticoccidial products, vaccines and other prevention strategies. *Vet. Q.* **2011**, *31*, 143–161. [CrossRef]
16. Pop, L.M.; Varga, E.; Coroian, M.; Nedișan, M.E.; Mircean, V.; Dumitrache, M.O.; Farczádi, L.; Fülöp, I.; Croitoru, M.D.; Fazakas, M.; et al. Efficacy of a commercial herbal formula in chicken experimental coccidiosis. *Parasites Vectors* **2019**, *12*, 343. [CrossRef]
17. Coroian, M.; Pop, L.M.; Popa, V.; Friss, Z.; Oprea, O.; Kalmár, Z.; Pintea, A.; Borșan, S.D.; Mircean, V.; Lobonțiu, I.; et al. Efficacy of Artemisia annua against Coccidiosis in Broiler Chickens: A Field Trial. *Microorganisms* **2022**, *10*, 2277. [CrossRef]
18. Jaramillo-Ortiz, J.M.; Burrell, C.; Adeyemi, O.; Werling, D.; Blake, D.P. First detection and characterisation of *Eimeria zaria* in Europe. *Vet. Parasitol.* **2023**, *324*, 110068. [CrossRef]
19. Blake, D.P.; Vrba, V.; Xia, D.; Jatau, I.D.; Spiro, S.; Nolan, M.J.; Underwood, G.; Tomley, F.M. Genetic and biological characterisation of three cryptic *Eimeria* operational taxonomic units that infect chickens (*Gallus gallus* domesticus). *Int. J. Parasitol.* **2021**, *51*, 621–634. [CrossRef]
20. Nicholls, J.; Obendorf, D.L. Application of a composite faecal egg count procedure in diagnostic parasitology. *Vet. Parasitol.* **1994**, *52*, 337–342. [CrossRef]
21. Györke, A.; Pop, L.; Cozma, V. Prevalence and distribution of *Eimeria* species in broiler chicken farms of different capacities. *Parasite* **2013**, *20*, 50. [CrossRef] [PubMed]
22. Schnitzler, B.E.; Thebo, P.L.; Mattsson, J.G.; Tomley, F.M.; Shirley, M.W. Development of a diagnostic PCR assay for the detection and discrimination of four pathogenic Eimeria species of the chicken. *Avian Pathol.* **1998**, *27*, 490–497. [CrossRef] [PubMed]
23. Schnitzler, B.E.; Thebo, P.L.; Tomley, F.M.; Uggla, A.; Shirley, M.W. PCR identification of chicken Eimeria: A simplified read-out. *Avian Pathol.* **1999**, *28*, 89–93. [CrossRef] [PubMed]
24. Haug, A.; Thebo, P.; Mattsson, J.G. A simplified protocol for molecular identification of Eimeria species in field samples. *Vet. Parasitol.* **2007**, *146*, 35–45. [CrossRef] [PubMed]
25. Blake, D.P.; Worthing, K.; Jenkins, M.C. Exploring *Eimeria* genomes to understand population biology: Recent progress and future opportunities. *Genes* **2020**, *11*, 1103. [CrossRef]
26. Feed&Additive. Global Poultry Industry and Trends. 2021. Available online: https://www.feedandadditive.com/global-poultry-industry-and-trends/ (accessed on 30 November 2023).
27. Butcher, G.D.; Miles, R.D. *Intestinal Parasites in Backyard Chicken Flocks*; University of Florida Cooperative Extension Service, Institute of Food and Agriculture Sciences, EDIS: Gainesville, FK, USA, 1995.
28. Tomza-Marciniak, A.; Pilarczyk, B.; Tobianska, B.; Tarasewicz, N. Gastrointestinal parasites of free-range chickens. *Ann. Parasitol.* **2014**, *60*, 305–308.
29. McDougald, L.R. Internal Parasites. In *Diseases of Poultry*, 13th ed.; Wiley-Blackwell: New York City, NY, USA, 2020; pp. 1117–1147.
30. Afolabi, O.J.; Simon-Oke, I.A.; Olasunkanmi, A.O. Intestinal parasites of domestic chicken (*Gallus gallus domesticus*) in Akure, Nigeria. *J. Biomed.* **2016**, *1*, e9771. [CrossRef]
31. Belete, A.; Addis, M.; Ayele, M. Review on major gastrointestinal parasites that affect chickens. *J. Biol. Agric. Healthc.* **2016**, *6*, 11–21.
32. Mesa-Pineda, C.; Navarro-Ruíz, J.L.; López-Osorio, S.; Chaparro-Gutiérrez, J.J.; Gómez-Osorio, L.M. Chicken coccidiosis: From the parasite lifecycle to control of the disease. *Front. Vet. Sci.* **2021**, *8*, 787653. [CrossRef]

MDPI AG
Grosspeteranlage 5
4052 Basel
Switzerland
Tel.: +41 61 683 77 34

Animals Editorial Office
E-mail: animals@mdpi.com
www.mdpi.com/journal/animals

www.ingramcontent.com/pod-product-compliance
Lightning Source LLC
Chambersburg PA
CBHW041344210926
51850CB00013B/2985
* 9 7 8 3 7 2 5 8 3 5 8 1 2 *